PROPERTY-PRESERVING
PETRI NET PROCESS ALGEBRA
IN SOFTWARE ENGINEERING

T0324829

PROPERTY-PRESERVING PETRI NET PROCESS ALGEBRA
IN SOFTWARE ENGINEERING

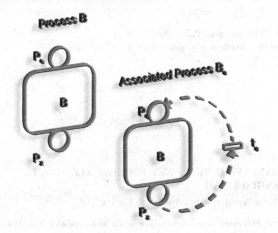

Hejiao Huang
Harbin Institute of Technology Shenzhen Graduate School, China

Li Jiao
Chinese Academy of Sciences, China

To-Yat Cheung
City University of Hong Kong, China

Wai Ming Mak
The Hongkong and Shanghai Banking Corporation Limited, China

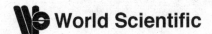 **World Scientific**

NEW JERSEY · LONDON · SINGAPORE · BEIJING · SHANGHAI · HONG KONG · TAIPEI · CHENNAI

Published by

World Scientific Publishing Co. Pte. Ltd.
5 Toh Tuck Link, Singapore 596224
USA office: 27 Warren Street, Suite 401-402, Hackensack, NJ 07601
UK office: 57 Shelton Street, Covent Garden, London WC2H 9HE

British Library Cataloguing-in-Publication Data
A catalogue record for this book is available from the British Library.

ISBN-13 978-981-4324-28-1
ISBN-10 981-4324-28-0

Printed in Singapore.

Preface

When you can measure what you are speaking about, and express it in numbers, you know something about it, but when you cannot measure it, when you cannot express it in numbers, your knowledge is of a meagre and unsatisfactory kind.

Lord Kelvin

The notion of 'software engineering' was first proposed in 1968 at a NATO conference to discuss software development problems. In particular, large software systems were not delivered on time, within budget and with user-desired functionality, and were also unreliable. As a result, a variety of new software design methods were developed in the following decades to bring about quality improvement in the software industry. Some representative design methods developed are Structured Analysis and Design, Object-Oriented Design and Component-Based Design.

In today's hypercompetitive software industry, software companies must continuously improve their products (with better product quality, outstanding features, cheaper price point and faster time to market) and even cannibalize them (i.e., replace popular products before competitors do so) in order to sustain their competitive advantage. This, in turn, calls for a systematic system design methodology that can significantly reduce design errors or eliminate them, and accurately and rigorously construct the desired system from components in a cost and time effective way.

This book presents a property preservation methodology for the component-based design of systems specified in Property-Preserving Petri Net Process Algebra (PPPA). A design approach aiming at property preservation is of paramount importance to Component-Based Design because of the intractability of the validation problem when the size of a system is very large. A property preservation methodology does not need a direct proof but is based on the ways of constructing the desired system from components. As a result, it can significantly reduce the substantial resources required for the subsequent analysis and verification.

Using this methodology, the basic components of a desired system are first specified as some primitive Petri net (PN) processes. These PN processes are then successively combined, refined or reduced using the appropriate operators of PPPA to construct the desired system. By making use of the property preservation theorems on the operators of PPPA, the designers can know immediately what properties are preserved in the desired system and how these properties are preserved. They do not need to go through the traditionally painful and complex (or sometimes intractable) analysis and verification of the desired system to check for the existence of both desirable and undesirable properties. It is this effective and efficient way of designing systems that sets PPPA apart from other traditional design methodologies not aiming at property preservation.

The primary goal of this book is to provide the readers with a proven basis for designing large and complex systems in an effective and efficient way. It shows the readers in a systematic manner on how to make use of PPPA to perform system design in various application domains, including manufacturing systems, multi-agent systems, job-shop scheduling systems and computer security policies. This text is particularly written for seniors, graduate students, researchers and practitioners of software engineering based on Petri nets. Although the understanding and the practical use of PPPA require only college set theory and matrix theory, readers with more experience in the field of Petri nets will definitely profit more from the material.

Acknowledgements

We are indebted to the following persons who have contributed to the successful development of this book. Prof. Helene Kirchner made some direct contribution to Chapter 11. Prof. Dingzhu Du helped with the improvement of the technical contents and presentation, and the recommendation of publishers. Our lovely students, Bo Fu, Taiping Lu, Nianxing Ji, Dong Liang, Hua Jiang, Yuelong Huang, Qichen Bai, Guoli Zhao, Jinling Liu, Chunyan Liu, Tixi He, Feng Shang, Qiang Zhou, Maozhe Xu, Qianqian Wu, Jia Dai, Jinqian Yuan, Bojun Fang, Zhijia Li, and so on, helped with the editing and drawing of figures, and the checking of errors.

Besides, we owe a debt of gratitude to those researchers whose theoretical results are referenced in this book. Although we have attempted to acknowledge their contributions where possible, we would like to apologize in advance for any oversights or inaccuracies in our acknowledgements.

Moreover, we would like to express our deepest gratitude to the Harbin Institute of Technology - Shenzhen Graduate School, Institute of Software Chinese Academy of Sciences, INRIA and the City University of Hong Kong for providing us with the logistic support throughout the preparation of the manuscripts.

Furthermore, we would like to express our sincere thanks to the financial support for this book, provided in part by the National Natural Science Foundation of China (with Grant No. 11071271 and No. 60970029), and by the Fundamental Research Funds for the Central Universities (HIT. NSRIF. 2009127).

Finally, we would like to show our greatest appreciation to our parents, our husbands and wives, our children, and our friends. They are definitely the source of unceasing love and support.

Hejiao Huang
Li Jiao
To-Yat Cheung
Wai Ming Mak

Contents

Chapter 1

Introduction

Petri nets are well-known for their graphical and analytical capabilities for the specification and verification of concurrent, asynchronous, distributed, parallel and nondeterministic systems. Various features contributing to such a success include graphical nature, the simplicity of the model and the firm mathematical foundation. They also provide modularity in design. Moreover, Petri nets have two main features particularly convenient for the formal specification and verification of system design:

1) *Petri-net representations are analytical and flexible*

    'Analytical' here means that both specification and verification can be conducted in terms of matrix algebra.

- Petri nets can provide logical and analytical representations for components that can be readily modified.
- The properties of components can be analytically defined and proved on the basis of their representations.
- All transformations, including composition, refinement and reduction, can be analytically performed on the ingredient components. Their functional purposes and characteristics can also be accurately and logically reflected.

2) *Many Petri-net-based techniques are available for verification*

- Many analytical techniques for verification are available, including reachability analysis, mathematical programming, characterizations and transformations, etc. (See Chapter 2 for a review.)
- There are abundant results concerning property-preserving transformations.

1

In the literature, a few formal techniques, such as logic, LOTOS, CSP and Z, are available for similar purposes. However, they possess only part of the above features and can provide only very limited and fragmentary solutions to the problems involved. Consider LOTOS, for example. LOTOS is handy for specifying operations on processes but is not versatile enough for conducting analytical verification directly. Its operations are only syntactically defined and are not analytically reflected in the representations themselves. To conduct a verification, for example, a LOTOS program has to be converted to another formal technique first.

1.1 This Book's Approach for Component-Based System Design

In a highly competitive society, in order to produce a product quickly, many software and manufacturing systems are built upon existing components. In the software world, for example, a workflow system may be composed of many software tools, workstations, agents, etc. In industry, a manufacturing system is often composed of several assembly lines, machining subsystems, robots, etc. Very often, these components have been in operation for a while. The issue is how to integrate them into a complex system.

The development of a system usually involves four phases: requirement analysis, design, implementation and maintenance. Based on the results of requirement analysis, the design phase produces *a design specification* that formally describes the system's operations, conditions, logic, etc. This design specification plays a core role throughout a system's development. Not only should it reflect the users' and the systems' requirements, it will also serve as the basis for performance analysis, implementation and maintenance. Hence, its correctness is of utmost importance. It should be carefully verified.

This book presents a component-based methodology for the creation and verification of design specifications. The methodology is formally presented as an algebra called Property-Preserving Petri Net Process Algebra (PPPA). Briefly, PPPA includes five classes of operators. The first two classes of operators, namely Extension and Composition, are mainly direct Petri net translations from LOTOS and can be applied only on PNPs. The last three classes, namely Refinement, Reduction and Place merging, originate mainly from the domain of Petri net transformations although the concept of refinement also exists in LOTOS. They can be applied on PNPs as well as other Petri net structures for the purposes of modification. A summary of these operators is given below:

1) *Extension* - This class of operators are used for bottom-up design wherein an option or element is added into a PNP in various ways. It includes: BYPASS, INTERATION, etc.

2) *Composition* - This class of operators are used for bottom-up design wherein several PNPs are integrated according to the semantics of a LOTOS operator, such as, ENABLE, CHOICE, SYNCHRONIZTION, DISABLE, etc.

3) *Refinement* - This class of operators are used for top-down design wherein an element of a component is replaced with a refined component. They can be used, for example, for modeling system calls or interfaces.

4) *Reduction* - This class of operators are used mainly for simplifying or reducing the structure and behavior of a component. They include, for example, replacing an entire sub-component with a single element, merging a set of places into one for solving a resource sharing problem, etc.

5) *Place merging* - This class of operators are used mainly for handling resources sharing problems. They include, merging an arbitrary set of places, merging a set of non-neighboring places, composition via merging pairs of places, composition via merging two single places.

PPPA solves three major problems as follows:

1) System components are specified as Petri-net-based processes (PNP), the elements of PPPA.

2) The operators of PPPA form the methods for integration. The five classes of operators represent most of the common methods for creating new processes from old ones. Together, they can create systems with many different structures. In fact, like LOTOS, they are more than adequate for operating distributed systems.

3) Each of the operators preserves a large number of properties. The PNP created after each operator satisfies these properties if the constituent components also do so. This is the main characteristic of our component-based approach for system design. This concept will be clarified in the subsequent chapters.

In a component-based approach for system design, one of the difficult problems is how to prove the correctness of the created components. Usually, the constituent components are supposed to be correct, i.e., possessing the desirable properties and being free from undesirable ones. However, the operators may destroy these properties or create new ones, resulting in an undesirable new component. Hence, every created component has to go a new process of verification. This indeed involves a tremendous amount of efforts! Our approach is to show that every operator of PPPA can preserve a large number of basic system properties. Hence, if the initial set of primitive components satisfies some of these properties, the created components will also 'automatically' satisfy them without the need of further verification. This greatly saves the efforts spent in verification.

1.2   Background Information and Main Features of PPPA

PPPA combines three major domains of technology from software practices and theories, namely, the component-based approach for system design, the property-preservation approach for system verification and a process algebra from formal techniques. Each of these three domains has had a long history of developments. It is the way they are integrated in this book that makes PPPA a rather extensive and unique methodology for component-based system design. This section first clarifies a few issues and confusions existing in these three domains. It then summarizes the main features of this methodology.

Various meanings of 'component':

The term 'component' has various meanings. In general, it simply means 'a unit for deployment'. Within the scope of software interface technology, it has the specialized meaning 'an autonomous unit for interface'. The main focus of study is on the interoperability and reusability [AF (1998); SZY (1998, 2000)] of the components rather than their logical integration. For a certain long period of time, the two terms 'object' and 'component' were often used interchangeably until a serious drawback in the development of object technology was identified. In fact, it has been found that most of the object libraries available in the markets have been designed mainly for solving problems at a low technical level. Consider the typical four-tiered service architecture for software interface (Figure 1.1). Previously, there were only three tiers - two at the top and one at the bottom. Those objects in the bottom tier do not cater directly or readily to the needs of the enterprise services (i.e., some user functions or business functions in the two top tiers) because incompatibility exists between their interfaces. This shortcoming has led to the addition of the third tier containing enterprise-oriented component processes. As a consequence of such developments, the term component often means 'enterprise processes' in the world of enterprise software.

However, it should be pointed out that this enterprise view of components is not adopted in this book. Instead, in this book, a component is defined as a Petri net with a special structure called Petri net process. Occasionally, a component may even be a more general Petri net.

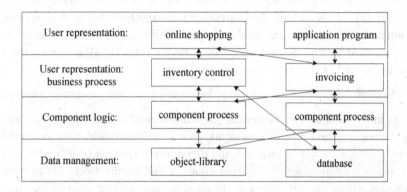

Fig. 1.1 Four-tier service architecture for component software interface.

Combination of process algebras, Petri nets and property-preservation techniques:

The term 'software process' appears mostly in the context of software development, such as code generation, interoperability and reusability of programs, etc. However, in this domain, a 'process' is mainly a unit of software with special organization. Not much has been reported about methods for the systematic construction or verification of software processes. Such methods are usually studied as part of a process algebra in the context of formal system theories. Many formal process algebras, such as CCS, COSY, LOTOS [BB (1987); LFH (1992)], etc., have been published in the literature. However, most of their results concentrate on the semantic aspects of the algebras, such as the semantic equivalence of two algebras, etc. The literature is full of property-preserving transformations on Petri nets. (See Chapter 2 for some references.) However, most of them are for individual irregular transformations which preserve the most basic system properties.

The results are too fragmentary to form an algebra for component-based system specification and verification. Furthermore, the idea of combining process algebra, Petri nets and property preservation techniques is not new either. Probably, the earliest work on combining Petri nets with algebraic processes was by Peter Lauer [JL (1992)]. Also, in the book Petri Net Algebra [BDK (2001)], the algebra Petri Box Calculus (PBC) is defined in terms of Petri nets. It also provides some results about the preservation of a few properties such as T-covers, uniqueness of entry points and S-invariants under some of its operators. However, the dominant concern of that book is on the semantic aspects of PBC, including its operational semantics and its semantic equivalence with other algebras. In brief, that book is about the semantic theory of an algebra rather than methods for system specification and verification. The first version of PPPA, other score of this book, was proposed by Zhu and Cheung [CZ (1994); CRM (1994)] as a structural translation of LOTOS processes to Petri nets.

Distinct features of the property-preserving petri net process algebra:

As mentioned above, within the three domains of technology, the results about property preserving transformations and their application to system development are too fragmentary and diversified to form a coordinated and systematic methodology for system creation and verification. Many well-known algebras are presented as formal theories with emphasis on their logical and semantic properties. In contrast, our property-preserving Petri Net Process Algebra (PPPA), the core of this book, provides a rather complete, extensive and application-oriented methodology for component-based system design. It has the following main features:

1) The algebra caters to the needs of real-life component-based system design: The algebra is defined over a set of elements called Petri net processes (PNP). A PNP is modeled after a real-life software process or manufacturing system process that has a unique entry place, a unique exit place and a specific set of places for handling resource sharing. The algebra includes most of the operators commonly used for the modification and integration of processes: extensions, compositions, refinements and reductions. Examples of operators include: Two PNPs are executed in a sequential order, in a synchronously or asynchronously interactive mode or in a mutually exclusive manner. Two PNPs share some common resources. One PNP interrupts another. A Petri net is refined by having some of its elements replaced with some other PNPs.

2) Verification is based on real-life system properties: Correctness of a system may have many meanings depending on the techniques used for specification and verification. There are many approaches for proving correctness. For example, it may be required to show that the system is semanitcally equivalent to another system that is known to be correct. In this book, correctness means 'possessing a set of properties specified by the user'. These properties reflect the nature of real-life systems, such as every function can eventually occur (liveness), absence of system overflow (boundedness) and every system must return to its initial state when terminated (proper termination).

3) A property-preservation approach for verification is built into the algebra: In a component-based approach for system design, the operands of every operator are supposed to be correct with respect to a set of properties. However, a general operator may destroy these properties or create new ones, resulting in an undesirable new component. A very important feature of PPPA is that all its operators are shown to be preserving many common properties. Hence, if the constituent components satisfy these properties, the created components will also satisfy them without the needs of further verification. Naturally, in this approach, the burden is shifted to designing correct primitive components at the beginning. Usually, however, these components are relatively simpler and their correctness can be verified more easily.

Another main advantage of PPPA is that the number of properties preserved by its operators is quite large. Hence, it may cater for a big variety of system requirements. Some of these properties are very common, such as liveness, boundedness and reversibility. Some of them are seldom directly used, such as siphon, RC-property, coverability by S-components, etc. However, they are often used to derive other properties.

While some general techniques for verification will be briefly reviewed in Chapter 2, it should be pointed out that this book is not concerned about the details of these techniques. Instead, it is devoted to showing the property-preservation nature of the operators of PPPA.

1.3 Organization of This Book

The contents of this book may be divided into two parts. Part I presents some background knowledge and the theory of PPPA. Part II illustrates the application of the algebra to some areas.

Part I. Background Knowledge and Theory of PPPA

This part of the book includes seven chapters. It provides some terminologies of Petri nets and some existing properties for understanding this book's main contribution: PPPA. Five types of opertions of PPPA and the property preservation results are introduced in this part.

- Chapter 1 gives an outline of this book. After pointing out the importance of a correct design specification in the development of a system, this chapter summarizes the main features of the property-preserving component-based methodology presented in this book.
- Chapter 2 defines some related terminologies for understanding the main theory of this book. It introduces the basic properties and special strcutures of Petri nets. A brief review on several major approaches for verification, such as reachability analysis, characterizations, mathematical programming, etc are also included in this chapter. At last, some related work for property-preserving transformations is presented and a discussion on the comparison our approach with them is given.
- Chapter 3 presents the elements, Petri net process (PNP), of PPPA and the related properties. The extension operators of PPPA, PREFIX, BYPASS, ITERATION, etc, are also introduced in this chapter.
- Chapter 4 presents the compositon operators of PPPA. It includes ENABLE, CHOICE, DISABLE, RESUME, etc.
- Chapter 5 presents the refinement operators of PPPA. It includes two kinds of refinements, one replacing a place with a PNP and another replacing a transition with a PNP.
- Chapter 6 presents the reduction operators of the process algebra. It includes various kinds of reductions, such as reducing a set of places, a path or a subnet to a single place.
- Chapter 7 presents the place merging operators for handlign resource sharing issues. It includes various kinds of place merging operators, such as merging an arbitrary set of places, merging a set of non-neighboring places, composition via merging pairs of places, composition via merging two single places, etc.

Part II. Application of Property-Preserving Petri Net Process Algebra (PPPA)

This part includes four chapters about the application of PPPA.

- Chapter 8 presents an example for illustrating the application of PPPA to manufacturing system design.

- Chapter 9 presents an example for illustrating the application of PPPA to multi-agent system design.
- Chapter 10 presents the application of PPPA to job-shop scheduling (JSS) system design. In this Chapter, the element of PPPA is colored Petri net process (CPNP). Based on PPPA, JSS system can be designed and the scheduling can be obtained for mininimizing the makespan.
- Chapter 11 presents an application of PPPA to secure access control system design. Based on PPPA and Petri net theory, the composition of access control policies is possible for preserving some security properties. Furthermore, the detection and resolution of four types of conflicts for policy composition can be handled properly.

Chapter 2

Fundamentals on PPPA and Related Work

As mentioned in Chapter 1, verification is a process of showing whether a system satisfies certain properties or not. Naturally, to make such a process feasible, the properties must first be defined in an analytical way.

In this chapter, after presenting some preliminaries of Petri nets, many properties and special structures of Petri nets are formally defined. The latter is needed as many properties are defined in terms of special structures of the Petri nets. Then several approaches for verifying these properties are given. At last, the related work of PPPA is presented and the differences between our technique and those in the literature are analyzed.

2.1 Preliminaries and Basic Properties of Petri Nets

[DES (1998a)], [DES (1998b)], [REI (1985)], [MUR (1989)], [STV (1998)]

Notation

Let $N(resp., N^+)$ be the set of non-negative (resp., positive) integers, Z be the set of integers and Q be the set of rational numbers. For a finite set A, the function $f : A \to N$(in notation, $f \in N^A$) is an $|A|$-vector of non-negative integers. Z^A and Q^A are defined in a similar way. For a finite subset $X = \{x_1, \ldots, x_m\} \subseteq Q^A$, X is said to be linearly independent if and only if $k_1 x_1 + \ldots + k_m x_m = 0$ implies $k_1 = \ldots = k_m = 0$, where $k_i \in Q$ for $i = 1, 2, \ldots, m$. Let P and T be two finite sets and V be a $|P| \times |T|$ matrix V whose row index is denoted by P and column index by T. For $P_s \subseteq P$ and $T_s \subseteq T$, $V[P_s, T_s]$ denotes the sub-matrix of V with rows in P_s and columns in T_s. In particular, row p (resp., column t) of V is written

11

as $V[p,T]$ (resp., $V[P,t]$) and $V[p,t]$ denotes the entry of V at row p and column t.

Definition 2.1. (net)
A *net* is a 4-tuple $N = (P, T, F, W)$ where,

1) P is a finite set of *places*;
2) T is a finite set of *transitions* such that $P \cap T = \phi$ and $P \cup T \neq \phi$;
3) $F \subseteq (P \times T) \cup (T \times P)$ is the *flow relation*; and
4) W is a *weight function* such that $W(x, y) \in N^+$ if $(x, y) \in F$ and $W(x, y) = 0$ if $(x, y) \notin F$.

A net can be represented as a directed bipartite graph with two kinds of nodes, *circles* and *bars*, representing the places and transitions, respectively. Each arc represents an element of F and connects two nodes of different kinds. The weight W of an arc is a positive integral value associated with the arc.

- A net $N = (P, T, F, W)$ is said to be *ordinary* if and only if $W = 0$ or 1, $\forall (x, y) \in F$. In this case, W will be omitted. N is said to be *pure* or *self-loop-free* if and only if $\forall x, y \in P \cup T : ((x, y) \in F \Rightarrow (y, x) \notin F)$.
- For a net N, a *path* $\sigma = x_1 x_2 \ldots x_n$ is a sequence such that $(x_i, x_{i+1}) \in F$, for $i \in \{1, ..., n-1\}$. σ is said to be *elementary* if, $\forall x_i, x_j$ in $\sigma : (i \neq j \Rightarrow x_i \neq x_j)$. In addition, $P(x_i, x_j)$ and $EP(x_i, x_j)$ denote the set of paths and the set of elementary paths from node x_i to node x_j, respectively.

Definition 2.2. (marking, token and Petri net)
The *marking* (or *state*) for a net $N = (P, T, F, W)$ is a function $M : P \rightarrow N$ such that $M(p)$ represents the number of *tokens* in place $p \in P$. A Petri net (N, M_0) is a net N with an initial marking M_0.

A marking M can be written as a $|P|$ - vector$(M_{p_1} \ldots M(p_{|P|}))$ or as a multi-set expression. For example, the vector $(1\ 2\ \ldots\ 0\ 3)$ and the multi-set expression $p_1 + 2p_2 + \ldots + 3p_{|P|}$ both represent the marking that has 1 token at place p_1, 2 tokens at place p_2, 3 tokens at place $p_{|P|}$. In the following of this book, for clearance, comma "," will be used to separate the elements of a vector, e.g., $(1, 2, \ldots, 0, 3)$. For convenience, these expressions will be used without confusion throughout this book.

Definition 2.3. (firing rule)
A transition $t \in T$ is *firable* (or *enabled*) at a *marking* M if and only if $\forall p \in P : (M(p) \geq W(p,t))$. *Firing* (or *executing*) transition t results in changing marking M to marking M', where $\forall p \in P : (M'(p) = M(p) - W(p,t) + W(t,p))$.

Definition 2.4. (firing sequence and reachability)
Let M, M' be markings, t be a transition, and σ be a sequence in a Petri net (N, M_0). In the following notations, N or M_0 may be omitted if there is no confusion in the context.

1) $M[N,t\rangle$ means that t is enabled at M.
2) $M[N,t\rangle M'$ means that M' is *reachable* from M by firing transition t.
3) $M[N,-\rangle M'$ means that M' is *reachable* from M by firing an unspecified transition
4) $M[N,\sigma\rangle$ means that *sequence σ is firable from* M.That is, for $\sigma = t_1 \ldots t_k \in T^*$, (the Kleene closure of T), there exist markings $M_1, M_2, \ldots, M_{k-1}$ and M_k such that $M_0 = M, M_{i-1}[N,t_i\rangle M_i, i = 1, 2, \ldots, k-1$ and $M_{k-1}[N,t_k\rangle$.
5) $M[N,\sigma\rangle M'$ means that M' is *reachable* from M by firing σ.
6) $M[N,*\rangle M'$ means that M' is *reachable* from M by firing an unspecified sequence.
7) $R(N,M)$ denotes the *reachability* set of N starting from M, i.e., the smallest set of markings such that:

    (a) $M \in R(N,M)$.
    (b) If $M' \in R(N,M)$ and $M'[N,t\rangle M''$ for some $t \in T$, then $M'' \in R(N,M)$.

Definition 2.5. (pre-incidence matrix, post-incidence matrix and incidence matrix)
The *pre-incidence matrix* PRE of a net N is a $|P| \times |T|$ matrix whose element at row p and column t is the weight $W(p,t)$ of the arc from place p to transition t. The *post-incidence matrix* POST of N is a $|P| \times |T|$ matrix whose element is the weight $W(t,p)$ of the arc from transition t to place p. $V = $ POST - PRE is called the *incidence matrix* of N.

Definition 2.6. (state equation and firing count vector)
For a Petri net (N, M_0), $M = M_0 + V\mu$, is called the *state equation*, where V is the incidence matrix of N and $\mu \in N^T$ is the *firing count vector* of

a firing sequence σ, i.e., $\mu[t]$ is the number of times transition t occurs in σ.

Definition 2.7. (pre-set, post-set, input set and output set)
For $x \in P \cup T$, $^{\bullet}x = \{y|(y,x) \in F\}$ and $x^{\bullet} = \{y|(x,y) \in F\}$ are called the *pre-set* (*input set*) and *post-set* (*output set*) of x, respectively. For a set $X \subseteq P \cup T$, $^{\bullet}X = \bigcup_{x \in X} {}^{\bullet}x$ and $X^{\bullet} = \bigcup_{x \in X} x^{\bullet}$.

Definition 2.8. (state machine, marked graph, free choice net, asymmetric choice net and augmented marked graph)
Let $N = (P, T, F)$ be a net.

1) N is called a *state machine* (*SM*) if and only if $\forall t \in T$: ($|^{\bullet}t| = |t^{\bullet}| = 1$).

2) N is called a *marked graph* (*MG*) if and only if $\forall p \in P$: ($|^{\bullet}p| = |p^{\bullet}| = 1$).

3) N is said to be *free choice* (*FC*) if and only if $\forall p_1, p_2 \in P$: ($p_1{}^{\bullet} \cap p_2{}^{\bullet} \neq \phi \Rightarrow p_1{}^{\bullet} = p_2{}^{\bullet}$).

4) N is said to be *asymmetric choice* (*AC*) if and only if $\forall p_1, p_2 \in P$: ($p_1{}^{\bullet} \cap p_2{}^{\bullet} \neq \phi \Rightarrow p_1{}^{\bullet} \subseteq p_2{}^{\bullet}$ or $p_2{}^{\bullet} \subseteq p_1{}^{\bullet}$).

5) N is called *augmented marked graph* (*AMG*) if its places consist of two disjoint subsets P and R such that the following conditions hold:

   (a) The net G obtained by removing all the resource places in R is a marked graph.

   (b) Each resource place $r \in R$ is associated with k pairs of transitions $(a_i, b_i), i = 1, 2, \ldots k$, where $a_i \neq a_j$ and $b_i \neq b_j$ $\forall i \neq j$ and $r^{\bullet} = \{a_i|i = 1, 2, \ldots, k\}$, $^{\bullet}r = \{b_i|i = 1, 2, \ldots, k\}$.

   (c) Every elementary cycle in G is marked by M_0 and every elementary path P_i connecting a_i to b_i is not marked by M_0.

   (d) $M_0(p) = 1, \forall p \in R$.

Notes:

- Every transition of an SM has exactly one incoming arc and one outgoing arc.
- Every place of an MG has exactly one incoming arc and one outgoing arc.
- A FC net [DE (1995); BES (1987)] allows controlled conflicts. Several transitions are in conflict if they have at least one input place in com-

mon. In a FC net, all conflicting transitions have the same set of input places. Hence, every marking enables either all or none of them. This makes it free to choose any of these transitions to fire.

Definition 2.9. (siphon and trap, ST-property and MST-property)
A set of places $D \subseteq P$ is called a *siphon* if ${}^\bullet D \subseteq D^\bullet$. D is called a *trap* if $D^\bullet \subseteq {}^\bullet D$. An ordinary Petri net N is said to satisfy ST-property (resp., MST-property) if every siphon contains a trap (resp., initially marked trap) in N.

During execution, a siphon remains token-free thereafter once it becomes free of tokens. On the contrary, a trap will always remain marked once it becomes marked. It can be easily shown that the union of two siphons (resp., traps) is also a siphon (resp., trap).

Definition 2.10. (cluster)
For a node $x \in P \cup T$, the *cluster* of x, denoted as $[x]$, is the smallest subset of $P \cup T$ satisfying three conditions: (1) $x \in [x]$; (2) if $p \in P \cap [x]$ then $p^\bullet \subseteq [x]$; and (3) if $t \in T \cap [x]$ then ${}^\bullet t \subseteq [x]$. The *set of clusters* of N is denoted as $C(N) = \{[x] \,|\, x \in P \cup T\}$.

It can be shown [DE (1995)] that $C(N)$ is a partition of the nodes of N.

The following part of this subsection presents the formal definitions of many properties of Petri nets (N, M_0). For convenience, we sometimes use the logical notations: $\forall$ means 'for every'; $\exists$ means 'there exists' or 'there exist'; : means 'such that'.

Definition 2.11. (boundedness, safeness and structural boundedness)
A place p is said to be *bounded* (resp., *k-bounded*) if and only if $\exists k \in \mathbb{N}^+$: $(\forall M \in \mathrm{R}(N, M_0), M(p) \leq k)$. (N, M_0) is said to be *bounded* (resp., *k-bounded*) if and only if every place of N is bounded (resp., *k-bounded*). A place is said to be *safe* if and only if it is 1-bounded. (N, M_0) is said to be *safe* if and only if all its places are 1-bounded. N is said to be *structurally bounded* if and only if it is bounded for any initial marking M_0.

Definition 2.12. (liveness and structural liveness)
A transition t is said to be *live* if and only if $\forall M \in \mathrm{R}(N, M_0)$, $\exists M' \in \mathrm{R}(N, M)$: $(M'[N, t\rangle)$, or equivalently, $\forall M \in \mathrm{R}(N, M_0)$, $(\exists \sigma \in T^*: (M[N, \sigma t\rangle))$. (N, M_0) is said to be *live* if and only if every transition of N is live. N is said to be *structurally live* if there exists a marking M_0 such that (N, M_0) is live.

Definition 2.13. (well-formedness)
A net N is said to be *well-formedness* if there exists a marking M_0 such that (N, M_0) is live and bounded.

Liveness requires the firability of every transition starting from any reachable marking, boundedness implies that the number of tokens existing in every place will not exceed a certain limit. Well-formedness is the structural guarantee for liveness and boundedness of a Petri net.

Definition 2.14. (reversibility)
(N, M_0) is said to be *reversible* if $\forall M \in R(N, M_0)$: $(M_0 \in R(N, M))$.

Reversibility implies re-initializability of the system. That is, the system can always return to its initial state from any reachable state.

Definition 2.15. (conservativeness)
N is said to be *conservative* if and only if there exists an integral $|P|$-vector $\alpha \geq 1$ such that $\alpha V = 0$. It is said to be *strictly conservative* if $\alpha = 1$.

For a conservative Petri net, a weighted sum of the numbers of tokens (over the set of all places) in any reachable marking remains constant.

Definition 2.16. (place invariant)
A *place invariant* (*P-invariant*) is an integral $|P|$-vector $\alpha \geq 0$ such that $\alpha V = 0$, i.e., $\forall t \in T$: $(\sum_{p \in {}^\bullet t} \alpha[p] = \sum_{p \in t^\bullet} \alpha[p])$.

A place invariant reflects a conservation property over a subset of places of P instead of the entire set P.

Definition 2.17. (consistency)
N is said to be *consistent* if and only if there exists a $|T|$-vector $\beta \geq 1$ such that $V\beta = 0$. It is said to be *strictly consistent* if $\beta = 1$.

N is consistent [MUR (1989)] if there exist a marking M_0 and a firing sequence σ starting from M_0 and returning back to M_0 such that every transition of N must occur at least once in σ.

Definition 2.18. (repetitiveness)
N is said to be *repetitive* if and only if there exists a $|T|$-vector $\beta \geq 1$ such that $V\beta \geq 0$.

N is repetitive [MUR (1989)] if there exists a marking M_0 and a firing sequence σ starting from M_0 such that every transition occurs infinitely many times in σ.

Definition 2.19. (transition invariant)
A *transition invariant* (*T-invariant*) is a non-negative integeral $|T|$-vector β such that $V\beta = 0$. (i.e., $\forall p \in P$: $(\sum_{t \in {}^\bullet p} \beta[t] = \sum_{t \in p^\bullet} \beta[t])$).

A transition invariant is a multi-set of transitions such that N will be at the same marking before and after firing these transitions in a specified order. However, in general, these transitions may not be all firable in any order. The value of the ith element of β represents the number of times that transition t_i participates in the transition invariant.

Definition 2.20. (semi-positive, positive place/transition invariant, support)
A (place or transition) invariant μ is said to be *semi-positive* if $\mu \geq 0$ and $\mu \neq 0$. An invariant μ is said to be *positive* if $\mu > 0$. The *support* of a semi-positive invariant μ, denoted by $||\mu||$, is the set of nodes x (i.e., places or transitions) satisfying $\mu[x] > 0$.

Definition 2.21. (connectedness and strong connectedness)
N is *connected* if and only if it is not composed of two disjoint and non-empty subnets. N is *strongly connected* if and only if, for every pair of nodes x and y, there exists a directed path from x to y.

Definition 2.22. (Rank-and-cluster property, i.e., RC-property)
N is said to satisfy the *rank-and-cluster property* if $\mathrm{Rank}(N) = |C(N)| - 1$, where $\mathrm{Rank}(N)$ is the rank of its incidence matrix and $|C(N)|$ is the number of clusters of N.

Definition 2.23. (Rank Theorem)
N is said to satisfy the *Rank Theorem* [DE (1995)] iff the following conditions hold: (1) N has a positive place invariant. (2) N has a positive transition invariant. (3) N satisfies the RC-Propery.

Definition 2.24. (SM-coverable and SM-cover)
A subnet $N_i = (P_i, T_i, F_i)$ of a net $N = (P, T, F)$ is said to be an *SM-component* of N if N_i is a strongly connected SM and $T_i = {}^\bullet P_i \cup P_i^\bullet$. N is said to be *SM-coverable* if and only if there exists a set of *SM-components* $\{N_1, \ldots, N_k\}$ such that $P = \cup_i P_i$, $T = \cup_i T_i$ and $F = \cup_i F_i$, where i runs from 1 to k. $\{N_1, \ldots, N_k\}$ is called an *SM-cover* of N and is said to be *minimal* if and only if none of its proper subsets is also an SM-cover of N.

Depending on the structure of a Petri net, the rank of its incidence matrix and the number of its clusters can provide a sufficient criterion and/or necessary criterion for its liveness and boundedness. One version of Rank Theorem was stated in [CCS (1991)] without a complete proof. Another version of the Rank Theorem was stated and proved in [ESP (1990)] for a class of SM-coverable free choice Petri nets. In [DES (1992)], the Rank Theorem in [ESP (1990)] was generalized to extended free choice nets with a complete proof. This is the first version that made use of clusters to formulate the Rank Theorem. In the subsequent years, the Rank Theorem was further generalized to ordinary Petri nets in [DES (1993); DE (1995)] and some special Place/Transition nets, such as equal conflict systems [TS (1994, 1996)] and deterministic systems of sequential processes [RTS (1995, 1996)].

In this book, the Rank Theorem proposed by Jörg Desel [DE (1995)] will be used because it is the most general one.

2.2 General Approaches for Verification

System analysis includes two major tasks: Verifying the system's properties and estimating its performances. This book is about the former.

When creating a system, a designer should determine for each component a set of desirable properties (i.e., properties that the various components of the system should satisfy) and a set of undesirable properties (i.e., properties the system should avoid). Some common properties, such as liveness, boundedness, etc. are probably desirable for most systems whereas properties such as overflow and deadlock should be avoided in most systems. In general, it is up to the designer to select these two sets of properties. To verify a component is to check whether it satisfies these two sets of properties or not.

For Petri nets, many methods for verification have been developed. The major methods can be classified into three approaches: reachability analysis, characterizations/problem conversion, and transformations.

2.2.1 *Verification via reachability analysis*

For Petri nets, a place is considered as a local state whereas a marking is considered as a global state. The main idea of reachability analysis is to generate all the reachable global states and, during the generation process, check whether the desirable properties are satisfied and the undesirable ones are absent at each of these global states. The process starts from the initial marking. In general, at any reachable marking, one fires all the firable transitions and thus generates all the 'next' reachable markings. At each of the generated markings, one checks whether the designated properties hold or not. For example, if it is found that no transitions can be fired at a marking, the system has a deadlock or terminates at that marking. An example (Figure 3.6) of reachability analysis is given in Chapter 3.

Reachability analysis is also called coverability analysis or occurance graph method. It is one of the most popular verification approaches for Petri net based models. The software CNP-Tool is usually applied to the simulation of this approach. However, its shortcoming is that, very often, a huge number of markings may have to be generated, causing the 'state explosion' problem. To reduce the effects of state explosion, many techniques for partial enumeration had been proposed. These techniques usually include a strategy for guiding the generation of some of global states. The main point is that the partial reachability graph generated by the strategy should still reflect the desirable and undesirable properties.

2.2.2   *Verification via characterizations or problem conversion*

Sometimes, direct verification (i.e., verifying a property or a statement according to its definition) is very difficult. Then, one may seek the assistance of some characterizations. A characterization is a relationship between several properties. When it is not known how to directly verify one of the properties involved in the characterization, one tries to verify the other properties instead. The characterization will then imply the validity of the remaining property. Usually, a characterization may appear in the form of an algebraic equation or inequality [BES (1987); DES (1998b); STC (1998)].

The characterization approach for verification can even be generalized to a problem conversion approach. The idea is to show that solving one problem is equivalent to solving another problem. In the literature of Petri nets, many problems have been converted to mathematical programming problems [DES (1998b); ESP (1990); STV (1998)].

Examples of characterizations:

The Petri net (N, M_0), where $N = (P, T, F, W)$, is involved in the following characterizations.

Characterization 2.1. [MUR (1989)]

A net is conservative if and only if there exists a $|P|$-vector $\alpha \geq 1$ such that $\alpha V = 0$.

Characterization 2.2. [MUR (1989)]

A net is structurally bounded if and only if there exists a $|P|$-vector $\alpha \geq 1$ such that $\alpha V \leq 0$.

Characterization 2.3. [MUR (1989)]

A net is consistent if and only if there exists a $|T|$-vector $\beta \geq 1$ such that $V\beta = 0$.

Characterization 2.4. [MUR (1989)]

A net is repetitive if and only if there exists a $|T|$-vector $\beta \geq 1$ such that $V\beta \leq 0$.

Characterization 2.5. (Sufficient condition for repetitiveness) [MUR (1989)]

A consistent Petri net is repetitive.

Characterization 2.6. (Sufficient condition for strong connectedness) [DE (1995)]

A conservative and consistent connected net is strongly connected.

Characterization 2.7. (Fundamental property of place invariants) [DE (1995)]

Let α be a place invariant of Petri net (N, M_0). If $M_0[N, *\rangle M$, then $\alpha M_0 = \alpha M$.

Characterization 2.8. (Fundamental property of invariants) [DE (1995)]

For every semi-positive invariant μ, $^\bullet||\mu|| = ||\mu||^\bullet$.

Characterization 2.9. (Fundamental property of minimal invariants) [DE (1995)]

Every semi-positive or positive invariant is the sum of some minimal invariants.

Characterization 2.10. (Necessary condition for liveness) [DE (1995)]

If a Petri net (N, M_0) is live, then $\alpha M_0 > 0$ for every semi-positive place invariant α of (N, M_0).

Characterization 2.11. (Sufficient condition for liveness) [DE (1995)]

Let N be an ordinary Petri net which is connected, conservative, consistent and satisfies the RC-property. (N, M_0) is live if $\alpha M_0 > 0$ for every semi-positive place invariant α of (N, M_0).

Characterization 2.12. (Necessary condition for well-formedness) [DE (1995)]

A well-formed Petri net (N, M_0) is consistent.

Characterization 2.13. (Rank Theorem) [DE (1995)]

Let N be an ordinary Petri net.

1) If N is connected, conservative, consistent and satisfies the RC-property, then N is well-formed.
2) If N is well-formed, then $\text{Rank}(N) < |\{{}^{\bullet}t | t \in T\}|$.

Characterization 2.14. (Commoner's Theorem for the well-formedness of an FC net) [DE (1995)]

Let N be an FC Petri net. N is well-formed if and only if N is connected, conservative, consistent and satisfies the RC-property.

Characterization 2.15. (Sufficient condition for SM-coverability) [DE (1995)]

Let N be an ordinary Petri net. If N is connected, conservative, consistent and satisfies the RC-property, then N is SM-coverable.

Characterization 2.16. (SM-components induce minimal place invariants) [DE (1995)]

Let $N_i = (P_i, T_i, F_i)$ be an SM-component of a Petri net N. Then, P_i (a multi-set representation of a vector) is a minimal place invariant of N.

Characterization 2.17. (Sufficient condition for the liveness of AC net) [JCL (2002); CX (1997)]

Let $N = (P, T, F, W)$ be an AC net. Then,

1) (N, M_0) is live if it satisfies MST-property and $W(x, y) = 1$ for any $(x, y) \in (P, T)$.
2) (N, M_0) is live if for each minimal siphon S, either it contains a marked trap or $F(S) > 0$, where $F(S) = \min \{M(S) | M = M_0 + V\sigma, M \geq 0, \sigma \geq 0\}$, V is the incidence matrix of N.

Characterization 2.18. (Sufficient condition for well-formedness) [DAI (1995)]

An ordinary and pure Petri net (N, M_0) is well-formed if N is SM-coverable and (N, M_0) satisfies MST-property.

Characterization 2.19. (Liveness of an FC net)

An FC net is live iff every siphon contains a marked trap.

Characterization 2.20. [CX (1997)]
An AMG (N, M_0) is live if it satisfies MST-property.

Characterization 2.21. [CX (1997)]
An AMG (N, M_0) is reversible if it is live.

Characterization 2.22. [CX (1997)]
An augmented marked graph (N, M_0) is live iff it does not contain any potential deadlock.

Characterization 2.23. [CX (1997)]
An augmented marked graph (N, M_0) is live and reversible if for every minimal siphon D containing at least one resource place, either it contains a marked trap or $F(D) > 0$, whére $F(D) = \min \{M(D)|M = M_0 + V\sigma, M \geq 0, \sigma \geq 0\}$, V is the incidence matrix of N.

Characterization 2.24. [HJC (2003)]
The resource places of an augmented marked graph are safe.

2.2.3 *Verification via property-preserving transformations*

Usually, a design may be subject to many transformations, such as composition, refinement, place-reduction, etc. A transformation may be used for system generation or system verification. For the former, a transformation creates a needed and 'permanent' modification on a design. For the latter, a transformation is purely temporary so that verification may proceed more easily under the transformed specification. Naturally, for both purposes, it is important that a transformation should not destroy or create those properties under investigation.

Some relevant issues concerning a property-preserving transformation are discussed below:

A. *The kinds of properties to be preserved* – Most works concentrated on the basic properties, such as liveness, boundedness and reversibility. Recently, the need of considering other properties arises for two reasons: Firstly, the other properties are also important features of some systems. Secondly, the characterization approach for verification has become more popular. A characterization is a relationship between several properties. When it is not known how to directly prove a certain property involved in a characterization, one tries to verify its other properties instead. Then, preservation of the other properties will imply preservation of the intended one. Many new characterizations concerning place-reducing transformations are derived throughout this paper.

B. *Forward preservation and backward preservation* – A transformation may preserve a property in two directions. Forward (resp., backward) preservation guarantees that a property of the original (resp., transformed) system is satisfied by the transformed (resp., original) system while being unable to guard against the creation of new and probably undesired properties in it. Backward preservation is particularly useful if a transformation serves purely verification purposes.

Although highly desirable, it is uncommon that a transformation can preserve a property in both directions. In fact, even for one-way preservation, additional conditions often have to be imposed.

C. *Preservation of multiple properties* – Very often, a system has several desirable properties. Then, for both system generation and verification, it is a challenge to discover a *single* transformation that can preserve all of them. Recent research aims at exploring for transformations which can preserve as many properties as possible [CHE (2002); HJC (2003); MAK (2001)].

2.3 Related Work and Discussion

This book concerns about property-preservation transformations for verification. Transformations on Petri nets may be roughly classified into three groups, namely refinement, reduction, and composition. A review on each group of transformations and a comparison with our approach are introduced in this section.

2.3.1 *Reviews on property-preserving refinements for Petri nets*

Valette [VAL (1979)] first proposed a method for stepwise refinement of transitions for synthesizing Petri nets. In the method, the refined transition is replaced with a well-formed block that is a safe Petri net with a single entry place and a single exit place. Properties of boundedness, safeness and liveness are preserved after the refinement with some conditional restrictions. Valette's-method can be considered as a generation of the results of [BA (1971)] which developed a theory for modeling the control structure of digital systems.

[MK (1980)] proposed six kinds of place and transition refinements for marked graphs that preserve boundedness, safeness and liveness. It also proposed the reductions that are exactly the converse of these refinements. The six refinement methods are briefly described as follows: Refining a place to a place-bordered path; Refining a place to a parallel structure; Refining a place by adding a transition to it such that the added transition and the refined place form a self-loop; Refinement for transitions is similar.

Suzuki and Murata [SM (1983)] generalized Valette's work and defined the concept of k-well-behaved. In [SM (1983)], a transition which is not $k +$ 1-enable can be replaced by a k-well-behaved Petri net while preserving the properties of boundedness, safeness and liveness. [SM (1983)] also proposed refinement of places by splitting the refined place p_0 into p_i-t_0-p_o structure first and refine transition t_0 later. Properties of boundedness, safeness and liveness can be preserved with place refinement.

[VOG (1987)] proposed a transition refinement technique that allows the refinement net be a *daughter-net*, a net with distributed *input places* and distributed *output places*. The refined transition t is replaced by the daughter net D such that the preset of t are identified with the input places of D and the postset of t are identified with the output places of D. This technique is more general than the above refinement methods but it required that the number of the identified places should be the same. Otherwise the refinement is impossible. Preservation of liveness and boundedness are considered.

[BC (1990); MUL (1985)] proposed a technique for the refinement of several transitions by one net. In the method, the refined transitions are chosen from each component net and refined with a common net. The common refinement net is in fact a block [VAL (1979)]. Preservation of liveness is considered.

Aalst [AAL (1997, 1998)] proposed a method for transition refinement. In this method, the refinement net is a WF-net (workflow net). A WF-net has the same net structure as our process in this thesis. In [AAL (1997)] and [AAL (1998)], liveness and boundedness of extending WF-net (associated process) can be preserved by proposing some transition refinement methods introduced in [MK (1980)].

Besides the normal place and transition refinement introduced above, action refinement is also considered in the literature. In the action refinement, all transitions are labeled by actions, and when we refine the net we replace every transition with some given label by a copy of the same refinement net. Instead, the work for action refinement pays attention on the preservation of semantic equivalence. E.g., [HEN (1987); AH (1988); NEL (1989); ACE (1990)] are first attempt to add an action refinement operator to process algebras like CCS; Action refinement of event structures in the sense of [NPW (1981)] and transition systems are studied in [GG (1989); GW (1989); GLA (1990); GG (1990); VOG (1990b)]; A split operator for Petri nets that splits every action into a sequence of two actions was introduced in [GV (1987)], while general action refinement for Petri nets are studied in [DEV (1988); BDK (1989); GG (1990); VOG (1990a); HCM (2004); JIA (2008)]. What makes the issue of action refinement so interesting is its connection to partial order semantics. [PRA (1986); CDP (1987)] suggested that partial order semantics is useful when considering action refinement. It is argued in [CDP (1987)] that partial order semantics is congruent with refinement, indeed.

Differences between our technique and those in the literature:

- *Our technique is applicable within the context of component-based system design whereas those in the literature are not* – In the process of component-based system design, one of the difficulties is to select a structure to represent the components and to define a set of operators on them so that every component created by any of the operators will be of the same structure. Under those refinement techniques in the literature, the refined Petri net may have a different structure from that of the original Petri net. In other words, after a refinement, the created Petri net may belong to a different class and hence those operators may not be applicable to them any more. That is, this does not fulfill the requirement of a systematic method for component-based system design. In our technique, a refinement will produce a Petri net of the same structure. It requires the refinement net (i.e., the net to be refined to) be a Petri net process. After replacing a single place or transition with another Petri net process, the resulting net keeps the same structure as the original one. This satisfies the basic requirement for component-based system design.
- Our refinements preserve nineteen system properties whereas those in the literature preserve up to three properties at most. Many do not even consider the issue of property preservation – As described in the papers [VAL (1979); BA (1971); MK (1980); AAL (1997, 1998); VOG (1987)], only properties liveness, safeness and boundedness are considered. For those papers concerning action refinement, property preservation was not considered at all.

- Our refinement net has a more general structure than those in the literature. Hence, our technique enhances and widens the application areas – Our refinement net is a Petri net process whereas the refinement nets in the literature are either a well-behaved block [VAL (1979); SM (1983)] or a WF-net [AAL (1997, 1998)] or a daughter net [VOG (1987)]. Although Petri net process is also a special Petri net, it is a generalization of k-well behaved block, daughter net and WF-net from the following points: a) In order for a refinement to work properly, the refinement net should be non-re-enterable. That is, once started, the refinement net cannot be initiated again until the current cycle has terminated. In the literature, this is ensured by requiring the refinement net to be well-formed or k-well-behaved and the refined transition not to be 2-enabled or $(k + 1)$-enabled in the refined net. In our model, this is ensured by requiring the Petri net process initiate and terminate properly and the refined transition not be 2-enabled. b) Daughter-net can be easily converted to the case of single entry (resp., exit) by creating a super entry (resp., exit) place and controlling the firing of its output (resp., input) places. The resulting structure is in fact a Petri net process. c) A WF-net is a special Petri net process such that the static marking is zero. As will be mentioned in Chapter 3, static marking is applicable for handling resource sharing problems. Hence, WF-net is non-applicable for handling this problem while our Petri net process can.

2.3.2 Reviews on property-preserving reductions for Petri nets

Research in reduction methods began with simple pattern modifications on Petri nets. Berthelot [BER (1985)] extended the work of reduction and verification of parallel programs proposed by Lipon [LIP (1981)], Kwong [KWO (1977)], and Kowalk and Valk [KV (1979)]. He presented several set of transformation rules for reduction and refinement of Petri nets. In his approach [BER (1985)], the reduction is done by eliminating arcs and by replacing subnets with places and transitions. The rules are classified into three categories: *place transformation*; *fusion of transitions* and *addition of nets*. In each category, there are some sub-rules. For example, place transformation includes three rules: simplification of redundant places, fusion of doubled places and fusion of equivalent places. Fusion of transition category is divided into three rules: postfusion, prefusion and lateral fusion. Addition of nets comprises four rules: addition of a derivative net, alternation of laterally fusible transitions, identical transition regulation and addition of non-restricting nets. In [BER (1985)], each transformation rule may be used if some application conditions are satisfied. Most application conditions are based on the structure of the net, but some depend on its behavior. It is easier to verify conditions on static structure than dynamic behavior but the former tends to require more detail.

Berthelot generalized the work of [BER (1985)] in another paper [BER (1986)]. He considered the first two categories in [BER (1985)] discussed above. In [BER (1986)], he used net's dynamic behavior as much as possible to describe the application conditions of the rules. These rules are general in the sense that they can include the structurally defined reduction rules as subsets. Berthelot [BER (1986)] also reviewed and organized a number of decomposition schemes to split a system into subsystems which can be analyzed separately.

Lee and Favrel [LF (1985)] developed a hierarchical reduction method for Petri nets. The method is based on a set of reduction rules and reduces the reducible subnet in a Petri net into the macro-nodes without changing the properties of liveness, boundedness and proper termination. In [LF (1985)], the application conditions are depended solely on the structure of the net and not directly on the dynamic behavior. In a later paper [LF (1987)], they extended the work of [LF (1985)] such that the reduction rules can be applied for generalized Petri nets, i.e., Petri nets with multiple arcs.

Desel and Esparza [DES (1990); ESP (1994)] considered the reduction rules in free choice (FC) nets. [DES (1990)] presented four kinds of reduction rules: P-reduction, T-reduction, R-reduction and A-reduction. The four rules are purely local and preserve the behavior properties in both directions. [DES (1990)] showed that a live and safe FC net without frozen tokens can be reduced either to a live and safe marked graph or a state machine. [ESP (1994)] provided a kit of reduction rules that preserve liveness and boundedness. The rules make it possible to reduce all and only live and bounded FC net to a circuit containing one place and one transition. Both [DES (1990)] and [ESP (1994)] specified that the reduction rules in their work provide not only a verification but also a synthesis technique. They can "reverse" the reduction rules to obtain the refinement methods for synthesizing. A well-known recent result is the preservation of well-formedness and Commoner's property under the merge of places within a free-choice net [ES (1990b, 1991b); DE (1995)] or an asymmetric-choice net [JCL (2002, 2004)].

Proth and Xie proposed seven transformations in their book [PX (1996)]. They are simplification of redundant places, merging twin places, merging equivalence places, simplification of implicit places, post-merging, pre-merging and side-merging. The seven rules preserve the following properties: boundedness, liveness, the system being deadlock free, reversibility, the existence of home state, structural liveness, structural boundedness, and the system being conservative, repetitive and consistent.

Sloan and Buy [SB (1996)] and Wang and Deng [WD (1998)] considered the reduction rules applied for Time Petri net (TPN). In [SB (1996)], they extend several rules provided by Berthelot [BER (1985, 1986)] for the reduction of ordinary Petri nets to work with time Petri nets. They introduced a notion of equivalence among TPNs and proved that the reduction rules yield equivalent nets. This notion of equivalence guarantees that crucial timing and concurrency properties are preserved. [WD (1998)] introduced five component-level reduction rules for TPN. Each of the rules simplifies the TPN while preserving the net's external observable timing properties.

Some researchers reduce Petri nets by applying place, transition or subnet merging. These operators [DE (1995); ESP (1994); JCL (2004); AC (1978); SM (1990); NV (1985); HJC (2005); JHC (2005, 2008)] are mainly applied for handling resource-sharing problems or synthesizing a complex system in system design. They will be described in more detail in the next chapter.

Differences between our technique and those in the literature:

- *Our technique is applicable within the context of component-based system design whereas those in the literature are not* – As mentioned above, in component-based system design, after applying a set of operators, the resulting component should belong to the same class as the original one. In our technique, a reduction applied on a Petri net will produce a Petri net of the same structure. Our technique achieves this purpose by preserving some structural properties of the Petri net. For example, it can preserve properties such as state machine, marked graph, free choice net, asymmetric choice net, Petri net process, etc.
- *Our reductions both forwardly and backwardly preserve nineteen properties whereas those in the literature did not consider the issues extensively. Many papers did not even touch the issue of property preservation* – For example, only Desel and Esparza [DES (1990); ESP (1994)] considered the two-directional preservation problem and their results can only be applied to free choice nets. This greatly limits the applica-

tion areas. Most papers in the literature just discussed forward preservation of system properties [LF (1985, 1987); PX (1996); SB (1996); WD (1998)]. Most of the papers discussed only a few important properties for preservations. For example, only liveness, bounededness and proper termination are considered in some papers [LF (1985, 1987); ESP (1994)]. In the earlier work [BER (1985, 1986)], property preservation is not discussed at all.

- *Our technique has some features in terms of the following four specification or verification problems:*

  a) *Modeling the system with general Petri nets* – The type of a Petri net used for modeling the system under design not only determines its scope of applications but also affects the process of verification. For example, in manufacturing engineering, most of the systems are specified as finite state machines or marked graphs [ZHO (1996)]. In use-case-based software system design, the use cases may be specified as case nets [CHE (2002)]. For each paper reviewed above, the reduction rules are applicable provided that some application conditions are satisfied. The application condition may either dependent on the structure [BER (1985, 1986); LF (1985, 1987); PX (1996); DES (1990); ESP (1994)] or on the dynamic behavior [BER (1985, 1986); SB (1996); WD (1998)] of the net. Our technique investigates general Petri nets. Hence, it can be applied in more general application areas.

  b) *Representing the resources and abstracted parts with subnets* – In the literature, a resource is uniquely represented as a place. Zhou's exclusions [ZHO (1996); ZV (1999)] and Chu's augmented marked graphs [CX (1997)] are formal descriptions of such representations. Also, it is assumed that a resource is switched from one user to another without any intermediate modification. In our technique, it is assumed that the given system is composed of connected or disconnected parts. (For the sake of flexibility, a part in this chapter has no fixed definition.) Each resource is originally represented by a set of places (called *resource-places* hereafter), one in each of the parts it is involved in. Also, a resource may go through some intermediate processing when switching from one user to another. This implies that the resource-places may form a connected subnet whose transitions represent the intermediate processes. As for abstraction, sequential systems are represented as directed paths and non-sequential systems having multiple entries and exits as state-machine subnets.

c) *Formulating resource sharing and subsystem abstraction as subnet-reducing transformations* – In all the models appearing in the literature, a resource is represented uniquely as a place. This chapter takes a synthesis approach. When a resource is shared with several parts or a part is abstracted into a single function, its representation (i.e., a subnet) will be merged into a single place or a single transition. Formally, this is a transformation that reduces a subnet to a single place or transition. Three transformations are formulated according to the structure of the shared resources or abstracted parts.

d) *Eliminating the requirement of system verification by a property-preservation technique* – In our approach, to verify a system is to check whether it possesses certain properties or not. For example, the deadlock and overflow errors are investigated as the liveness and boundedness properties of the system's Petri net representation. In the literature dealing purely with resource-sharing or system abstraction, rarely any 'specific and systematic' methods for verification have been reported. Most of the time, just general techniques are used. By viewing these two problems as transformations, this chapter proposes a property-preserving approach

2.3.3   *Reviews on property-preserving compositions for Petri nets*

As mentioned above, many researchers proposed the reduction of Petri net by applying place or transition merging operators. For example, 1-way merge of a set of non-neighboring places was studied in [AC (1978)]. P-invariants are preserved under such merging operators. [SM (1990)] proposed the constraints for the preservation of liveness. Narahari et al [NV (1985)] investigated the following properties of the merged system: absence of deadlocks, boundedness and conservativeness. The theorems for P-invariants allows more than one set of places to be merged at each synthesis step, and allows simplified computation of the P-invariants of the merged net when the P-invariants of the subnets are known. Another well-known result concerning this method is the preservation of well-formedness and Commoner's property under the merge of places within an FC net [DE (1995); ESP (1994)] or an asymmetric choice net [JCL (2004)]. By the similar way, Cheung [CHE (2002)] considered the problem of merging the places of two marked graphs. He has proposed a condition called cycle-inclusion property for checking the liveness, boundedness and reversibility of the integrated net. This condition was proved to be equivalent to the ST-property. Recently, Huang, et al. [HJC (2003)] extended this approach to augmented marked graphs and provided a different method for checking these properties.

Krogh and Beck [BEC (1985); KB (1986)] developed a method for synthesizing live and safe Petri nets. The methods start with a collection of simple elementary circuits that correspond to the basic activity cycles in the system with the assumption that the system is designed to perform a repetitive operation. Petri nets are then constructed from the circuits by merging the coincident place-bordered paths or the transition-bordered paths. The Petri net obtained in this way is live and safe with respect to any initial marking for which there is exactly one token in each of the P-invariants of the system. Koh and DiCesare [KD (1991)] extended the work of Krogh and Beck to bounded and generalized Petri nets.

Souissi and Memmi [SM (1990)] proposed a method for sharing a subnet called communication medium from two dependent bounded and live ordinary Petri net components. The method consists of three parts: composition by sharing places, structural composition by sequential processes and transition merging and composition by sharing a well-formed block [VAL (1979)]. This method preserves liveness and boundedness.

Jeng [JD (1995); JEN (1997)] proposed a method for the merging of transitions and transition subnets from two Resource Control Nets (RCN). A RCN is a strongly connected state machine that has one initially marked resource place and other initially unmarked operation places. In the context of resource sharing system, a RCN models a resource control process. Interaction between two RCNs is represented as the common transitions and common transition subnets. These shared transition or transition subnets represent the common activity or synchronization of the resource control processes. Petri net constructed by this way are conservative and bounded. Liveness has to be checked after each composition operator was applied.

Proth et al [PWX (1997)] proposed a method for linking a set of Controllable-Output (CO) nets by inserting some so called interface places. A CO net is an ordinary Petri net where the extremity nodes are all transitions. It is live, reversible and unbounded for any initial marking. Petri net constructed in this way preserve liveness, consistency and reversibility. Boundedness can also be considered if controlled adequately. CO net is mainly used for modeling manufacturing systems.

Kindler [KIN (1997)] introduced a method for integrating components by merging the corresponding interface places of each component. He presented a compositional semantics for components. The main feature of the compositional semantics is that composition of components corresponds to conjunction. This feature makes the compositional semantics applicable in combination with a temporal logic, which then allows to reason about systems in a compositional way.

Differences between our technique and those in the literature:

- *Our technique is applicable within the context of component-based system design while those in the literature are not* – Similar to refinement and reduction techniques discussed above, our place-merging operators can also guarantee the resulting component be of the same class as the original one.
- *Our place-merging approach is applicable to general Petri nets and nineteen system properties are considered whereas those in the literature are applicable for special Petri nets and not so many properties are considered for preservation* – As described in the above review, most papers [DE (1995); ESP (1994); JCL (2004); CHE (2002); HJC (2003); BEC (1985); KB (1986); JD (1995); JEN (1997); PWX (1997)] consider the operators in special Petri nets. The properties considered are limited to liveness [BEC (1985); KB (1986); SM (1990); PWX (1997)], boundedness [NV (1985); SM (1990); JEN (1997)], reversibility [PWX (1997)], conservativeness [NV (1985); JD (1995); JEN (1997)], consistency [PWX (1997)] and P-invariants [AC (1978)].
- *Our technique uses both the pre-condition and post-condition strategies for verifying the resulting systems whereas those in the literature did not* – For some special nets such as augmented marked graph, both the pre-condition and post-condition strategies can be used for verifying the resulting systems. At the same time, they can be applied iteratively since the resulting net is still an AMG. In some papers reviewed, either the post-conditions [BEC (1985); CHE (2002); KB (1986); JD (1995); JEN (1997)] or pre-conditions [DE (1995); ESP (1994); JCL (2004); AC (1978); SM (1990); NV (1985); SM (1990); PWX (1997)] are proposed for verifying the desirable properties. The operators cannot be iteratively applied because the structural properties cannot be preserved properly.
- Our technique includes conditions for property preservation in some well-known special Petri nets whereas those in the literature consider

their own specifically-defined nets – In our technique, conditions are proposed for property preservation in the Petri nets such as AMG, FC, AC, etc. such that the designer has more choices for modeling the systems. In the literature, subnet-merging operators are applicable to the special nets such as Resource Control Nets [JD (1995); JEN (1997)] and Controllable Output nets [PWX (1997)]. These special nets cannot be applied in a large application area because they are not known by the designer and have no enough well-defined properties.

- Our technique is applicable not only for synthesis but also for handling resource-sharing problems in system design whereas those in the literature are mainly applied to synthesizing systems – In the literature, only [JD (1995); JEN (1997)] considered resource sharing problem while most papers focus on synthesizing systems by applying the operators. Handling resource-sharing problem is not their main concern.

Chapter 3

Petri Net Processes and Extension Operators

The core of this book is the Property-Preserving Petri Net Process Algebra (PPPA). An algebra consists of a set of operators to be applied over a set of elements. This chapter first describes in detail its elements - Petri net processes (PNP) and then the first kind of operators: Extension operators.

3.1 The Structure of a Petri Net Process

Petri net processes (PNP), the elements of PPPA, form a closed set. That is, the result of applying any of the operators of PPPA on a PNP or two PNPs is another PNP.

Definition 3.1. (Petri net process PNP, associated process) (Figure 3.1) A Petri net process (or just process when there is no confusion in the context) is a 3-tuple $B = (N, p_e, p_x)$, where

- $N = (P, T, F)$ is an ordinary, connected net;
- p_e (called the entry place) is the only place $p \in P$ such that $^\bullet p = \phi$;
- p_x (called the exit place) is the only place $p \in P$ such that $p^\bullet = \phi$.

The associated process B_a of B is created from B by adding to N an associated transition t_a and two associated arcs (p_x, t_a) and (t_a, p_e). That is, B_a is the 3-tuple (N_a, p_e, p_x), where $N_a = (P, T \cup \{t_a\}, F \cup \{(p_x, t_a), (t_a, p_e)\})$.

The introduction of the associated process B_a is mainly for the convenience in describing many properties of B. In fact, though B is connected by definition, B is never strongly connected because there is no path from p_x to p_e. Hence, B can never satisfy those properties that require strong connectedness, such as consistency, SM-coverability, etc. On the other hand,

35

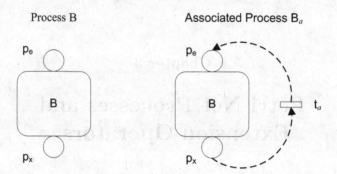

Fig. 3.1   A Petri net process B and its associated process B_a.

in many applications, it is sufficient to require just B_a, rather than B, to satisfy certain properties. In brief, while B is used for the actual modeling of a real-life process, B_a is used to study some of the properties of B.

Notation (to be used throughout the book):

- If B is a PNP, B_a always represents its associated process.
- If V is the incidence matrix of B, the incidence matrix V_a of B_a is given below,where $p_e - p_x$ is a multi-set representation of the $|P|$-vector $(1, 0, \cdots, 0, -1)$.

$$
\begin{array}{cc}
 T & t_a \\
P\ (V & p_e - p_x)
\end{array}
$$

- When nor explicitly stating otherwise, p_e and p_x are supposed to be in the top row and bottom row of V, respectively.

3.2   Structural Properties of Petri Net Processes

This section presents some structural properties of a PNP B and the relationships of these properties between B and B_a.

Definition 3.2. (structural 'almost' properties of a Petri net process)
Let B_a be the associated process of Petri net process B. B is said to be almost consistent (resp., almost repetitive, almost a marked graph (MG), almost SM-coverable) if B_a is consistent (resp., repetitive, an MG, SM-coverable).

The cyclomatic number Z defined below had its origin from graph theory. In software engineering, it is a commonly used method for measuring the software complexity of programs whose logical flow can be represented as a flowgraph (also called program graph). Z also denotes the maximum number of independent entry-to-exit paths that can cover the entire flowgraph. The following formula is for the cyclomatic flowgraphs that have a single connected component.

Definition 3.3. (cyclomatic complexity [PRE (1997)])
Let $G = (U, E)$ be a connected flowgraph that has q sink nodes. The cyclomatic complexity of G is defined as $Z(G) = |E| - |U| + q + 1$, where U is the set of nodes and E is the set of edges.

Zhu and Cheung [CZ (1994); CRM (1994)] extended Definition 3.3 from flowgraphs to Petri nets processes. The following theorem describes the relationships of many properties between B and B_a. Some of them are quite trivial. They are listed here for completeness.

Theorem 3.1. *(preservation of structural properties between Process B and its associated process B_a)*
Let B be a marked process and B_a be its associated process. Then, the following characterizations hold:

1) *B is an SM iff B_a is an SM.*
2) *B is almost an MG iff B_a is an MG.*
3) *B is an FC net iff B_a is an FC.*
4) *B is an AC net iff B_a is an AC net.*
5) *B is conservative if B_a is conservative. If B is conservative and in the incidence matrix $V_a = V_a(P, T \cup \{t_a\})$ of B_a, the column $V_a(P, t_a)$ is a linear combination of the columns of $V_a(P, T)$, then B_a is conservative.*
6) *B is structurally bounded if B_a is structurally bounded. If B is structurally bounded and in the incidence matrix $V_a = V_a(P, T \cup \{t_a\})$ of B_a, the column $V_a(P, t_a)$ is a linear combination of the columns of $V_a(P, T)$, then B_a is structurally bounded.*
7) *B is almost consistent iff B_a is consistent.*
8) *B is almost repetitive iff B_a is repetitive.*
9) *In general, $Rank(B_a) - 1 \leq Rank(B) \leq Rank(B_a)$. In particular, $Rank(B) = Rank(B_a)$ if, in the incidence matrix $V_a = V_a(P, T \cup \{t_a\})$ of B_a, the column $V_a(P, t_a)$ is a linear combination of the columns of $V_a(P, T)$.*
10) *$|C(B)| = |C(B_a)|$.*

11) B satisfies the RC-property if B_a satisfies the RC-property. B_a satisfies the RC-property if B satisfies the RC-property and $V_a(P, t_a)$ is a linear combination of some columns of $V_a(P, T)$.

12) B almost has a minimal SM-cover iff B_a has a minimal SM-cover.

13) A siphon of B_a is also a siphon of B. A siphon D of B is also a siphon of B_a if $(p_e \notin D)$ or $(p_e \in D$ and $p_x \in D)$.

14) A trap of B_a is also a trap of B. A trap S of B is also a trap of B_a if $(p_x \notin S)$ or $(p_e \in S$ and $p_x \in S)$.

15) $Z(B) = Z(B_a)$.

Proof.

1-4) According to the structures of B and B_a, adding or deleting transition t_a and the two arcs does not make B_a or B violate the definitions of SM, MG, FC and AC nets.

5) Since B_a is conservative, by Characterization 2.1, $\exists \alpha = (x_1, \cdots, x_{|P|}) \geq 1$ such that $\alpha V_a = 0$. Hence, $\alpha V = (\alpha V_a[P,T]) = 0$. It follows from Characterization 2.1 that B is conservative. If B is conservative, then $\exists \alpha = (x_1, \cdots, x_{|P|}) \geq 1$ such that $\alpha V = 0$. Furthermore, $\alpha V_a[P, t_a] = 0$ if in the incidence matrix $V_a = V_a(P, T \cup \{t_a\})$ of B_a, the column $V_a(P, t_a)$ is a linear combination of the columns of $V_a(P, T)$. Hence, B_a is conservative.

6) Since B_a is structurally bounded, by Characterization 2.2, $\exists \alpha = (x_1, \cdots, x_{|P|}) \geq 1$ such that $\alpha V_a \leq 0$. Then, $\alpha V = (\alpha V_a[P,T]) \leq 0$. It follows that B is structurally bounded. If B is structurally bounded, then $\exists \alpha = (x_1, \cdots, x_{|P|}) \geq 1$ such that $\alpha V \leq 0$. Furthermore, $\alpha V_a[P, t_a] \leq 0$ if in the incidence matrix $V_a = V_a(P, T \cup \{t_a\})$ of B_a, the column $V_a(P, t_a)$ is a linear combination of the columns of $V_a(P, T)$. Hence, B_a is structurally bounded.

7-8) These follow from definitions.

9) The first part is obvious. If $V_a(P, t_a)$ is a linear combination of the columns of $V_a(P, T)$, then $V_a(P, t_a)$ can be reduced to zero by Gaussian eliminations. It follows that $Rank(B_a) = Rank(B)$.

10) In B_a, since the only new transition t_a is absorbed into the cluster $[p_x]$, no cluster is created or destroyed. Hence, $|C(B_a)| = |C(B)|$.

11) Suppose B_a satisfies the RC-property. Then, by 10), $Rank(B_a) = |C(B)| - 1 \leq |T|$, $Rank(B_a) \leq |T| - 1$. This implies that the column $V_a(P, t_a)$ is a linear combination of some columns of $V_a(P, T)$. Then,

by 9), $Rank(B_a) = Rank(B)$. It follows that $Rank(B) = |C(B)| - 1$. Next, suppose $Rank(B) = |C(B)| - 1$. It follows from 9) and 10) that $Rank(B_a) = |C(B_a)| - 1$.

12) This follows from definition.

13) $\forall D \subseteq P$, we have: $(^\bullet D \text{ in } B) = (^\bullet D \text{ in } B_a) - \{t_a\}$ and $(D^\bullet \text{ in } B) = (D^\bullet \text{ in } B_a) - \{t_a\}$. Hence, if D is a siphon of B_a, then $(^\bullet D \text{ in } B) = (^\bullet D \text{ in } B_a) - \{t_a\} \subseteq (D^\bullet \text{ in } B_a) - \{t_a\} = (D^\bullet \text{ in } B)$. Conversely, if D is a siphon of B, then $(^\bullet D \text{ in } B_a) - \{t_a\} = (^\bullet D \text{ in } B) \subseteq (D^\bullet \text{ in } B) = (D^\bullet \text{ in } B_a) - \{t_a\}$. If $p_e \notin D$, then $(t_a \notin {}^\bullet D)$ in B_a and $(^\bullet D \text{ in } B_a) = (^\bullet D \text{ in } B_a) - \{t_a\} \subseteq (D^\bullet \text{ in } B_a) - \{t_a\} \subseteq (D^\bullet \text{ in } B_a)$. If $p_e \in D$ and $p_x \in D$, then $(^\bullet D \text{ in } B_a) = (^\bullet D \text{ in } B) \cup \{t_a\} \subseteq (D^\bullet \text{ in } B) \cup \{t_a\} = (D^\bullet \text{ in } B_a)$. For both cases, D is a siphon of B_a.

14) $\forall S \subseteq P$, we have: $(^\bullet S \text{ in } B) = (^\bullet S \text{ in } B_a) - \{t_a\}$ and $(S^\bullet \text{ in } B) = (S^\bullet \text{ in } B_a) - \{t_a\}$. Hence, if S is a trap of B_a, then $(S^\bullet \text{ in } B) = (S^\bullet \text{ in } B_a) - \{t_a\} \subseteq (^\bullet S \text{ in } B_a) - \{t_a\} = (^\bullet S \text{ in } B)$. If S is a trap of B, then $(S^\bullet \text{ in } B_a) - \{t_a\} = (S^\bullet \text{ in } B) \subseteq (^\bullet S \text{ in } B) = (^\bullet S \text{ in } B_a) - \{t_a\}$. If $p_x \notin S$, then $t_a \notin S^\bullet$ in B_a and $(S^\bullet \text{ in } B_a) = (S^\bullet \text{ in } B_a) - \{t_a\} \subseteq (^\bullet S \text{ in } B_a) - \{t_a\} \subseteq (^\bullet S \text{ in } B_a)$. If $p_e \in S$ and $p_x \in S$, then $(S^\bullet \text{ in } B_a) = (S^\bullet \text{ in } B) \cup \{t_a\} \subseteq (^\bullet S \text{ in } B) \cup \{t_a\} = (^\bullet S \text{ in } B_a)$. For both cases, S is a trap of B_a.

15) Since B is a flowgraph with one sink and B_a is a flowgraph without any sink nodes, we have: $Z(B_a) = (|F| + 2) - (|P \cup T| + 1) + 0 + 1 = |F| - |P \cup T| + 1 + 1 = Z(B)$.

$\qquad\qquad\qquad\qquad\qquad\qquad\qquad\qquad\qquad\qquad\qquad\qquad\qquad\quad$ $\square$

3.3 Behavioral Properties of Petri Net Processes

This section presents some properties of a PNP that depend on its markings.

Definition 3.4. (special markings and states of a Petri net process)
For a Petri net process $B = (N, p_e, p_x)$, three special markings can be defined: A static marking with the form $M_s = (0, M_c, 0)$, where M_c is a control sub-marking defined over the set of places $P - \{p_e, p_x\}$, an entry marking of the form $M_e = (1, M_c, 0)$ and an exit marking of the form $M_x = (0, M_c, 1)$.

Before execution starts, a process B is supposed to be in the static state M_s. It can start execution at the entry state M_e when a token has been entered into p_e. The exit state M_x is used for some other purposes to be described later.

Definition 3.5. (static marking and firing rule of a marked Petri net process)

A Petri net process $B = (N, p_e, p_x)$ is said to be marked if it is associated with a static marking M_s and an entry marking M_e that satisfy the following two rules:

- Rule for Deadness of Static Marking: Before firing starts, B is marked by a dead $M_s = (0, M_c, 0)$. That is, in (B, M_s), $\forall t \in T$ cannot be fired at M_s.
- Rule for Proper Initiation: Firing can start only at the entry state M_e. That is, the only possible initial marking of B has the form M_e.

Definition 3.6. (behavioral 'almost' properties of a Petri net process)

(B, M_e) is said to be almost live (resp., almost bounded, almost reversible) if (B_a, M_e) is live (resp., bounded, reversible).

The following theorem states the relationship of boundedness between a marked process and its associated process.

Definition 3.7. (proper termination and soundness)

Let B be a process marked with static marking $M_s = (0, M_c, 0)$. Let $M_e = (1, M_c, 0)$ be the entry marking and $M_x = (0, M_c, 1)$ be the exit marking. We say that (B, M_e) terminates properly if $\forall M \in R(B, M_e)$: $M(p_x) \geq 1 \Rightarrow M = M_x.(B, M_e)$ is said to be sound iff:

1) (B, M_e) is almost reversible (i.e., $M_e[B,^* \rangle M \Rightarrow M[B,^* \rangle M_x)$.
2) (B, M_e) terminates properly.
3) (B, M_e) has no dead transitions.(i.e., $\forall t \in T$, $\exists M$ such that $M_e[B,^* \rangle M[B, t \rangle$.)

Note: When it is clear from the context what the entry marking M_e is, we may say 'B terminates properly' and 'B is sound' instead of '(B, M_e) terminates properly' and '(B, M_e) is sound', respectively.

Discussion on Definitions 3.4 - 3.7:

1) Uniqueness of the entry place and exit place: By definition, a PNP has only one entry place and one exit place. This uniqueness assumption is mainly for the consistency in modeling and convenience in creating composite processes under the various operators. For modeling real-life problems, the case of multiple entries (resp., exits) can be easily converted to the case of single entry (resp., exit) by creating a super

entry (resp., exit) place and controlling the firing of its output (resp., input) places.

2) Role of the exit place p_x: It should be pointed out that the exit place p_x of a PNP plays a slightly different role from its counterparts in other process-based languages such as LOTOS [LFH (1992)]. These languages usually have two kinds of terminal points, EXITs and STOPs, representing successful and unsuccessful terminations, respectively. In a Petri net model like PPPA, p_x is simply a sink place, indicating the location where the control flow may leave the process after one cycle of executions. In Petri nets, there is no such thing of unsuccessful firing of a transition. During actual modeling, for example, a STOP may be designed as a deadlock.

3) M_c represents a token distribution assigned to the set of places $P - \{p_e, p_x\}$ before execution of B starts. Those places having tokens serve certain 'controlling' purposes. For example, they may represent some system resources that are available before B starts its execution. The association of a static marking $M_s = (0, M_c, 0)$ with B, where $M_c \neq 0$, is a special feature of PPPA. It greatly enlarges the scope of application of PPPA. Other Petri-net-based process models, including those proposed by Cheung [CRM (1994)] and Aalst [AAL (1997, 2000)], are special cases, where $M_c = 0$, of PPPA. This condition imposes a severe constraint on the application of the model to many systems of which resource sharing is an important feature. They include workflow systems, multi-agents systems and job-shop scheduling systems (introduced in Chapter 9 and Chapter 10 of this book).

4) The rule for Deadness of Static Marking is a constraint on the static marking M_s. Together with the Rule for Proper Initiation, it implies two facts: a) A process can start execution only after a token has been deposited into its entry place p_e. b) Without this deposit of tokens, (B, M_s) cannot 'self-start'. This reflects the realistic requirement that a process cannot start by itself. In order to start, it must be called by another process or by itself (i.e., in recursion).

5) The Rule for Proper Initiation is imposed as part of the firing rules for a PNP. It does not guarantee that (B, M_e) can always be initiated. It only says that, if B can ever start firing, it must start at M_e and not at any other marking.

6) In general, proper termination by itself does not guarantee that a process can always terminate. It just requires a process to be at the exit state M_x whenever a token has been deposited into p_x. Proper ter-

mination models the well-known 'memoryless' property of a software process that it should return to its initial 'ready' state after having completed a cycle of execution.

7) Together with the Rule for Deadness of Static Marking, proper termination guarantees that no transitions can be fired when p_x gets a token. This follows from the fact, whenever p_x gets a token, the system reaches M_x which is obviously a dead marking because M_s is dead.

8) The rule for Deadness of static marking, together with the properties of proper initiation and proper termination, guarantees that a PNP is non-reenterable. This means that, once having been initiated, a PNP cannot be initiated again until its previous execution cycle has been completed. In general, to avoid mixing two independent execution cycles of a PNP, one either has to use colored Petri nets or control the procedure of entering into the process. We adopt the second approach.

Theorem 3.2. *Let (B, M_e) be a marked process and (B_a, M_e) be its associated process, where $M_e = (1, M_c, 0)$ is the entry marking of both B and B_a. Then, (B, M_e) is bounded if (B_a, M_e) is bounded. In general, (B_a, M_e) may be unbounded even if (B, M_e) is bounded. In particular, (B_a, M_e) is bounded if (B, M_e) is bounded and terminates properly.*

Proof. If (B_a, M_e) is bounded, it is obvious that (B, M_e) is also bounded. Suppose (B, M_e) is bounded. If (B, M_e) terminates properly, then every new cycle of execution of (B_a, M_e) is the same as (B, M_e) and (B_a, M_e) is therefore bounded. Otherwise, it is possible that there exists a firing sequence σ such that $M_e[B_a, \sigma\rangle M$ and $M \geq M_e$. Then, repeated executions of σ will make (B_a, M_e) unbounded. □

Lemma 3.1. *Let (B, M_e) be a marked process and (B_a, M_e) be its associated process. If (B, M_e) terminates properly, then $M_e[B_a, {}^*\rangle M$ implies $M_e[B, {}^*\rangle M$.*

Proof. Suppose $M_e[B_a, \sigma_a\rangle M.\sigma_a$ can be split into the form $\sigma'\sigma$, where σ' is either null or has t_a as its last transition and σ does not contain any t_a. Due to proper termination, we have $M_e[B_a, \sigma'\rangle M_e[B, \sigma\rangle M$. □

The following corollary characterizes the property soundness of a process B.

Corollary 3.1. *Let (B, M_e) be a marked process and (B_a, M_e) be its associated process, where $M_e = (1, M_c, 0)$ is the entry marking of both B and*

B_a. Then, (B_a, M_e) *is live, bounded and reversible if* (B, M_e) *is sound. Conversely,* (B, M_e) *is sound if* $M_c = 0$ *and* (B_a, M_e) *is live, bounded and reversible.*

Proof. If (B_a, M_e) is not bounded, then there exist two markings M_1 and M_2 reachable from M_e such that $M_2 \geq M_1$ and $M_2 \neq M_1$. By Lemma 3.1 and Condition (2) of the soundness assumption on (B, M_e), we have $M_1[B, \sigma\rangle M_x$. It follows that $M_2[B, \sigma\rangle M$, where $M = M_x + M_2 - M_1 \geq M_x$ but $M \neq M_x$, contradicting with the assumption that (B, M_e) terminates properly. Next, reversibility of (B_a, M_e) follows from Condition (1) of the soundness assumption on (B, M_e) and the fact that $M_x[B_a, t_a\rangle M_e$. Lastly, consider any $t \in T \cup \{t_a\}$ and any $M \in R(B_a, M_e)$. By Condition (1) of the soundness assumption, t_a is obviously live. Hence, we have to consider only the case where $t \in T$. Since (B_a, M_e) is reversible, there exists $\sigma \in (T \cup \{t_a\})^*$ such that $M[B_a, \sigma\rangle M_e$. Then, by Condition (3) of the soundness assumption, there exists M' such that $M_e[B,^*\rangle M'[B, t\rangle$. It follows that $M[B_a, \sigma\rangle M_e[B,^*\rangle M'[B, t\rangle$. That is, t is live in (B_a, M_e). $\square$

The converse follows from a known result for workflow nets [AAL (1997)].

Note that, without the condition $M_c = 0$, (B, M_e) may not be sound even if (B_a, M_e) is live, bounded and reversible. For example, it is easy to show that (B_a, M_e) in Figure 3.2 is live, bounded and reversible, however, (B, M_e) is not sound since it does not terminate properly and has a dead transition t_2. In fact, after firing $t_1 t_3$, B terminates at the marking $(0, 0, 1, 0, 0, 1)$.

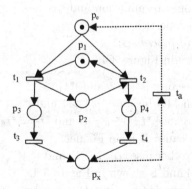

Fig. 3.2   An unsound (B, M_e) with a live, bounded and reversible (B_a, M_e).

3.4   An Example

Example 3.1. (Illustration of properties of a process) (Figure 3.3)
Process B specifies a workflow system for handling insurance claims related
to car damage [AAL (1997)]. Table 3.1 contains an interpretation of the
places and transitions of B.

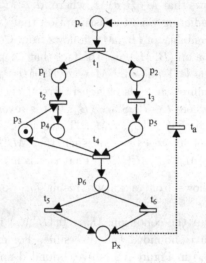

Fig. 3.3   Process B specifying the workflow system for handling insurance claims.

An example similar to Figure 3.3 was used in [AAL (1997)], where
only boundedness, proper termination and liveness were considered. This
example shows more properties.
Structural *properties of process B*:
Let V and V_a be shown in Figure 3.4.

1) B is not an SM because $|t_1^\bullet| > 1$ and $|{}^\bullet t_4| > 1$.
2) B is not almost an MG because $|p_6^\bullet| > 1$ and $|{}^\bullet p_x| > 1$.
3) B is an FC net because ${}^\bullet t_5 \cap {}^\bullet t_6 \neq \phi$ and ${}^\bullet t_5 = {}^\bullet t_6 = \{p_6\}$.
4) B is an AC net because B is an FC net.
5) B is conservative because $\alpha \geq 1$ and $\alpha V = 0$, where $\alpha = (2,1,1,1,2,1,2,2)$ and is shown in Figure 3.4.
6) B is structurally bounded because B is conservative.
7) B is almost consistent because $\beta \geq 1$ and $V_a \beta = 0$, where $\beta = (2,2,2,2,1,1,2)$.

Table 3.1 Interpretation of places and transitions of the workflow system (Figure 3.3).

Place/Transition	Interpretation
p_e	Ready to handle an insurance claim
p_1	Ready to check the insurance policy information
p_2	Ready to contact the garage
p_3	Insurance policy statement
p_4	Waiting for the return of the insurance policy information
p_5	Waiting for the car information from the garage
p_6	Ready to pay for the claimant or reject the claim
p_x	Ready to close the claim case
t_1	Process an insurance claim
t_2	Check the insurance policy information
t_3	Contact the garage for car information
t_4	Determine whether the claim is justified
t_5	Pay the claimant for damage
t_6	Send a letter of rejection to the claimant
t_a	Associated transition

$$
\begin{array}{c}
\quad\quad t_1 \ \ t_2 \ \ t_3 \ \ t_4 \ \ t_5 \ \ t_6 \ \ t_a \\
\begin{array}{c}
p_e \\ p_1 \\ p_2 \\ p_3 \\ p_4 \\ p_5 \\ p_6 \\ p_x
\end{array}
\left(
\begin{array}{ccccccc}
-1 & 0 & 0 & 0 & 0 & 0 & 1 \\
1 & -1 & 0 & 0 & 0 & 0 & 0 \\
1 & 0 & -1 & 0 & 0 & 0 & 0 \\
0 & -1 & 0 & 1 & 0 & 0 & 0 \\
0 & 1 & 0 & -1 & 0 & 0 & 0 \\
0 & 0 & 1 & -1 & 0 & 0 & 0 \\
0 & 0 & 0 & 1 & -1 & -1 & 0 \\
0 & 0 & 0 & 0 & 1 & 1 & -1
\end{array}
\right)
\end{array}
$$

Fig. 3.4 Incidence matrix V_a of B_a (V of B is the part of V_a without column t_a).

8) B is almost repetitive because B is almost consistent.

9) Let t denote the column vector $V[P, t]$. Since $t_6 = t_5$ and t_1, t_2, t_3, t_4 and t_5 are linearly independent, we have $Rank(B) = 5$.

10) Set of clusters $C(B) = \{[p_e], [p_1], [p_2], [t_5], [p_6], [p_x]\}$, where $[p_e] = \{p_e, t_1\}, [p_1] = \{p_1, p_3, t_2\}, [p_2] = \{p_2, t_3\}, [t_4] = \{p_4, p_5, t_4\}, [p_6] = \{p_6, t_5, t_6\}$ and $[p_x] = \{p_x\}$. Hence, $|C(B)| = 6$.

11) B satisfies the RC-property because $Rank(B) = 5 = |C(B)| - 1$.

12) B is almost SM-coverable with three minimal SM-components shown in Figure 3.5:

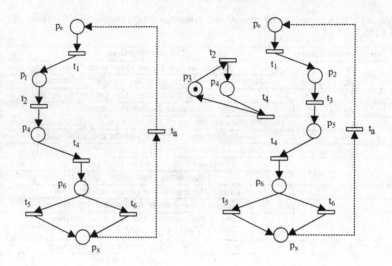

Fig. 3.5   Three minimal SM-components of process B.

13) Examples of siphons: $\{p_e, p_1, p_4, p_6, p_x\}$, $\{p_e, p_2, p_5, p_6, p_x\}$ and $\{p_3, p_4\}$.

14) Examples of traps: $\{p_4, p_5, p_6, p_x\}$, $\{p_e, p_1, p_4, p_6, p_x\}$ and $\{p_3, p_4\}$.

15) $Z(B) = 4$. Four linearly independent entry-to-exit paths which cover all arcs: $\{(p_e, t_1), (t_1, p_1), (p_1, t_2), (t_2, p_4), (p_4, t_4), (t_4, p_6), (p_6, t_5), (t_5, p_x)\}$, $\{(p_e, t_1), (t_1, p_1), (p_1, t_2), (t_2, p_4), (p_4, t_4), (t_4, p_6), (p_6, t_6), (t_6, p_x)\}$, $\{(p_e, t_1), (t_1, p_2), (p_2, t_3), (t_3, p_5), (p_5, t_4), (t_4, p_3), (p_3, t_2), t_2, p_4), (p_4, t_4),$ $(t_4, p_6), (p_6, t_6), (t_6, p_x)\}$ and $\{(p_e, t_1), (t_1, p_2), (p_2, t_3), (t_3, p_5), (p_5, t_4),$ $(t_4, p_6), (p_6, t_5), (t_5, p_x)\}$.

Behavioral properties of process (B, Me):

16) (B, M_e) has no deadlock transitions.

17) (B, M_e) is safe because the maximum number of tokens in every place does not exceed 1 for all reachable markings (Figure 3.6).

18) (B, M_e) terminates properly because any M where $M(p_x) \geq 1, M = M_x = (0, 0, 0, 1, 0, 0, 0, 1)$.

19) (B, M_e) is almost live as shown in reachability tree (Figure 3.6).

20) (B, M_e) is almost reversible as shown in reachability tree (Figure 3.6).

3.5   Notations for the Description of PPPA

In the remaining sections, three extension operators of the PPPA, namely, ACTION PREFIX, BYPASS and ITERATION, are defined and shown to

$M_e=(1\ 0\ 0\ 1\ 0\ 0\ 0\ 0)$

$M_x=(0\ 0\ 0\ 1\ 0\ 0\ 0\ 1)$

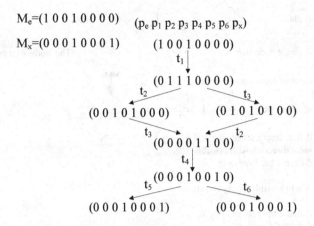

Fig. 3.6   Reachability tree of process (B_a, M_e) (Reachability tree of (B, M_e) is the same as (B_a, M_e) except without the bottom level).

be preserving many properties. The notations for these operators are shown in Table 3.2.

Throughout this chapter, the notations $N = (P, T, F), B = (N, p_e, p_x)$ and $M_e = (1, M_c, 0)$ represent elements of a given PNP; whereas the notations $N' = (P', T', F'), B' = (N', p'_e, p'_x)$ and $M'_e = (1, M'_c, 0)$ represent the corresponding elements of the composite PNP obtained by an extension operator.

Table 3.2   Three extension operators of PPPA.

Type	Name of operator	Notation
Extension	ACTION PREFIX	$b\ ;\ B$
	BYPASS (not a LOTOS operator)	$b|B$
	ITERATION	B'

The self-explanatory graphical notations in Figure 3.7 will be used in the remaining part of this book for the description of the operators of PPPA. The bottom part of the figure on the right, for example, means that a set of input transitions ends at the same place p_x.

3.6   The Operator Action Prefix

The operator ACTION PREFIX extends a process B by adding an action b in front of B.

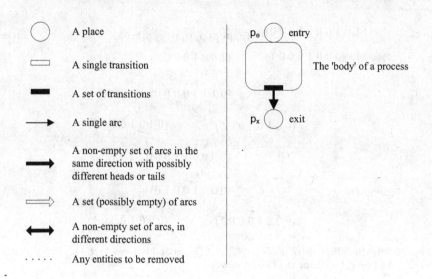

Fig. 3.7   Graphical notations for the PPPA.

Definition 3.8. (ACTION PREFIX, Figure 3.8)
For a marked process (B, M_e) and a transition $b \notin T$ representing a non-null action, the *ACTION PREFIX extension* to (B, M_e), in notation $(b \; ; \; B, M'_e)$, is defined as follows:

$P' = P \cup \{p'_e\}, p'_x = p_x$

$T' = T \cup \{b\}$

$F' = F \cup \{(p'_e, b), (b, p_e)\}$

$M'_e = (1, 0, M_c, 0).$

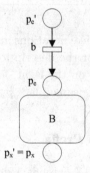

Fig. 3.8   b ; B: Extension to B by the operator ACTION PREFIX.

It is obvious that $(b \; ; \; B, M'_e)$ is also a marked process. The incidence matrix V' of $(b \; ; \; B)$ is formed from the incidence matrix V of B as follows.

$$\begin{array}{c} b \quad T \\ \begin{array}{c} p'_e \\ P \end{array} \begin{pmatrix} -1 & 0 \\ p_e & V \end{pmatrix} \end{array}$$

Theorem 3.3. *(property preservation under ACTION PREFIX $b \; ; \; B$) Let (B, M_e) and $(b \; ; \; B, M'_e)$ be the two marked PNPs involved in Definition 3.8. Then, the following propositions hold:*

1) *If B is an SM, so is $b \; ; \; B$.*
2) *If B is almost an MG, so is $b \; ; \; B$.*
3) *If B is an FC net, so is $b \; ; \; B$.*
4) *If B is an AC net, so is $b \; ; \; B$.*
5) *If B is conservative, so is $b \; ; \; B$.*
6) *If B is structurally bounded, so is $b \; ; \; B$.*
7) *If B is almost consistent, so is $b \; ; \; B$.*
8) *If B is almost repetitive, so is $b \; ; \; B$.*
9) *$Rank(b \; ; \; B) = Rank(B) + 1$.*
10) *$|C(b \; ; \; B)| = |C(B)| + 1$.*
11) *If B satisfies the RC-property, so does $b \; ; \; B$.*
12) *If B almost has a minimal SM-cover, so does $b \; ; \; B$.*
13) *Suppose D is a siphon of B. Then, a) D is a siphon of $b \; ; \; B$ iff $p_e \notin D$; b) $D \cup \{p'_e\}$ is a siphon of $b \; ; \; B$ if $p_e \in D$.*
14) *Every trap of B is also a trap of $b \; ; \; B$.*
15) *$Z(b \; ; \; B) = Z(B)$.*
16) *If (B, M_e) has no dead transitions, then $(b \; ; \; B, M'_e)$ does not have either.*
17) *If (B, M_e) is k-bounded, so is $(b \; ; \; B, M'_e)$.*
18) *If (B, M_e) terminates properly, so does $(b \; ; \; B, M'_e)$.*
19) *If (B, M_e) is almost live, so is $(b \; ; \; B, M'_e)$.*
20) *If (B, M_e) is almost reversible, so is $(b \; ; \; B, M'_e)$.*

Proof.

1-4) According to the structures of B and $b \; ; \; B$, adding transition b and the arcs does not violate the definitions of an SM, MG, FC net and AC net.

5) Since B is conservative, by Characterization 2.1, $\exists \alpha = (x_1, \cdots, x_{|P|}) \geq 1$ such that $\alpha V = 0$. Let $\alpha' = (x_1, \alpha)$ and V' be

shown in Figure 3.9. Then, $\alpha' \geq 1$ and $\alpha'b = -x_1 + x_1 = 0$ implies $\alpha'V' = (0, \alpha V) = 0$. It follows from Characterization 2.1 that $b; B$ is conservative.

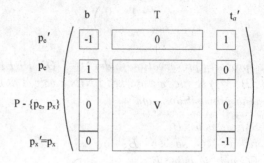

Fig. 3.9   Incidence matrix V_a' of $(b \; ; B)_a$. (V' of $b \; ; B$ is the part of V_a' without column t_a').

6) Suppose B is structurally bounded. The same argument as in 5) shows that $\exists \alpha' \geq 1$ such that $\alpha'V' \leq 0$. It follows from Characterization **??** that $b; B$ is structurally bounded.

7) Since B is almost consistent, by Characterization 2.3, $\exists \beta_a = (\beta, z) \geq 1$, where $\beta = (y_1, \cdots, y_{|T|})$, such that $V_a\beta_a = 0$. Since $t_a = p_e - p_x$, this equality can be rewritten as $V_a\beta_a = (V, t_a)(\beta, z) = V\beta + (p_e - p_x)z = 0$. Let $\beta_a' = (z, \beta_a)$. Then, $\beta_a' \geq 1$ and $V_a'\beta_a' = (-z + z, p_e z + V\beta - p_x z) = (0, V_a\beta_a) = 0$. It follows from Characterization 2.3 that $b \; ; B$ is almost consistent.

8) Suppose B is almost repetitive. The same argument as in 7) shows that $\exists \beta_a' \geq 1$ such that $V_a'\beta_a' \geq 0$. It follows from Characterization 2.4 that $b \; ; B$ is almost repetitive.

9) It is obvious that, in V', the leftmost column vector $V'[P', b]$ is linearly independent from the other columns. Hence, $Rank(b \; ; B) = Rank(B) + 1$.

10) According to the structure of $b \; ; B$, $[b] = \{p_e', b\}$ is the only new cluster created and all clusters of B are still clusters of $b \; ; B$. Hence, $|C(b \; ; B)| = |C(B)| + 1$.

11) This follows from 9) and 10).

12) In order to be strongly connected, any SM-component of B_a containing p_e must also contain t_a and p_x. Suppose K is a minimal SM-cover of B_a. For every $S \in K$ which contains p_e and p_x, extend S by including p_e', b and the arcs (p_e', b) and (b, p_e) in it (with some trivial changes in labels). Obviously, this results in a minimal SM-cover for $(b \; ; B)_a$.

(Note: The main idea is to absorb the new elements into those SM-components that contain p_e and p_x.)

13) a) If $p_e \notin D$, then $^\bullet D$ in $(b \; ; \; B) = \,^\bullet D$ in B and $D^\bullet$ in $(b \; ; \; B) = D^\bullet$ in B. Since $(^\bullet D \subseteq D^\bullet)$ in B, it follows that $(^\bullet D \subseteq D^\bullet)$ in $(b \; ; \; B)$. If $p_e \in D$, then $^\bullet D$ in $(b \; ; \; B) = (^\bullet D$ in $B) \cup \{b\}$ and $D^\bullet$ in $(b \; ; \; B) = D^\bullet$ in B. Hence, $(^\bullet D \not\subset D^\bullet)$ in $(b \; ; B)$ even if $(^\bullet \subseteq D^\bullet)$ in B. b) Suppose $p_e \in D$. Let $D' = D \cup \{p'_e\}$. Since $(D')^\bullet$ in $(b \; ; \; B) = (D^\bullet$ in $B) \cup \{b\}$ and $^\bullet D'$ in $(b \; ; \; B) = (^\bullet D$ in $B) \cup \{b\}$, we have $(^\bullet D' \subseteq (D')^\bullet)$ in $(b \; ; \; B)$ if $(^\bullet D \subseteq D^\bullet)$ in B.

14) According to the structures of B and $(b \; ; \; B)$, $\forall S \subseteq P$: $S^\bullet$ in $(b \; ; \; B) = S^\bullet$ in B and $^\bullet S$ in $B \subseteq \,^\bullet S$ in $(b \; ; \; B)$. In particular, for a trap S of B, $S^\bullet$ in $(b \; ; \; B) = S^\bullet$ in $B \subseteq \,^\bullet S$ in $B \subseteq \,^\bullet S$ in $(b \; ; \; B)$.

15) [CZ (1994)]. $Z(b \; ; \; B) = (|F| + 2) - (|P| + |T| + 2) + 1 + 1 = Z(B)$.

16) Obviously, b is not dead. The behavior of $b \; ; \; B$ is the same as B after firing b.

17) Any $M' \in R(b \; ; \; B, M'_e)$ can only be of two forms: a) $M' = M'_e$, where M'_e is a constant. b) $M' = (0, m)$, where $m \in R(B, M_e)$ is k-bounded by assumption.

18) Note that $M'_e[b \; ; \; B, b\rangle(0, M_e)$ and that the behaviour of $(b \; ; \; B, M'_e)$ is identical with (B, M_e) after firing b. Since (B, M_e) terminates properly at M_s, $(b \; ; \; B, M'_e)$ terminates properly at $M'_s = (0, M_s)$.

19) All we have to show is that the two transitions b and t'_a are live because all the other transitions behave in the same way in B_a or $(b \; ; \; B)_a$. Since t'_a plays the same role in $(b \; ; \; B)_a$ as t_a in B_a, t'_a is also live and thus b is obviously also live.

20) Any $M' \in R((b \; ; \; B)_a, M'_e)$ can only have two forms: a) $M' = (0, m)$, where $m \in R(b \; ; \; B_a, M'_e)$. Obviously, it is also true that $m \in R(B_a, M_e)$. Since B_a is reversible, $\exists \sigma \in T^*_a$: $(m[B_a, \sigma\rangle M_e)$, where $\sigma = \sigma_1 t_a \sigma_2 t_a \cdots \sigma_r t_a$. (Note: Since B_a may not terminate properly, it may have to fire t_a several times before reaching M_e.) Hence, $M'[(b \; ; \; B)_a, \sigma'\rangle M'_e$, where $\sigma' = \sigma_1 t'_a b \sigma_2 t'_a b \cdots \sigma_r t'_a$. b) $M' = (1, m)$. This case becomes case a) after firing bk times at M'. $\quad\square$

3.7 The Operator BYPASS $b|B$

The operator BYPASS extends a process by adding an option of skipping the process entirely. In fact, BYPASS is a special case of the operator Choice. It is not a LOTOS operator but is added into PPPA for convenience in modeling.

Definition 3.9. (BYPASS, Figure 3.10)

For a marked process (B, M_e) and a transition $b \notin T$ representing the bypass action, the *BYPASS extension* to (B, M_e), in notation $(b|B, M'_e)$, is defined as follows:

$P' = P$, $p'_e = p_e$ and $p'_x = p_x$

$T' = T \cup \{b\}$

$F' = F \cup \{(p'_e, b), (b, p'_x)\}$

$M'_e = M_e = (1, M_c, 0)$.

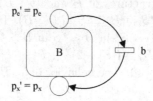

Fig. 3.10 $b|B$: Extension of B by BYPASS.

It is obvious that $(b|B, M'_e)$ is also a marked process. The incidence matrix V' of $b|B$ is formed from the incidence matrix V of B as follows.

$$\begin{matrix} & b & T \\ P & (-P_e + P_x & V) \end{matrix}$$

Theorem 3.4. (*property preservation under BYPASS $b|B$*)

Let (B, M_e) and $(b|B, M'_e)$ be the two marked PNPs involved in Definition 3.9. Then, the following propositions hold:

1) If B is an SM, so is $b|B$.

2) $b|B$ is not almost an MG, whether or not B is so.

3) Suppose B is an FC net. $b|B$ is an FC net iff $^\bullet(p^\bullet_e) = \{p_e\}$.

4) If B is an AC net, then so is $b|B$ if $\forall p \in P$: $(p^\bullet \cap p^\bullet_e \neq \phi \Rightarrow p^\bullet \subseteq p^\bullet_e)$ in B.

5) a) If B is conservative, then $b|B$ is conservative provided that $M_e[B, *)M_x$. b) If B is almost conservative, then $b|B$ is almost conservative.

6) a) If B is structurally bounded, then $b|B$ is structurally bounded. b) If B is almost structurally bounded, $b|B$ is almost structurally bounded.

7) If B is almost consistent, so is $b|B$.

8) If B is almost repetitive, so is $b|B$.

9) *Rank (b|B) = Rank(B) iff B_a has a T-invariant including t_a. Otherwise, Rank(b|B) = Rank(B) + 1.*

10) *$|C(b|B)| = |C(B)|$.*

11) *If B satisfies the RC-property, b|B satisfies the RC property provided that B_a has a T-invariant including t_a.*

12) *If B almost has a minimal SM-cover, so does b|B.*

13) *Suppose D is a siphon of B, Then, D is a siphon of b|B iff ($p_e \in D$) or ($p_e \notin D$ and $p_x \notin D$).*

14) *Suppose S is a trap of B. Then, S is a trap of b|B iff ($p_x \in S$) or ($p_e \notin S$ and $p_x \notin S$).*

15) *$Z(b|B) = Z(B) + 1$.*

16) *If (B, M_e) has no dead transitions, then $(b|B, M'_e)$ does not have either.*

17) *If (B, M_e) is $k-$bounded, so is $(b|B, M'_e)$.*

18) *If (B, M_e) terminates properly, so does $(b|B, M'_e)$.*

19) *If (B, M_e) is almost live, so is $(b|B, M'_e)$.*

20) *If (B, M_e) is almost reversible, so is $(b|B, M'_e)$.*

Proof.

1-3) In $b|B$, since $|{}^\bullet b| = |b^\bullet| = 1$ and $|p_e^\bullet| > 1$, $b|B$ is an SM but not an MG. Also, it does not violate the definition of a FC net only if ${}^\bullet(p_e^\bullet) = \{p_e\}$. If $\exists t \in p_e^\bullet$: (${}^\bullet t \neq \{p_e\}$), then ${}^\bullet b \cap {}^\bullet t \neq \phi$ but ${}^\bullet b = \{p_e\} \neq {}^\bullet t$.

4) Since $\forall p \in P$: $(p^\bullet \cap p_e^\bullet \neq \phi \Rightarrow p^\bullet \subseteq p_e^\bullet)$ in B, $((p^\bullet \cap p_e^\bullet \neq \phi)$ in $b|B) \Rightarrow ((p^\bullet$ in $b|B) \subseteq (p_e^\bullet$ in $B) \cup \{b\}) \Rightarrow ((p^\bullet \subseteq p_e^\bullet)$ in $b|B)$. Hence, $b|B$ is also an AC net.

5) a) Suppose B is conservative. By Characterization 2.1, $\exists \alpha = (x_1, \cdots, x_{|P|}) \geq 1$ such that $\alpha V = 0$. If $M_e[B, *\rangle M_x$, then $\alpha M_e = \alpha M_x$. This implies $x_1 = x_{|P|}$. Let $\alpha' = \alpha$ and V' be shown in Figure 3.11. Then, $\alpha' \geq 1$ and $\alpha'V' = (\alpha'V'[P, b], \alpha'V'[P, T]) = (-x_1 + x_{|P|}, \alpha V) = 0$. It follows from Characterization 2.1 that $b|B$ is conservative.

b) Suppose B is almost conservative. Then, $\exists \alpha(x_1, \cdots, x_{|P|}) \geq 1$ such that $\alpha V_a = 0$. This implies $x_1 = x_{|P|}$. Let $\alpha' = \alpha$ and V' be shown as in Figure 3.11. Then, $\alpha' \geq 1$ and $\alpha'V'_a = (\alpha'V'[P, b], \alpha'V_a) = (-x_1 + x_{|P|}, \alpha V_a) = 0$. It follows from Characterization 2.1 that $b|B$ is almost conservative.

6) a) Suppose B is structurally bounded. Then $\forall M_0$, the reachable marking of (B, M_0) is bounded. Consider, $(b|B, M_0)$, either $M_0[b|B, b\rangle M_x$

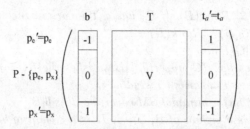

Fig. 3.11   Incidence matrix V'_a of $(b|B)_a$. (V' of $b|B$ is the part of V'_a without column t'_a).

or $M_0[b|B, *\rangle M$, where $M \in R(B, M_0)$. Hence, $(b|B, M_0)$ is bounded and $b|B$ is structurally bounded. b)Similar to a).

7) Since B is almost consistent, $\exists \beta = (y_1, \cdots, y_{|T|}, z) \geq 1$ such that $V_a\beta = 0$. Let $\beta' = (z, \beta[T], 2z)$ and V'_a be shown in Figure 3.11. Then, $\beta' \geq 1$. Since $V'_a[P', b] = -V'_a[P', t'_a]$, $V'_a\beta' = zV'_a[P', b] + V_a\beta + zV'_a[P', t'_a] = 0$. It follows from Characterization 2.3 that $b|B$ is almost consistent.

8) Suppose B is almost repetitive. The same argument as in 7) shows that $V_a\beta \geq 0$ leads to $V'_a\beta'_a \geq 0$ for $\beta'_a \geq 1$. It follows from Characterization 2.4 that $b|B$ is almost repetitive.

9) In V'_a (Figure 3.11), since the leftmost and rightmost columns are negative of each other, $\text{Rank}(b|B) = \text{Rank}(B_a)$. If B_a has a T-invariant including t_a, $\exists \beta$ where $\beta[t_a] \neq 0$ such that $V_a\beta = 0$. This means that the rightmost column $V_a[P, t_a]$ is linearly dependent on the columns of T. Hence, $\text{Rank}(B_a) = \text{Rank}(B)$. Otherwise, obviously $\text{Rank}(B_a) = \text{Rank}(B) + 1$.

10) Since $b \in (p'_e)^\bullet$, b is absorbed into the cluster $[p_e]$ of B to form $[p'_e]$ for $b|B$. Also, there are no other changes to the clusters of B. Hence, $|C(b|B)| = |C(B)|$.

11) This follows from 9) and 10).

12) By definition, any SM-component of B_a containing p_e must also contain t_a and p_x. Suppose K is a minimal SM-cover of B_a. For every $S \in K$ which contains p_e and p_x, extend S by adding b and the arcs (p'_e, b) and (b, p'_x) into it (with some trivial changes in labels). Obviously, this results in a minimal SM-cover for $(b|B)_a$.

13) Consider three cases: a) $p_e \in D$. Since $(b \in p^\bullet_e)$ in $b|B$, we have $D^\bullet$ in $b|B = (D^\bullet$ in $B) \cup \{b\}$. Since $(^\bullet D \subseteq D^\bullet)$ in B, we have $^\bullet D$ in $b|B \subseteq (^\bullet D$ in $B) \cup \{b\} \subseteq (D^\bullet$ in $B) \cup \{b\} = D^\bullet$ in $b|B$. b) $p_e \notin D$ and $p_x \in D$. Then, $^\bullet D$ in $b|B = (^\bullet D$ in $B) \cup \{b\}$ but $b \notin D^\bullet$ in $b|B$.

Hence, $(^\bullet D \not\subseteq D^\bullet)$ in $b|B$ even if $(^\bullet D \subseteq D^\bullet)$ in B. c) $p_e \notin D$ and $p_x \notin D$. Then, $^\bullet D$ in $b|B = {}^\bullet D$ in $B \subseteq D^\bullet$ in $B = D^\bullet$ in $b|B$. Hence, $(^\bullet D \subseteq D^\bullet)$ in $b|B$ provided that $(^\bullet D \subseteq D^\bullet)$ in B.

14) Consider three cases: a) $p_x \in S$. Since $(b \in {}^\bullet p_x)$ in $b|B$, we have $^\bullet S$ in $b|B = (^\bullet S$ in $B) \cup \{b\}$. Since $(S^\bullet \subseteq {}^\bullet S)$ in B, we have $S^\bullet$ in$b|B \subseteq (S^\bullet$ in $B)\cup\{b\} \subseteq (^\bullet S$ in $B)\cup\{b\} = {}^\bullet S$ in $b|B$. b) $p_e \in S$and $p_x \notin S$. Then, $(b \notin {}^\bullet S)$ in $b|B$ and $S^\bullet$ in $b|B = (S^\bullet$ in $B) \cup \{b\}$. Hence, $(S^\bullet \not\subseteq {}^\bullet S)$ in $b|B$ even if $(S^\bullet \subseteq {}^\bullet S)$ in B. c) $p_e \notin S$ and $p_x \notin S$. Then, $S^\bullet$ in $b|B = S^\bullet$ in $B \subseteq {}^\bullet S$ in $B = {}^\bullet S$ in $b|B$.

15) $Z(b|B) = |F'| - |P' \cup T'| + 2 = (|F| + 2) - (|P| + |T| + 1) + 2 = Z(B) + 1$.

16) Any transition dead in $b|B$ is also dead in B.

17) Since $R(b|B, M_e') = R(B, M_e) \cup \{M_x'\}$, k-boundedness of $(b|B, M_e')$ follows from the k-boundedness of $R(B, M_e)$ and M_x'.

18) If b is fired, $M_e'[b|B, b\rangle M_x'$. That is, $(b|B, M_e')$ terminates properly if b is fired. The behaviour of $(b|B, M_e')$ is identical with (B, M_e) if firing occurs within B. Proper termination of $(b|B, M_e')$ follows from proper termination of (B, M_e).

19) Since t_a' plays the same role in $(b|B)_a$ as t_a plays in B_a, liveness of any transition other than b in $((b|B)_a, M_e')$ follows from its liveness in (B_a, M_e). b can be shown to be live as follows. Since t_a' is live, for any $M' \in R((b|B)_a, M_e')$, $\exists M''$: $(M'[(b|B)_a, *\rangle M''[(b|B)_a, t_a'\rangle)$. Obviously, then $M''[(b|B)_a, t_a'b\rangle$.

20) Since $R(b|B)_a, M_e') = \{M_x'\} \cup R(B_a, M_e)$, reversibility of $((b|B)_a, M_e')$ follows from $M_x'[(b|B))_a, t_a'\rangle M_e'$ and reversibility of (B_a, M_e). $\qquad\square$

3.8 The Operator ITERATION B'

The operator ITERATION models the recursion operation that exists in almost any programming language. It allows a process to be executed again.

Definition 3.10. (ITERATION, Figure 3.12)
For a marked process (B, M_e), an *ITERATION* of B, in notation (B', M_e'), is defined as follows:

$P' = P \cup \{p_e', p_x'\}$

$T' = T \cup \{\varepsilon_1, \varepsilon_2, \varepsilon_3\}$

$F' = F \cup \{(p_e', \varepsilon_1), (\varepsilon_1, p_e), (\varepsilon_2, p_e), (p_x, \varepsilon_2), (p_x, \varepsilon_3), (\varepsilon_3, p_x')\}$

$M_e' = (1, 0, M_c, 0, 0)$.

It is obvious that (B', M_e') is also a marked process. The incidence matrix V' of B' is formed from the incidence matrix V of B as follows.

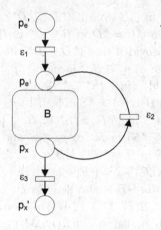

Fig. 3.12 B': ITERATION.

$$
\begin{array}{c}
\quad\quad \varepsilon_1 \quad\quad \varepsilon_2 \quad\quad \varepsilon_3 \quad\quad T \\
\begin{array}{c} p'_e \\ P \\ P'_x \end{array}
\left(
\begin{array}{cccc}
-1 & 0 & 0 & 0 \\
p_e & p_e - p_x & -p_x & V \\
0 & 0 & 1 & 0
\end{array}
\right)
\end{array}
$$

Theorem 3.5. *(property preservation under ITERATION B')*
Let $(B,\ M_e)$ and $(B',\ M'_e)$ be the two marked PNPs involved in Definition 3.10. Then, the following propositions hold:

1) *If B is an SM, so is B'.*
2) *B' is not almost an MG, whether or not B is so.*
3) *If B is an FC net, so is B'.*
4) *If B is an AC net, so is B'.*
5) *If B is conservative, B' is conservative provided that $M_e[B, *\rangle M_x$.*
6) *a) If B is structurally bounded, B' is structurally bounded provided that $M_e[B, *\rangle M_x$. b) If B is almost structurally bounded, B' is almost structurally bounded.*
7) *If B is almost consistent, so is B'.*
8) *If B is almost repetitive, so is B'.*
9) *In general, $Rank(B) + 3 \geq Rank(B') \geq Rank(B) + 2$.*
   $Rank(B') = Rank(B) + 2$ provided that B_a has a T-invariant including t_a.
10) *$|C(B')| = |C(B)| + 2$.*

11) *If B satisfies the RC-property, B' satisfies the RC-property provided that B_a has a T-invariant including t_a.*

12) *If B almost has a minimal SM-cover, so does B'.*

13) *Suppose D is a siphon of B. Then, two cases hold: a) D is a siphon of B' iff $p_e \notin D$. b) $D \cup \{p'_e\}$ is a siphon of B' if p_e, $p_x \in D$.*

14) *Suppose S is a trap of B. Then, two cases hold: a) S is a trap of B' iff $p_x \notin S$. b) S is a trap of B' if p_e, $p_x \in S$.*

15) $Z(B') = Z(B) + 1$.

16) *If (B, M'_e) has no dead transitions, then (B', M'_e) does not have either.*

17) *If (B, M_e) is k-bounded, then (B', M'_e) is k-bounded provided that (B, M_e) terminates properly. In particular, (B', M'_e) is 1-bounded if (B, M_e) is almost 1-bounded.*

18) *If (B, M_e) terminates properly, so does (B', M'_e).*

19) *If (B, M_e) is almost live, so is (B', M'_e).*

20) *If (B, M_e) is almost reversible, so is (B', M'_e).*

Proof.

1-4) According to the structure of B', adding transitions ε_1, ε_2, ε_3 and the arcs does not violate the definitions of SM, FC and AC nets because of $p^\bullet_x = \phi$ in B. However, B' cannot almost be an MG because $|{}^\bullet p_e| = |\{\varepsilon_1, \varepsilon_2\}| > 1$.

5) Since B is conservative, $\exists \alpha = (x_1, \cdots, x_{|P|}) \geq 1$ such that $\alpha V = 0$. Since $M_e[B, *\rangle M_x$, $\alpha M_e = \alpha M_x$. This implies $x_1 = x_{|P|}$. Let $\alpha' = (x_1, \alpha, x_{|P|})$ and V' be shown in Figure 3.13. Then, $\alpha' \geq 1$ and $\alpha' V' = (\alpha' V'[P', \varepsilon_1], \alpha' V'[P', \varepsilon_2], \alpha' V'[P', \varepsilon_2], \alpha' V'[P', T]) = (-x_1 + x_1, x_1 - x_{|P|}, -x_{|P|} + x_{|P|}, \alpha V) = 0$. It follows from Characterization 2.1 that B' is conservative.

6) a) Suppose B is structurally bounded. The same argument as in 5) shows that $\exists \alpha' \geq 1$ such that $\alpha' V' \leq 0$. It follows from Characterization 2.2 that B' is structurally bounded.

b) Suppose B is almost structurally bounded. Then, $\exists \alpha = (x_1, \cdots, x_{|P|}) \geq 1$ such that $\alpha V_a \leq 0$. This implies $x_1 = x_{|P|}$ since $x_1 \leq x_{|P|}$ and $x_{|P|} \leq x_1$. Let $\alpha' = (x_1, \alpha, x_{|P|})$ and V' be shown as in Figure 3.13. Then, $\alpha' \geq 1$ and $\alpha' V' = (\alpha' V'[P', \varepsilon_1], \alpha' V'[P', \varepsilon_2], \alpha' V'[P', \varepsilon_3], \alpha' V'[P', T], \alpha' V'[P', t_a]) = (-x_1 + x_1, x_1 - x_{|P|}, -x_{|P|} + x_{|P|}, \alpha V, x_1 - x_{|P|}) \leq 0$. It follows from Characterization 2.2 that B' is almost structurally bounded.

7) Since B is almost consistent, $\exists \beta_a = (\beta, z) \geq 1$ such that $V_a \beta_a = 0$. Let $\beta'_a = (z, z, z, 2\beta, z)$. Then, $\beta'_a \geq 1$ and $V'_a \beta'_a = (-z + z, p_e z + $

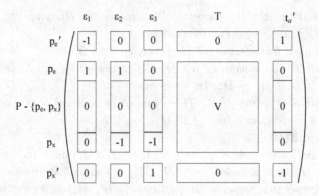

Fig. 3.13   Incidence matrix V_a' of B_a' (V' of B' is the part of V_a' without column t_a').

$(p_e - p_x)z - p_x z + 2V\beta + 0z, z - z) = (0, 2V\beta + 2(p_e - p_x)z, 0) = (0, 2V_a\beta_a, 0) = 0$. It follows from Characterization 2.3 that B' is almost consistent.

8) Suppose B is almost repetitive. The same argument as in 7) shows that $\exists \beta_a' \geq 1$ such that $V_a'\beta_a' \geq 0$. It follows from Characterization 2.4 that B' is almost repetitive.

9) In V', it is obvious that columns ε_1 and ε_3 are linearly independent and that they are also linearly independent from the columns of $V'[P', T]$. Also, column ε_2 is linearly independent from columns ε_1 and ε_3. Hence, in general, $Rank(B)+3 \geq Rank(B') \geq Rank(B)+2$. However, if column ε_2 is linearly dependent on the columns of $V'[P', T]$, then $Rank(B') = Rank(B) + 2$. Furthermore, if B_a has a T-invariant including t_a, $\exists \beta$, where $\beta[t_a] \neq 0$, such that $V_a\beta = 0$. This means that, in V_a, column t_a is linearly dependent on the columns of $V[P, T]$. Since $\varepsilon_2 = (0, t_a, 0)$, this implies that, in V', ε_2 is linearly dependent on the columns of $V'[P', T]$.

10) Among the added elements, ε_2 and ε_3 are absorbed into the existing cluster $[p_x]$. Two new clusters $[p_e'] = \{p_e', \varepsilon_1\}$ and $[p_x'] = \{p_x'\}$ are created and no existing clusters are destroyed. Hence, $|C(B')| = |C(B)| + 2$.

11) This follows from 9) and 10).

12) By definition, any SM-component of B_a containing p_e must also contain t_a and p_x. Suppose K is a minimal SM-cover of B_a. For any $S \in K$ that contains p_e and p_x, change S to S' by deleting t_a, (p_x, t_a), (t_a, p_e) and adding p_e', ε_1, ε_2, ε_3, p_x', t_a', (p_e', ε_1), (ε_1, p_e),

(p_x, ε_2), (ε_2, p_e), (p_x, ε_3), (ε_3, p_x'), (p_x', t_a') and (t_a', p_e'). Obviously, S' is a minimal SM-cover for B_a'.

13) a) Consider two cases: Case 1) $p_e \in D$. Then, ${}^\bullet D$ in $B' = ({}^\bullet D$ in $B) \cup \{\varepsilon_1, \varepsilon_2\}$. Since $(\varepsilon_1 \notin D^\bullet)$ in B', $({}^\bullet D \not\subseteq D^\bullet)$ in B' even if $({}^\bullet D \subseteq D^\bullet)$ in B. Case 2) $p_e \notin D$. Then, ${}^\bullet D$ in $B' = {}^\bullet D$ in B. Since $D^\bullet$ in $B \subseteq D^\bullet$ in B', ${}^\bullet D$ in $B' = {}^\bullet D$ in $B \subseteq D^\bullet$ in $B \subseteq D^\bullet$ in B'. b) Let $D' = D \cup \{p_e'\}$. Since $p_e, p_x \in D$, ${}^\bullet(D')$ in $B' = ({}^\bullet D$ in $B) \cup \{\varepsilon_1, \varepsilon_2\} \subseteq (D^\bullet$ in $B) \cup \{\varepsilon_1, \varepsilon_2, \varepsilon_3\} = (D')^\bullet$ in B'.

14) a) Consider two cases: Case 1) $p_x \in S$. Then, $S^\bullet$ in $B' = (S^\bullet$ in $B) \cup \{\varepsilon_2, \varepsilon_3\}$. Since $(\varepsilon_3 \notin {}^\bullet S)$ in B', $(S^\bullet \not\subseteq {}^\bullet S)$ in B' even if $(S^\bullet \subseteq {}^\bullet S)$ in B. Case 2) $p_x \notin S$. Then, $S^\bullet$ in $B' = S^\bullet$ in B. Since ${}^\bullet S$ in $B \subseteq {}^\bullet S$ in B', $S^\bullet$ in $B' = S^\bullet$ in $B \subseteq {}^\bullet S$ in $B \subseteq {}^\bullet S$ in B'. b) Let $S' = S \cup \{p_x'\}$. Since $p_e, p_x \in S$, $(S')^\bullet$ in $B' = (S^\bullet$ in $B) \cup \{\varepsilon_2, \varepsilon_3\} \subseteq ({}^\bullet S$ in $B) \cup \{\varepsilon_1, \varepsilon_2, \varepsilon_3\} = {}^\bullet S'$ in B'.

15) $Z(B') = |F'| - |P'| \cup T'| + 2 = (|F| + 6) - (|P| + 2 + |T| + 3) + 2 = (|F| - |P \cup T| + 2) + 1 = Z(B) + 1$.

16) The result is obvious.

17) Because of proper termination, (B, M_e) will reach the exit marking M_x after each iteration. Hence, k-boundedness of (B_a', M_e') follows from k-boundedness of (B_a, M_e). It is obvious that (B', M_e') is 1-bounded if (B, M_e) is almost 1-bounded.

18) Any $M \in R(B', M_e')$ can be expressed in the form $(M(p_e'), M(p_e), m,$
$M(p_x), M(p_x'))$. If $M(p_x) \geq 1$, M must be of the form $(0, 0, M_c, 1, 0)$ for three reasons: a) B terminates properly. b) B cannot be reinitiated without firing ε_2. c) p_x' cannot get a token without firing ε_3. Since p_x' can get a token only when p_x has a token, any $M \in R(B', M_e')$, where $M(p_x') \geq 1$, must be of the form $(0, 0, M_c, 0, 1)$.

19-20) Let B_ε be obtained from B_a by replacing t_a with ε_2. Since ε_2 can play the same role as the associated transition t_a, (B_ε, M_e) is the same as (B_a, M_e). Hence, liveness of (B_ε, M_e) follows from liveness of (B_a, M_e). Then, almost liveness (resp., almost reversibility) of (B', M_e') follows by similar argument as used in Theorem 3.3. $\qquad\square$

Chapter 4

Composition Operators of PPPA

In this chapter, six composition operators of PPPA, namely, EN-ABLE, CHOICE, INTERLEAVE, PARALLEL, DISABLE and DISABLE-RESUME, are defined and shown to be preserving many properties.

Table 4.1   Six operators for composition in PPPA.

Type	Name of operator	Notation
Composition	ENABLE (Sequential Composition)	$B_1 >> B_2$
	CHOICE	$B_1 [] B_2$
	Parallelism: INTERLEAVE	$B_1 \|\|\| B_2$
	PARALLEL	$B_1 \| G \| B_2$
	DISABLE	$B_1 [> B_2$
	DISABLE-RESUME	$B_1 [r > B_2$

Throughout this chapter, the notations $B_i = (N_i, p_{ie}, p_{ix})$, $N_i = (P_i, T_i, F_i)$ and $M_{ie} = (1, M_{ic}, 0)$ represent the elements of a given PNP, where $i = 1$ and 2; whereas the notations $B = (N, p_e, p_x)$, $N = (P, T, F)$, M_e and M_c represent the corresponding elements of the composite PNP obtained by an operator.

4.1   The Operator ENABLE

ENABLE models the sequential execution of two processes as follows: B_1 is executed first and B_2 is executed after successful termination of B_1. However, if B_1 has a deadlock or infinite loop within itself, B_2 will never be activated.

Definition 4.1. (composition by the operator ENABLE, Figure 4.1)
For two marked processes $(B_i, M_{ie}), i = 1, 2$, their marked composition by ENABLE, in notation $(B_1 >> B_2, M_e)$, is defined as follows:

61

$P = P_1 \cup P_2$, with $p_e = p_{1e}$ and $p_x = p_{2x}$
$T = T_1 \cup T_2 \cup \{\varepsilon\}$
$F = F_1 \cup F_2 \cup \{(p_{1x}, \varepsilon), (\varepsilon, p_{2e})\}$
$M_e = (1, M_{1c}, 0, 0, M_{2c}, 0)$

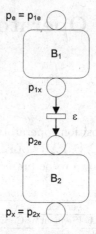

Fig. 4.1 $B_1 >> B_2$: Sequential composition by the operator *ENABLE*.

It is obvious that $(B_1 >> B_2, M_e)$ is also a marked process. The incidence matrix V of $B_1 >> B_2$ is formed from the incidence matrices V_1 of B_1 and V_2 of B_2 as follows:

$$\begin{array}{c} \\ P_1 \\ P_2 \end{array}\begin{array}{ccc} \varepsilon & T_1 & T_2 \\ \left(\begin{array}{ccc} -p_{1x} & V_1 & 0 \\ p_{2e} & 0 & V_2 \end{array}\right) \end{array}$$

Theorem 4.1. *(property preservation under ENABLE $B_1 >> B_2$)*
Let (B_1, M_{1e}), (B_2, M_{2e}) and $(B_1 >> B_2, M_e)$ be the three marked PNPs involved in Definition 4.1. Then, the following propositions hold:

1) If both B_1 and B_2 are SMs, so is $B_1 >> B_2$.
2) If both B_1 and B_2 are almost MGs, so is $B_1 >> B_2$.
3) If both B_1 and B_2 are FC nets, so is $B_1 >> B_2$.
4) If both B_1 and B_2 are AC nets, so is $B_1 >> B_2$.
5) If both B_1 and B_2 are conservative, so is $B_1 >> B_2$.
6) If both B_1 and B_2 are structurally bounded, so is $B_1 >> B_2$.
7) If both B_1 and B_2 are almost consistent, so is $B_1 >> B_2$.
8) If both B_1 and B_2 are almost repetitive, so is $B_1 >> B_2$.

9) If $V[P_1, \varepsilon]$ is independent from the columns of $V[P_1, T_1]$ or $V[P_2, \varepsilon]$ is linearly independent from the columns of $V[P_2, T_2]$, then $Rank(B_1 >> B_2) = Rank(B_1) + Rank(B_2) + 1$. Otherwise, $Rank(B_1 >> B_2) = Rank(B_1) + Rank(B_2)$.

10) $|C(B_1 >> B_2)| = |C(B_1)| + |C(B_2)|$.

11) If both B_1 and B_2 satisfy the RC-property, then either $B_1 >> B_2$ satisfies the RC-property or $Rank(B_1 >> B_2) = |C(B_1 >> B_2)| - 2$.

12) If both B_1 and B_2 almost have a minimal SM-cover, so does $B_1 >> B_2$.

13) Suppose D_i is a siphon of B_i, $i = 1, 2$. Then,

   (a) D_1 is a siphon of $B_1 >> B_2$;
   (b) D_2 is a siphon of $B_1 >> B_2$ iff $p_{2e} \notin D_2$;
   (c) $D_1 \bigcup D_2$ is a siphon of $B_1 >> B_2$ if $p_{1x} \in D_1$ and $p_{2e} \in D_2$

14) Suppose S_i is a trap of B_i, $i = 1, 2$. Then,

   (a) S_2 is a trap of $B_1 >> B_2$;
   (b) S_1 is a trap of $B_1 >> B_2$ iff $p_{1x} \notin S_1$;
   (c) $S_1 \bigcup S_2$ is a trap of $B_1 >> B_2$ if $p_{1x} \in S_1$ and $p_{2e} \in S_2$.

15) $Z(B_1 >> B_2) = Z(B_1) + Z(B_2) - 1$.

16) If both (B_1, M_{1e}) and (B_2, M_{2e}) have no dead transitions, then $(B_1 >> B_2, M_e)$ does not have either.

17) If both (B_1, M_{1e}) and (B_2, M_{2e}) are bounded and (B_1, M_{1e}) terminates properly, then $(B_1 >> B_2, M_e)$ is bounded. In particular, if both (B_1, M_{1e}) and (B_2, M_{2e}) are 1-bounded, so is $(B_1 >> B_2, M_e)$.

18) If both (B_1, M_{1e}) and (B_2, M_{2e}) terminate properly, so does $(B_1 >> B_2, M_e)$.

19) If both (B_1, M_{1e}) and (B_2, M_{2e}) are almost live and terminate properly, then $(B_1 >> B_2, M_e)$ is almost live.

20) If both (B_1, M_{1e}) and (B_2, M_{2e}) are almost reversible and terminate properly, then $(B_1 >> B_2, M_e)$ is almost reversible.

Proof.

1-4) According to the structure of $B_1 >> B_2$, adding transition ε and the arcs does not violate the definitions of SMs, MGs, FC nets and AC nets.

5) Since B_1 and B_2 are conservative, by Characterization 2.1, $\exists \alpha_1 = (x_1, \ldots, x_{|P_1|}) \geq 1$ and $\alpha_2 = (y_1, \ldots, y_{|P_2|}) \geq 1$ such that $\alpha_1 V_1 = 0$ and $\alpha_2 V_2 = 0$. Let $\alpha = (y_1 \alpha_1, x_{|P_1|} \alpha_2)$ and V be shown in Figure 4.2. Then, $\alpha \geq 1$ and $\alpha V = (\alpha V[P, \varepsilon], \alpha V[P, T_1], \alpha V[P, T_2]) = (-y_1 x_{|P_1|} +$

$x_{|P_1|}y_1$, $y_1\alpha_l V_1$, $x_{|P_1|}\alpha_2 V_2) = 0$. It follows from Characterization 2.1 that $B_1 >> B_2$ is conservative.

6) Suppose B_1 and B_2 are structurally bounded. The same argument as in 5) shows that $\exists \alpha \geq 1$ such that $\alpha V \leq 0$. It follows from Characterization 2.2 that $B_1 >> B_2$ is structurally bounded.

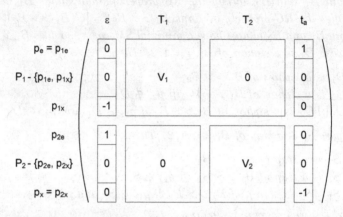

Fig. 4.2   Incidence matrix V_a of $(B_1 >> B_2)_a$ (V of $B_1 >> B_2$ is the part of V_a without column t_a).

7) Since B_1 and B_2 are almost consistent, by Characterization 2.3, $\exists \beta_1 = (x_1, \ldots, x_{|T_1|}, z_1) \geq 1$ and $\beta_2 = (y_1, \ldots, y_{|T_2|}, z_2) \geq 1$ such that $V_{1a}\beta_1 = 0$ and $V_{2a}\beta_2 = 0$. Let $\beta = (z_2 z_1, z_2 \beta_1[T_1], z_1 \beta_2[T_2], z_1 z_2)$ and V_a be shown in Figure 4.2. Then, $\beta \geq 1$. Since $(t_{1a}, t_{2a}) = V[P, \varepsilon] + V[P, t_a]$, $V_a\beta = (V_a[P_1, T_a]\beta, V_a[P_2, T_a]\beta) = (z_2 V_{1a}\beta_1, z_1 V_{2a}\beta_2) = 0$. It follows from Characterization 2.3 that $B_1 >> B_2$ is almost consistent.

8) Suppose B_1 and B_2 are almost repetitive. The same argument as in 7) shows that $\exists \beta \geq 1$ such that $V_a\beta \geq 0$. It follows from Characterization 2.4 that $B_1 >> B_2$ is almost repetitive.

9) As shown in Figure 4.2, V becomes diagonal after ignoring column ε. Hence, $Rank(B_1) + Rank(B_2) \leq Rank(B_1 >> B_2) \leq Rank(B_1) + Rank(B_2) + 1$. If the column $V[P_1, \varepsilon]$ is linearly independent from the columns of $V[P_1, T_1]$ or the column $V[P_2, \varepsilon]$ is linearly independent from the columns of $V[P_2, T_2]$, then $Rank(B_1 >> B_2) = Rank(B_1) + Rank(B_2) + 1$. Otherwise, $Rank(B_1 >> B_2) = Rank(B_1) + Rank(B_2)$.

10) Since the only added transition ε is absorbed into the cluster $[p_{1x}]$ of

B_1, no clusters have been created or destroyed. Hence, $|C(B_1 >> B_2)| = |C(B_1)| + |C(B_2)|$.

11) As shown in Figure 4.2, by 9) and 10), if the column $V[P, \varepsilon]$ is linearly independent from the columns of $V[P, T_1 \bigcup T_2]$, then $Rank(B_1 >> B_2) = Rank(B_1) + Rank(B_2) + 1 = |C(B_1)| - 1 + |C(B_2)| - 1 + 1 = |C(B_1 >> B_2)| - 1$. If the column $V[P, \varepsilon]$ is linearly dependent from the columns of $V[P, T_1 \bigcup T_2]$, then $Rank(B_1 >> B_2) = Rank(B_1) + Rank(B_2) = |C(B_1)| - 1 + |C(B_2)| - 1 = |C(B_1 >> B_2)| - 2$.

12) By definition, for $i = 1, 2$, any SM-component of B_{ia} must either contain p_{ie}, t_{ia} and p_{ix}, or none of them. Suppose K_1 and K_2 are minimal SM-covers of B_{1a} and B_{2a}, respectively. For any $S_i \in K_1$ that contains p_{1e} and p_{1x}, and any $S_j \in K_2$ that contains p_{2e} and p_{2x}, create S_{ij} by combining S_i and S_j, adding ε and replacing arcs (p_{1x}, t_{1a}) and (t_{2a}, p_{2e}) with the arcs (p_{1x}, ε) and (ε, p_{2e}), respectively. Obviously, S_{ij} is an SM-component of $(B_1 >> B_2)_a$. Then, $K_1 \bigcup K_2$, after deleting all such S_i and S_j, and including such S_{ij} is a minimal SM-cover for $(B_1 >> B_2)_a$.

13) (a) $^\bullet D_1$ in $(B_1 >> B_2) = {}^\bullet D_1$ in $B_1 \subseteq D_1^\bullet$ in $B_1 \subseteq D_1^\bullet$ in $B_1 >> B_2$.

(b) If $p_{2e} \notin D_2$, ${}^\bullet D_2$ in $(B_1 >> B_2) = {}^\bullet D_2$ in B_2 and $D_2^\bullet$ in $(B_1 >> B_2) = D_2^\bullet$ in B_2. Since $({}^\bullet D_2 \subseteq D_2^\bullet)$ in B_2, $({}^\bullet D_2 \subseteq D_2^\bullet)$ in $(B_1 >> B_2)$. If $p_{2e} \in D_2$, then $\varepsilon \in {}^\bullet D_2$ but $\varepsilon \notin D_2^\bullet$ in $(B_1 >> B_2)$. Hence, $({}^\bullet D_2 \subseteq D_2^\bullet)$ in $(B_1 >> B_2)$.

(c) If $p_{1x} \in D_1$ and $p_{2e} \in D_2$, ${}^\bullet(D_1 \bigcup D_2)$ in $(B_1 >> B_2) = \{\varepsilon\} \bigcup ({}^\bullet D_1 \text{in } B_1) \bigcup ({}^\bullet D_2 in B_2)$ and $(D_1 \bigcup D_2)^\bullet$ in $(B_1 >> B_2) = \{\varepsilon\} \bigcup (D_1^\bullet$ in $B_1) \bigcup (D_2^\bullet$ in $B_2)$. Since $({}^\bullet D_1 \subseteq D_1^\bullet)$ in B_1 and $({}^\bullet D_2 \subseteq D_2^\bullet)$ in B_2, it follows that $({}^\bullet(D_1 \bigcup D_2) \subseteq (D_1 \bigcup D_2)^\bullet)$ in $(B_1 >> B_2)$.

14) (a) $S_2^\bullet$ in $(B_1 >> B_2) = S_2^\bullet$ in $B_2 \subseteq {}^\bullet S_2$ in $B_2 \subseteq {}^\bullet S_2 in (B_1 >> B_2)$.

(b) If $p_{1x} \notin S_1$, ${}^\bullet S_1$ in $(B_1 >> B_2) = {}^\bullet S_1$ in B_1 and $S_1^\bullet$ in $(B_1 >> B_2) = S_1^\bullet$ in B_1. Since $(S_1^\bullet \subseteq {}^\bullet S_1)$ in B_1, $(S_1^\bullet \subseteq {}^\bullet S_1)$ in $(B_1 >> B_2)$. If $(S_1^\bullet \subseteq {}^\bullet S_1)$ in $(B_1 >> B_2)$, then $p_{1x} \notin S_1$ (because otherwise $\varepsilon \in S_1^\bullet$ but $\varepsilon \notin {}^\bullet S_1 in (B_1 >> B_2)$).

(c) Same as 13)(c).

15) [CZ (1994)] $Z(B_1 >> B_2) = |F| - |P \cup T| + 2 = (|F_1| + |F_2| + 2) - (|P_1| + |P_2| + |T_1| + |T_2| + 1) + 2 = |F_1| - (|P_1| + |T_1|) + 2 + |F_2| - (|P_2| + |T_2|) + 2 - 1 = Z(B_1) + Z(B_2) - 1$.

16) Since (B_1, M_{1e}) has no dead transitions, $\exists M' \in R(B_1 >> B_2, M_e)$ such that $M'(p_{1x}) = 1$ and $M'(P_2) = M_{2c}$. Hence, ε is friable at M'. After firing ε at M', one marking M'' will be reached such that

$M''(P_2) = M_{2e}$. Since (B_2, M_{2e}) also has no dead transitions, $(B_1 >> B_2, M_e)$ also has no dead transitions.

17) Within B_1, since (B_1, M_{1e}) terminates properly, M_{1x} is the only possible marking reached when B_1 terminates. Since $M_{1x}[B_1 >> B_2, \varepsilon\rangle$ and the behavior of $B_1 >> B_2$ before ε is fired (resp., ε has been fired) is the same as that of B_1 (resp., B_2), boundedness of $(B_1 >> B_2, M_e)$ follows from boundedness of (B_1, M_{1e}) and (B_2, M_{2e}), respectively. It is obvious that $(B_1 >> B_2, M_e)$ is 1-bounded if (B_1, M_{1e}) and (B_2, M_{2e}) are 1-bounded.

18) Let $M \in R(B_1 >> B_2, M_e)$. Since (B_1, M_{1e}) terminates properly, M must be in the form $M = M_{1s} + m_2$, where $m_2 \in R(B_2, M_{2e})$ and $m_2 \geq M_{2x}$. Since (B_2, M_{2e}) terminates properly, $m_2 \geq p_{2x} \Rightarrow m_2 = M_{2x} = M_{2s} + p_{2x}$. Hence, $M = M_{1s} + M_{2s} + p_x$.

19) Since both (B_1, M_{1e}) and (B_2, M_{2e}) terminate properly, any $M \in R((B_1 >> B_2)_a, M_e)$: $M(p_e) \leq 1$, $M(p_{1x}) \leq 1$, $M(p_{2e}) \leq 1$ and $M(p_x) \leq 1$. Since both (B_1, M_{1e}) and (B_2, M_{2e}) are almost live, it is obvious that $(B_1 >> B_2, M_e)$ is also almost live.

20) Similar to the proof of 19).

$\square$

4.2   The Operator CHOICE

CHOICE models the selection for execution between two processes B_1 and B_2.

Definition 4.2. (composition by the operator CHOICE, Figure 4.3)
For two marked processes (B_i, M_{ie}), $i = 1, 2$, their composition by *CHOICE*, in notation $(B_1 [\,] B_2, M_e)$, is defined as follows:

$P = P_1 \cup P_2 \cup \{p_e, p_x\} - \{p_{1x}, p_{2x}\}$
$T = T_1 \cup T_2 \cup \{\varepsilon_1, \varepsilon_2\}$
$F = F_1' \cup F_2' \cup \{(p_e, \varepsilon_1), (p_e, \varepsilon_2), (\varepsilon_1, p_{1e}), (\varepsilon_2, p_{2e})\}$, where
$F_i' = F_i - \{(t, p_{ix}) : t \in {}^\bullet p_{ix}\} \cup \{(t, p_x) : t \in {}^\bullet p_{ix}\}$, $i = 1, 2$.
$M_e = (1, 0, M_{1c}, 0, M_{2c}, 0)$.

It is obvious that $(B_1 [\,] B_2, M_e)$ is also a marked process. The incidence matrix V of $B_1 [\,] B_2$ is formed from the incidence matrices V_1 of B_1 and V_2 of B_2 as follows:

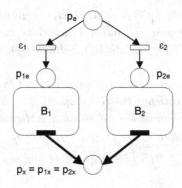

Fig. 4.3 $B_1[\ ]B_2$: Composition by the operator CHOICE.

$$
\begin{array}{c}
\begin{array}{cccc}
\varepsilon_1 & \varepsilon_2 & T_1 & T_2
\end{array} \\
\begin{array}{c}
p_e \\
P_1 - \{p_{1x}\} \\
P_2 - \{p_{2x}\} \\
p_x
\end{array}
\left(
\begin{array}{cccc}
-1 & -1 & 0 & 0 \\
p_{1e} & 0 & V_{11} & 0 \\
0 & p_{2e} & 0 & V_{21} \\
0 & 0 & V_{12} & V_{22}
\end{array}
\right)
\end{array}
$$

Explanation:

(a) $V_{11} = V_1[P_1 - \{p_{1x}\}, T_1]$ and $V_{21} = V_2[P_2 - \{p_{2x}\}, T_2]$.
(b) $V_{12} = V_1[p_{1x}, T_1]$ and $V_{22} = V_2[p_{2x}, T_2]$.

Theorem 4.2. *(property preservation under CHOICE $B_1\ [\ ]\ B_2$))*
Let (B_1, M_{1e}), (B_2, M_{2e}) and $(B_1\ [\ ]\ B_2, M_e)$ be the three marked PNPs involved in Definition 4.2. Then, the following propositions hold:

1) *If both B_1 and B_2 are SMs, so is $B_1[\ ]\ B_2$.*
2) *$B_1[\ ]\ B_2$ is not almost an MG, whether or not B_1 and B_2 are so.*
3) *If both B_1 and B_2 are FC nets, so is $B_1\ [\ ]\ B_2$.*
4) *If both B_1 and B_2 are AC nets, so is $B_1\ [\ ]\ B_2$.*
5) *(a) If both B_1 and B_2 are conservative, then $B_1\ [\ ]\ B_2$ is conservative provided that $M_{1e}[B_1,\ ^*\rangle M_{1x}$ and $M_{2e}[B_2,\ ^*\rangle M_{2x}$.*
   (b) If B_1 and B_2 are almost conservative, then $B_1\ [\ ]\ B_2$ is almost conservative.
6) *If both B_1 and B_2 are structurally bounded, so is $B_1\ [\ ]\ B_2$.*
7) *If both B_1 and B_2 are almost consistent, so is $B_1[\ ]\ B_2$.*
8) *If both B_1 and B_2 are almost repetitive, so is $B_1\ [\ ]\ B_2$.*

9) *In general,* $Rank(B_1 \;[\;]\; B_2) = Rank(B_1) + Rank(B_2) + k$, $k = 1$ *or*
    2.In particular, suppose B_{ia} *has a T-invariant including* t_{ia}, $i = 1, 2$.
    Then, $Rank(B_1[\;]\; B_2) = Rank(B_1) + Rank(B_2) + 1$.

10) $|C(B_1[\;]\; B_2)| = |C(B_1)| + |C(B_2)|$.

11) *If both* B_1 *and* B_2 *satisfy the RC-property and the condition stated in
    9), then* $B_1[\;]\; B_2$ *satisfies the RC-property.*

12) *If both* B_1 *and* B_2 *almost have a minimal SM-cover, so does* $B_1[\;]\; B_2$.

13) *Suppose* D *is a siphon of* B_1 *(resp.,* B_2*). Then,*

    (a) *D is a siphon of* $B_1 \;[\;]\; B_2$ *iff* $p_{1e} \notin D$ *and* $p_{1x} \notin D$ *(resp.,* $p_{2e} \notin D$
        and $p_{2x} \notin D$*);*

    (b) *$D \cup \{p_e\}$ is a siphon of* $B_1 \;[\;]\; B_2$ *if* $p_{1e} \in D$ *and* $p_{1x} \notin D$ *(resp.,*
        $p_{2e} \in D$ *and* $p_{2x} \notin D$*).*

14) *Suppose* S *is a trap of* B_1 *or* B_2. *Then,* S *is a trap of* $B_1[\;]\; B_2$.

15) $Z(B_1[\;]\; B_2) = Z(B_1) + Z(B_2)$.

16) *If both* (B_1, M_{1e}) *and* (B_2, M_{2e}) *have no dead transitions, then
    $(B_1 \;[\;]\; B_2, M_e)$ does not have either.*

17) *If both* (B_1, M_{1e}) *and* (B_2, M_{2e}) *are bounded, so is* $(B_1[\;]\; B_2, M_e)$.

18) *If both* (B_1, M_{1e}) *and* (B_2, M_{2e}) *terminate properly, so does* $(B_1[\;]\;
    B_2, M_e)$.

19) *If both* (B_1, M_{1e}) *and* (B_2, M_{2e}) *are almost live and terminate prop-
    erly, then* $(B_1[\;]\; B_2, M_e)$ *is almost live.*

20) *If both* (B_1, M_{1e}) *and* (B_2, M_{2e}) *are almost reversible and terminate
    properly, then* $(B_1 \;[\;]\; B_2, M_e)$ *is almost reversible.*

Proof.

1-4) According to the structure of $B_1 \;[\;]\; B_2$, the addition of p_e, ε_1, ε_2 and
     the arcs and the fusion of p_{1x} and p_{2x} do not violate the definitions
     of SM, FC and AC nets. However, they do violate the definition of
     MG because $|p_e^\bullet| > 1$ and $|{}^\bullet p_x| > 1$.

5) (a) Since B_1 and B_2 are conservative, by Characterization 2.1, $\exists \alpha_1$
    $= (x_1, \ldots, x_{|P_1|}) \geq 1$ and $\alpha_2 = (y_1, \ldots, y_{|P_2|}) \geq 1$ such that $\alpha_1 V_1 =$
    0 and $\alpha_2 V_2 = 0$. Since $M_{1e}[B_1, *\rangle M_{1x}$ and $M_{2e}[B_2, *\rangle M_{2x}$, $\alpha_1 M_{1e} =$
    $\alpha_1 M_{1x}$ and $\alpha_2 M_{2e} = \alpha_2 M_{2x}$. This implies $x_1 = x_{|P_1|}$ and $y_1 = y_{|P_2|}$.
    Let $\alpha = (x_1 y_1, y_1 \alpha_1 [P_1 - \{p_{1x}\}], x_1 \alpha_2 [P_2 - \{p_{2x}\}], x_{|P_1|} y_{|P_2|})$ and V
    be shown in Figure 4.4. Then, $\alpha \geq 1$ and $\alpha V = (\alpha V[P, \varepsilon_1], \alpha V[P,$
    $\varepsilon_2], \alpha V[P, T_1], \alpha V[P, T_2]) = (-x_1 y_1 + y_1 x_1, -x_1 y_1 + x_1 y_1, y_1 \alpha_1 V_1,$
    $x_1 \alpha_2 V_2) = 0$. It follows from Characterization 2.1 that $B_1 \;[\;]\; B_2$ is
    conservative.

(b) In fact, since B_1 and B_2 are almost conservative, $\exists \alpha_1 = (x_1, \ldots, x_{|P_1|}) \geq 1$ and $\alpha_2 = (y_1, \ldots, y_{|P_2|}) \geq 1$ such that $\alpha_1 V_{a1} = 0$ and $\alpha_2 V_{a2} = 0$. This implies that $x_1 = x_{|P_1|}$ and $y_1 = y_{|P_2|}$. Let $\alpha = (x_1 y_1, y_1 \alpha_1 [P_1 - \{p_{1x}\}], x_1 \alpha_2 [P_2 - \{p_{2x}\}], x_{|P_1|} y_{|P_2|})$ and V be shown in Figure 4.4. Then, $\alpha \geq 1$ and $\alpha V_a = (\alpha V_a [P, \varepsilon_1]$; $\alpha V_a [P, \varepsilon_2], \alpha V_a [P, T_1], \alpha V_a [P, T_2], \alpha V_a [P, t_a]) = (-x_1 y_1 + y_1 x_1, -x_1 y_1 + x_1 y_1, y_1 \alpha_1 V_1, x_1 \alpha_2 V_2, x_1 y_1 - x_{|P_1|} y_{|P_2|}) = 0$. It follows that $B_1 [\ ] B_2$ is almost conservative.

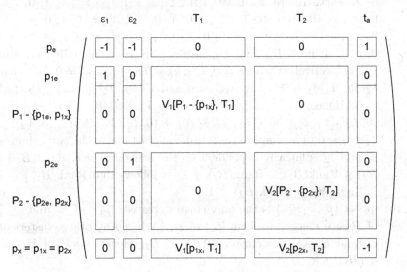

Fig. 4.4   Incidence matrix V_a of $(B_1 [\ ] B_2) a$ (V of $B_1 [\ ] B_2$ is the part of V_a without column t_a).

6) Since both B_1 and B_2 are structurally bounded, by Characterization 2.2, $\exists \alpha_1 = (x_1, \ldots, x_{|P_1|}) \geq 1$ and $\alpha_2 = (y_1, \ldots, y_{|P_2|}) \geq 1$ such that $\alpha_1 V_1 \leq 0$ and $\alpha_2 V_2 \leq 0$. Let $\alpha = (x_1 y_1, y_1 \alpha_1 [P_1 - \{p_{1x}\}], x_1 \alpha_2 [P_2 - \{p_{2x}\}], 1)$. Then, $\alpha \geq 1$ and $\alpha V \leq (-x_1 y_1 + y_1 x_1, -x_1 y_1 + x_1 y_1, y_1 \alpha_1 V_1, x_1 \alpha_2 V_2) \leq 0$. It follows from Characterization 2.2 that $B_1 [\ ] B_2$ is structurally bounded.

7) Since B_1 and B_2 are almost consistent, by Characterization 2.3, $\exists \beta_1 = (x_1, \ldots, x_{|T_1|}, z_1) \geq 1$ and $\beta_2 = (y_1, \ldots, y_{|T_2|}, z_2) \geq 1$ such that $V_{1a} \beta_1 = 0$ and $V_{2a} \beta_2 = 0$. Let $\beta = (z_1, z_2, \beta_1 [T_1], \beta_2 [T_2], z_1 + z_2)$ and V_a be shown in Figure 4.4. Then, $\beta \geq 1$ and $V_a \beta = (V_a [p_e, T_a] \beta, V_a [P_1 - \{p_{1x}\}, T_a] \beta, V_a [P_2 - \{p_{2x}\}, T_a] \beta, V_a [p_x, T_a] \beta) = (-z_1 - z_2 + z_1 + z_2, V_{1a} [P_1 - \{p_{1x}\}, T_{1a}] \beta_1, V_{2a} [P_2 - \{p_{2x}\}, T_{2a}] \beta_2, V_{1a} [p_{1x},$

$T_{1a}]\beta_1 + V_{2a}[p_{2x}, T_{2a}]\beta_2) = 0$. It follows from Characterization 2.3 that $B_1 [ ] B_2$ is almost consistent.

8) Suppose B_1 and B_2 are almost repetitive. The same argument as in 7) shows that $\exists \beta \geq 1$ such that $V_a\beta \geq 0$. It follows from Characterization 2.4 that $B_1 [ ] B_2$ is almost repetitive.

9) In V, since the columns ε_1 and ε_2 are both linearly independent from the columns of $T_1 \cup T_2$, $\text{Rank}(B_1) + \text{Rank}(B_2) + 1 = \text{Rank}(V[P, T_1]) + \text{Rank}(V[P, T_2]) + 1 \leq \text{Rank}(B_1[ ] B_2) \leq \text{Rank}(B_1) + \text{Rank}(B_2) + 2$. Furthermore, we have: $V[P, \varepsilon_1] - V[P, \varepsilon_2] = (-1, 1, 0, \ldots, 0, 0, \ldots, 0) - (-1, 0, 0, \ldots, 0, 0, 1, 0, \ldots, 0) = (0, 1, 0, \ldots, 0, 0, \ldots, -1) - (0, 0, 0, \ldots, 0, 0, 1, 0, \ldots, -1)^T = (0, t_{1a}, 0) - (0, 0, t_{2a})$. If both B_{1a} and B_{2a} have a T-invariant including t_{1a} and t_{2a}, respectively, then $\exists k \neq 0$, $(\beta_1, k) \neq 0$ and $(\beta_2, k) \neq 0$ such that $V_1[P_1, T_1]\beta_1 + kV_1[P_1, t_{1a}] = 0$ and $V_2[P_2, T_2]\beta_2 + kV_2[P_2, t_{2a}] = 0$. Hence, $kV[P, \varepsilon_1] - kV[P, \varepsilon_2] + V[P, T_1]\beta_1 - V[P, T_2]\beta_2 = (0, kV_1[P_1, t_{1a}] + V_1[P_1, T_1]\beta_1, 0) + (0, 0, kV_2[P_2, t_{2a}] + V_2[P_2, T_2]\beta_2) = 0$. This implies that, in V, at least one of the two columns ε_1 and ε_2 is linearly dependent on the others. Hence, $\text{Rank}(B_1 [ ] B_2) \leq \text{Rank}(B_1) + \text{Rank}(B_2) + 1$. It follows that $\text{Rank}(B_1[ ] B_2) = \text{Rank}(B_1) + \text{Rank}(B_2) + 1$.

10) $[p_e] = \{p_e, \varepsilon_1, \varepsilon_2\}$ is the only cluster created. $[p_{1x}]$ of B_1 and $[p_{2x}]$ of B_2 are merged to form $[p_x]$ of $B_1 [ ] B_2$. Any other clusters of B_1 and B_2 are also clusters of $B_1 [ ] B_2$. Hence, $|C(B_1[ ] B_2)| = |C(B_1)| + |C(B_2)|$.

11) This follows from 9) and 10).

12) By definition, any SM-component of B_{ia} containing p_{ie} must also contain t_{ia} and p_{ix}, where i = 1, 2. Suppose K_1 and K_2 are minimal SM-covers of B_{1a} and B_{2a}, respectively. For any $S_i \in K_1$ which contains p_{1e} and p_{1x} and any $S_j \in K_2$ which contains p_{2e} and p_{2x}, create S_{ij} by combining S_i and S_j, adding p_e, ε_1, ε_2 and the arcs (p_e, ε_1), (ε_1, p_{1e}), (p_e, ε_2) and (ε_2, p_{2e}) and merging (p_{ix}, t_{ia}), (t_{ia}, p_{ie}), i = 1, 2, to (p_x, t_a) and (t_a, p_e), respectively. Obviously, S_{ij} is an SM-component of $(B_1 [ ] B_2)_a$. Then, $K_1 \cup K_2$, after deleting all such S_i and S_j, and adding such S_{ij} is a minimal SM-cover for $(B_1 [ ] B_2)_a$.

13) We shall prove for the case $D \subseteq P_1$. For $D \subseteq P_2$, the proof is similar.

   (a) If $p_{1e} \notin D$ and $p_{1x} \notin D$, then $^\bullet D$ in $(B_1 [ ] B_2) = ^\bullet D$ in $B_1 \subseteq D^\bullet$ in $B_1 \subseteq D^\bullet$ in $(B_1 [ ] B_2)$. If $(^\bullet D \subseteq D^\bullet)$ in $(B_1 [ ] B_2)$, then

$p_{1e} \notin D$ in B_1 (because otherwise $\varepsilon_1 \in {}^\bullet D$ but $\varepsilon_1 \notin D^\bullet$ in $(B_1$ [ ] $B_2)$) and $p_{1x} \notin D$ in B_1 (because, otherwise, ${}^\bullet D$ in $(B_1$ [ ] $B_2)= ({}^\bullet D$ in $B_1) \cup^\bullet p_{2x} \not\subset D^\bullet$ in $B_1 = D^\bullet$ in $(B_1$ [ ] $B_2)$.

(b) If $p_{1e} \in D$ and $p_{1x} \notin D$, then ${}^\bullet(D \cup \{p_e\})$ in $(B_1$ [ ] $B_2) = \{\varepsilon_1\}$ $\cup ({}^\bullet D$ in $B_1)$ and $(D \cup \{p_e\})^\bullet$ in $(B_1$ [ ] $B_2) = \{\varepsilon_1, \varepsilon_2\} \cup (D^\bullet$ in $B_1)$. Since $({}^\bullet D \subseteq D^\bullet)$ in B_1, it follows that $({}^\bullet(D \cup \{p_e\}) \subseteq (D \cup \{p_e\})^\bullet)$ in $(B_1$ [ ] $B_2)$.

14) According to the structure of B_1 [ ] B_2, $\forall S \subseteq P_i$ for $i = 1$ or 2: $(S^\bullet$ in $B_i = S^\bullet$ in $(B_1$ [ ] $B_2)$ and ${}^\bullet S$ in $B_i \subseteq {}^\bullet S$ in $(B_1$ [ ] $B_2))$. Therefore, S is a trap of B_1 [ ] B_2 if S is a trap of B_1 or B_2.

15) [CZ (1994)] $Z(B_1$ [ ] $B_2) = (|F_1|+|F_2| + 4) - (|P_1|+|P_2| + |T_1|+|T_2| +2) + 2 = Z(B_1) + Z(B_2)$.

16) The result is obvious.

17-18) Since the behavior of $(B_1$ [ ] $B_2, M_e)$ is the same as (B_i, M_{ie}) after firing ε_i, boundedness (resp., proper termination) of $(B_1$ [ ] $B_2, M_e)$ follows from the boundedness (resp., proper termination) of (B_1, M_{1e}) and (B_2, M_{2e}).

19) Since both (B_1, M_{1e}) and (B_2, M_{2e}) terminate properly, any marking $M \in R((B_1$ [ ] $B_2)_a, M_e)$ can have only three forms: (1) (after ε_1 is fired) $M = (0, m_1, 0, M_{2c}, M(p_x))$, where $(m_1, M(p_x)) \in R(B_{1a}, M_{1e})$. Consider two possible locations of an arbitrary transition t. (a) $t \in T_1 \cup \{\varepsilon_1, t_a\}$. If $t = t_a$ or ε_1, replace t with t_{1a}. Then, $\exists \sigma_1 \in T_{1a}^*$: $((m_1, M(p_x))[B_{1a}, \sigma_1 t\rangle)$, where $\sigma_1 = \sigma_{11}$ or $\sigma_1 = \sigma_{11}t_{1a}\sigma_{12}$ and $\sigma_{11}, \sigma_{12} \in T_1^*$. If $\sigma_1 = \sigma_{11}$, then $M[(B_1$ [ ] $B_2)_a, \sigma t\rangle$, where $\sigma = \sigma_{11}$ for $t \in T_1 \cup \{t_a\}$ and $\sigma = \sigma_{11}t_a$ for $t = \varepsilon_1$. If $\sigma_1 = \sigma_{11}t_{1a}\sigma_{12}$, then $M[(B_1$ [ ] $B_2)_a, \sigma t\rangle$, where σ is the same as σ_1 except that every t_{1a} is replaced with $t_a \varepsilon_1$. (b) $t \in T_2 \cup \{\varepsilon_2, t_a\}$. Since (B_1, M_{1e}) is almost live, $\exists \sigma_1 \in T_1^*$: $((m_1 M(p_x))[B_{1a}, \sigma_1 t_{1a}\rangle)$. Then, $M[(B_1$ [ ] $B_2)_a, \sigma_1 t_a \varepsilon_2 \rangle (0, 0, M_{1c}, m_2, M(p_x))$, where $M_{2e} = (m_2, M(p_x))$. This becomes Case (2) to be discussed below. (2) (after ε_2 is fired) $M = (0, 0, M_{1c}, m_2, (p_x))$, where $(m_2, M(p_x)) \in R(B_{2a}, M_{2e})$. Similar argument as in (1) shows that t is live. (3) $M = (1, 0, M_{1c}, 0, M_{2c}, 0)$. Similar argument as in (1) and (2) shows that t is live.

20) Any $M \in R((B_1$ [ ] $B_2)_a, M_e)$ can only have three forms: (1) $M = (0, m_1, 0, M_{2c}, M(p_x))$, where $(m_1, M(p_x)) \in R(B_{1a}, M_{1e})$. If (B_1, M_{1e}) terminates properly, then $\exists \sigma_1 \in T_1^*$: $((m_1, M(p_x))[B_{1a}, \sigma_1 t_{1a}\rangle M_{1e})$. Hence, $M[(B_1$ [ ] $B_2)_a, \sigma_1 t_a\rangle M_e$. (2) $M = (0, 0,$

$M_{1c}, m_2, M(p_x))$, where $(m_2, M(p_x)) \in R(B_{2a}, M_{2e})$. If (B_2, M_{2e}) terminates properly, then $\exists \sigma_2 \in T_2^*$: $((m_2, M(p_x))[B_{2a}, \sigma_2 t_{2a}) M_{2e})$. Hence, $M[(B_1 \;[\;]\; B_2)_a, \sigma_2 t_a) M_e$. (3)$M = (1, 0, M_{1c}, 0, M_{2c}, 0)$. Obviously, $M[(B_1 \;[\;]\; B_2)_a, \lambda) M_e$, where λ is the null sequence of transitions. Similarly, we can prove that $(B_1[\;] B_2, M_e)$ is almost reversible if both (B_1, M_{1e}) and (B_2, M_{2e}) terminate properly or are almost 1-bounded.

$\square$

4.3   The Operator INTERLEAVE

INTERLEAVE models the concurrent but non-interfering execution of two processes B_1 and B_2. They are synchronized at the entrance and at the exit.

Definition 4.3. (composition by the operator INTERLEAVE, Figure 4.5) For two marked processes (B_i, M_{ie}), $i = 1, 2$, their composition by *INTERLEAVE*, in notation $(B_1|||B_2, M_e)$ is defined as follows:

$P = P_1 \cup P_2 \cup \{p_e, p_x\} - \{p_{1x}, p_{2x}\}$

$T = T_1 \cup T_2 \cup \{\varepsilon_1, \varepsilon_2\}$

$F = F_1 \cup F_2 \cup \{(p_e, \varepsilon_1), (\varepsilon_1, p_{1e}), (\varepsilon_1, p_{2e}), (p_{1x}, \varepsilon_2), ((p_{2x}, \varepsilon_2), (\varepsilon_2, p_x)\}$

$M_e = (1, 0, M_{1c}, 0, 0, M_{2c}, 0, 0)$.

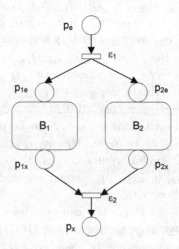

Fig. 4.5 $B_1|||B_2$: Composition by the operator INTERLEAVE.

It is obvious that $(B_1|||B_2, M_e)$ is also a marked process. The incidence matrix V of $B_1|||$ B_2 is formed from the incidence matrices V_1 of B_1 and V_2 of B_2 as follows:

$$
\begin{array}{c}
\begin{array}{cccc} \varepsilon_1 & \varepsilon_2 & T_1 & T_2 \end{array} \\
\begin{array}{c} p_e \\ P_1 \\ P_2 \\ p_x \end{array}
\left(\begin{array}{cccc}
-1 & 0 & 0 & 0 \\
p_{1e} & -p_{1x} & V_1 & 0 \\
p_{2e} & -p_{2x} & 0 & V_2 \\
0 & 1 & 0 & 0
\end{array}\right)
\end{array}
$$

Theorem 4.3. *(property preservation under INTERLEAVE $B_1 |||B_2$)*
Let (B_1, M_{1e}), (B_2, M_{2e}) and $(B_1|||B_2, M_e)$ be the three marked PNPs involved in Definition 4.3. Then, the following propositions hold.

1) $B_1|||B_2$ *is not an SM, whether or not B_1 and B_2 are so.*
2) *If both B_1 and B_2 are almost MGs, so is $B_1|||B_2$.*
3) *If both B_1 and B_2 are FC nets, so is $B_1|||B_2$.*
4) *If both B_1 and B_2 are AC nets, so is $B_1|||B_2$.*
5) *If both B_1 and B_2 are conservative, so is $B_1|||B_2$.*
6) *If both B_1 and B_2 are structurally bounded, so is $B_1|||B_2$.*
7) *If both B_1 and B_2 are almost consistent, so is $B_1|||B_2$.*
8) *If both B_1 and B_2 are almost repetitive, so is $B_1|||B_2$.*
9) *$Rank(B_1|||B_2) = Rank(B_1) + Rank(B_2) + 2$.*
10) *$|C(B_1|||B_2)| = |C(B_1)| + |C(B_2)| + 1$.*
11) *If both B_1 and B_2 satisfy the RC-property, so does $B_1|||B_2$.*
12) *If both B_1 and B_2 both almost have a minimal SM-cover, so does $B_1|||B_2$.*
13) *Suppose D is a siphon of B_1 (resp., B_2). Then,*

    (a) *D is a siphon of $B_1|||B_2$ iff $p_{1e} \notin D$ (resp., $p_{2e} \notin D$);*
    (b) *$D \cup \{p_e\}$ is a siphon of $B_1|||B_2$ provided that $p_{1e} \in D$ (resp., $p_{2e} \in D$).*

14) *Suppose S is a trap of B_1 (resp., B_2). Then,*

    (a) *S is a trap of $B_1|||B_2$ iff $p_{1x} \notin S$ (resp., $p_{2x} \notin S$);*
    (b) *$S \cup \{p_x\}$ is a trap of $B_1|||B_2$ provided that $p_{1x} \in S$ (resp., $p_{2x} \in S$).*

15) *$Z(B_1|||B_2) = Z(B_1) + Z(B_2)$.*
16) *If both (B_1, M_{1e}) and (B_2, M_{2e}) have no dead transitions, then $(B_1|||B_2, M_e)$ does not have either.*

17) *If both (B_1, M_{1e}) and (B_2, M_{2e}) are bounded, so is $(B_1|||B_2, M_e)$.*

18) *If both (B_1, M_{1e}) and (B_2, M_{2e}) terminate properly, so does $(B_1|||B_2, M_e)$.*

19) *If both (B_1, M_{1e}) and (B_2, M_{2e}) are almost live and terminate properly, then $(B_1|||B_2, M_e)$ is almost live.*

20) *If both (B_1, M_{1e}) and (B_2, M_{2e}) are almost reversible and terminate properly, then $(B_1|||B_2, M_e)$ is almost reversible.*

Proof.

1-4) According to the structure of $B_1|||B_2$, the addition of p_e, p_x, ε_1, ε_2 and the arcs does not violate the definitions of almost MG, FC and AC nets. However, $B_1|||B_2$ is no longer an SM because $|\varepsilon_1^\bullet| > 1$ and $|^\bullet\varepsilon_2| > 1$.

5) Since B_1 and B_2 are conservative, by Characterization 2.1, $\exists \alpha_1 = (x_1, \ldots, x_{|P_1|}) \geq 1$ and $\alpha_2 = (y_1, \ldots, y_{|P_2|}) \geq 1$ such that $\alpha_1 V_1 = 0$ and $\alpha_2 V_2 = 0$. Let $\alpha = (x_1 + y_1, \alpha_1, \alpha_2, x_{|P_1|} + y_{|P_2|})$ and V be shown in Figure 4.6. Then, $\alpha \geq 1$ and $\alpha V = (\alpha V[P, \varepsilon_1], \alpha V[P, \varepsilon_2], \alpha V[P, T_1], \alpha V[P, T_2]) = (-x_1 - y_1 + x_1 + y_1, -x_{|P_1|} - y_{|P_2|} + x_{|P_1|} + y_{|P_2|}, \alpha_1 V_1, \alpha_2 V_2) = 0$. It follows from Characterization 2.1 that $B_1|||B_2$ is conservative.

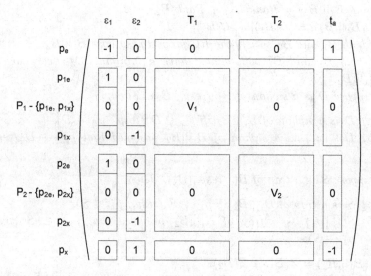

Fig. 4.6 Incidence matrix V_a of $(B_1|||B_2)_a$ (V of $B_1|||B_2$ is the part of V_a without column t_a).

6) Suppose B_1 and B_2 are structurally bounded. The same argument as in 5) shows that $\exists \alpha \geq 1$ such that $\alpha V \leq 0$. It follows from Characterization 2.2 that $B_1|||B_2$ is structurally bounded.

7) Since B_1 and B_2 are almost consistent, by Characterization 2.3, $\exists \beta_1 = (x_1, \ldots, x_{|T1|}, z_1) \geq 1$ and $\beta_2 = (y_1, \ldots, y_{|T2|}, z_2) \geq 1$ such that $V_{1a}\beta_1 = 0$ and $V_{2a}\beta_2 = 0$. Let $\beta = (z_1 z_2, z_1 z_2, z_2 \beta_1[T_1], z_1 \beta_2[T_2], z_1 z_2)$ and V_a be shown in Figure 4.6. Then, $\beta \geq 1$. Since $V[P_i, \varepsilon_1] + V[P_i, \varepsilon_2] = V_{ia}[P_i, t_{ia}]$ for $i = 1, 2$, we have $V_a\beta = (V_a[p_e, T_a]\beta, V_a[P_1, T_a]\beta, V_a[P_2, T_a]\beta, V_a[p_x, T_a]\beta) = (-z_1 z_2 + z_1 z_2, z_2 V_{1a}\beta_1, z_1 V_{2a}\beta_2, z_1 z_2 - z_1 z_2) = 0$. It follows from Characterization 2.3 that $B_1|||B_2$ is almost consistent.

8) Suppose B_1 and B_2 are almost repetitive. The same argument as in 7) shows that $\exists \beta \geq 1$ such that $V_a\beta \geq 0$. It follows from Characterization 2.4 that $B_1|||B_2$ is almost repetitive.

9) It is obvious that column ε_1 (resp., ε_2) is linearly independent from the columns of $V[P, \{\varepsilon_2\} \cup T_1 \cup T_2]$ (resp., $V[P, \{\varepsilon_1\} \cup T_1 \cup T_2]$). Therefore, Rank$(B_1 |||B_2)$ = Rank(B_1) + Rank(B_2) + 2.

10) For $B_1|||B_2$, two new clusters $[p_e] = \{p_e, \varepsilon_1\}$ and $[p_x] = \{p_x\}$ are created, whereas $[p_{1x}] = \{p_{1x}\}$ of B_1 and $[p_{2x}] = \{p_{2x}\}$ of B_2 are merged to form cluster $[\varepsilon_2] = \{p_{1x}, p_{2x}, \varepsilon_2\}$ of $B_1|||B_2$. All other clusters of B_1 and B_2 are also clusters of $B_1||| B_2$. Hence, $|C(B_1|||B_2)| = |C(B_1)| + |C(B_2)| + 1$.

11) This follows from 9) and 10).

12) By definition, any SM-component of B_{ia} containing p_{ie} must also contain t_{ia} and p_{ix}, where $i = 1, 2$. Suppose K_1 and K_2 are minimal SM-covers of B_{1a} and B_{2a}, respectively. For any $S_i \in K_1$ which contains p_{1e} and p_{1x}, extend S_i by adding $p_e, \varepsilon_1, \varepsilon_2, p_x, t_a$ and the arcs $(p_e, \varepsilon_1), (\varepsilon_1, p_{1e}), (p_{1x}, \varepsilon_2), (\varepsilon_2, p_x) (p_x, t_a)$ and (t_a, p_e) and deleting $t_{1a}, (p_{1x}, t_{1a})$ and (t_{1a}, p_{1e}). Do similarly for any $S_j \in K_2$ which contains p_{2e} and p_{2x}. Obviously, these result in a minimal SM-cover of $(B_1|||B_2)_a$.

13) We shall prove for the case $D \subseteq P_1$. For $D \subseteq P_2$, the proof is similar.

   (a) $p_{1e} \notin D$. Then, $^\bullet D$ in $(B_1|||B_2) = {}^\bullet D$ in $B_1 \subseteq D^\bullet$ in $B_1 \subseteq D^\bullet$ in $(B_1|||B_2)$. If $p_{1e} \in D$, then $\varepsilon_1 \in {}^\bullet D$ but $\varepsilon_1 \notin D^\bullet$ in $(B_1|||B_2)$. Hence, $(^\bullet D \not\subset D^\bullet)$ in $(B_1|||B_2)$ even if $(^\bullet D \subseteq D^\bullet)$ in B_1.

   (b) $p_{1e} \in D$. Then, $^\bullet(D \cup \{p_e\})$ in $(B_1|||B_2) = \{\varepsilon_1\} \cup (^\bullet D$ in $B_1) \subseteq \{\varepsilon_1\} \cup (D^\bullet$ in $B_1) \subseteq (D \cup \{p_e\})^\bullet$ in $(B_1|||B_2)$.

14) We shall prove for the case $S \subseteq P_1$. For $S \subseteq P_2$, the proof is similar.

(a) $p_{1x} \notin S$. Then, $S^\bullet$ in $(B_1|||B_2)' = S^\bullet$ in $B_1 \subseteq {}^\bullet S$ in $B_1 \subseteq {}^\bullet S$ in $(B_1|||B_2)$. If $p_{1x} \in S$, then $\varepsilon_2 \in S^\bullet$ but $\varepsilon_2 \notin {}^\bullet S$ in $(B_1|||B_2)$. Hence, $(S^\bullet \not\subseteq {}^\bullet S)$in $(B_1|||B_2)$ even if $({}^\bullet S \subseteq S^\bullet)$ in B_1.

(b) $p_{1x} \in S$. Then, $(S \cup \{p_x\})^\bullet$ in $(B_1|||B_2) = \{\varepsilon_2\} \cup (S^\bullet$ in $B_1) \subseteq \{\varepsilon_2\} \cup ({}^\bullet S$ in $B_1) \subseteq {}^\bullet(S \cup \{p_x\}))$ in $(B_1|||B_2)$.

15) $Z(B_1|||B_2) = |F| - |P \cup T| + 2 = |F_1| + |F_2| + 6 - (|P_1| + |P_2| + 2 + |T_1| + |T_2| + 2) + 2 = |F_1| - (|P_1| + |T_1|) + 2 + |F_2| - (|P_2| + |T_2|) + 2 = Z(B_1) + Z(B_2)$.

16) Obviously ε_1 is friable at M_e. After p_{1e} and p_{2e} get one token, the behavior of any transition t in $T_1 \cup T_2$ is the same as either a t in (B_1, M_{1e}) or a t in (B_2, M_{2e}). Hence, t is not dead in $(B_1|||B_2, M_e)$. Furthermore, since some $t \in {}^\bullet p_{1x}$ is not dead, $\exists M' \in R(B_1|||B_2, M_e)$ such that $M'(p_{1x}) = 1$. Hence, ε_2 is also not dead in $(B_1||| B_2, M_e)$.

17) Suppose $R(B_i, M_{ie})$ is k_i-bounded for $i = 1, 2$. Consider any $M \in R(B_1|||B_2, M_e)$. Obviously, $M(p_e) \leq 1$. For $p \in P_i$, $i = 1$ or 2, $\exists M_i \in R(B_i, M_{ie})$ such that $M(p) \leq M_i(p) \leq k_i$. Lastly, since p_x obtains its tokens and p_{ix}, $i = 1, 2$, lose their tokens only by firing ε_2, $M(p_x)$ cannot be unbounded. Otherwise, $M(p_{1x})$ and $M(p_{2x})$ will also be unbounded if we withhold the firing of ε_2.

18) Consider any $M \in R(B_1||| B_2, M_e)$ such that $M(p_x) \geq 1$. Note that p_x can get a token only by firing $\varepsilon_2$when there is at least one token in both p_{1x} and p_{2x}. Since both B_i terminates property, just before firing ε_2, M must be of the form $(0, 0, M_{1c}, 0, M_{2c}, 1, 0)$. Then, after firing ε_2, one can only have $M = M_x$.

19) Since (B_1, M_{1e}) and (B_2, M_{2e}) terminate properly, any marking $M \in R((B_1|||B_2)_a, M_e)$ can only have three forms: (1)$M = (0, m_1, m_2, 0)$, where $m_1 \in R(B_{1a}, M_{1e})$ and $m_2 \in R(B_{2a}, M_{2e})$. Consider two possible locations of an arbitrary transition t. (a) $t \in T_1 \cup \{\varepsilon_1, \varepsilon_2, t_a\}$. Since (B_1, M_{1e}) and (B_2, M_{2e}) are almost live and terminate properly, $\exists \sigma_1 \in T_{1a}^*$: $(m_1[B_{1a}, \sigma_1 t\rangle)$, where $\sigma_1 = \sigma_{11}$ or $\sigma_1 = \sigma_{11} t_{1a} \sigma_{12}$, and $\sigma_{11}, \sigma_{12} \in T_1^*$. Also, $\exists \sigma_{21} \in T_2^*$: $(m_2[B_{2a}, \sigma_{21} t_{2a}\rangle)$. If $\sigma_1 = \sigma_{11}$, then $M[(B_1|||B_2)_a, \sigma t\rangle$, where $\sigma = \sigma_{11}$ for $t \in T_1$, $\sigma = \sigma_{21} \sigma_{11}$ for $t = \varepsilon_2$, $\sigma = \sigma_{21} \sigma_{11} \varepsilon_2$ for $t = t_a$, and $\sigma = \sigma_{21} \sigma_{11} \varepsilon_2 t_a$ for $t = \varepsilon_1$. If $\sigma_1 = \sigma_{11} t_{1a} \sigma_{12}$, then $M[(B_1||| B_2)_a, \sigma t\rangle$, where $\sigma = \sigma_{21} \sigma_{11} \varepsilon_2 t_a \varepsilon_1 \sigma_{12}$ for $t \in T_1$, $\sigma = \sigma_{21} \sigma_{11}$ for $t = \varepsilon_2$, $\sigma = \sigma_{21} \sigma_{11} \varepsilon_2$ for $t = t_a$ and $\sigma = \sigma_{21} \sigma_{11} \varepsilon_2 t_a$ for $t = \varepsilon_1$. That is, t is live. (b) $t \in T_2 \cup \{\varepsilon_1, \varepsilon_2, t_a\}$. Similar argument as in (a) can prove that t is also live. (2)$M = (0, 0, M_{1c}, 0, 0, M_{2c}, 0, 1)$. This case becomes case (1) after firing $t_a \varepsilon_1$ at

M. (3)$M = (1, 0, M_{1c}, 0, 0, M_{2c}, 0, 0)$. This case becomes case (1) after firing ε_1 at M.

20) Since (B_1, M_{1e}) and (B_2, M_{2e}) terminate properly, any marking $M \in R((B_1|||B_2)_a, M_e)$ can only have three forms: (1)$M = (0, m_1, m_2, 0)$, where $m_1 \in R(B_{1a}, M_{1e})$ and $m_2 \in R(B_{2a}, M_{2e})$. For $i = 1, 2$, since (B_i, M_{ie}) is almost reversible and terminates properly, $\exists \sigma_i \in T_i^*$: $(m_i[B_{ia}, \sigma_i\rangle M_{ix}[B_{ia}, t_{ia}\rangle M_{ie})$. Hence, $M[(B_1|||B_2)_a, \sigma_1\sigma_2\varepsilon_2 t_a\rangle M_e$. (2)$M = (0, 0, M_{1c}, 0, 0, M_{2c}, 0, 1)$. This case becomes case (1) after firing $t_a\varepsilon_1$ at M. (3)$M = M_e = (1, 0, M_{1c}, 0, 0, M_{2c}, 0, 0)$. Obviously, $M[(B_1|||B_2)_a, \lambda\rangle M_e$, where λ is the null sequence of transitions. □

4.4 The Operator PARALLEL

PARALLEL with synchronization models the concurrent execution of two processes B$_1$ and B$_2$ with synchronizations at entrance at some transitions and at their exit points.

Definition 4.4. (composition by the operator PARALLEL, Figure 4.7) For two marked processes (B_i, M_{ie}), $i = 1, 2$, that have a 'common' set of transitions $G = T_1 \cap T_2$ to be synchronized, their composition by *PARALLEL* with synchronization, in notation $(B_1|G| B_2, M_e)$, is defined as follows:

$P = P_1 \cup P_2 \cup \{p_e, p_x\}$

$T = T_1 \cup T_2 \cup \{\varepsilon_1, \varepsilon_2\}$, where $T_1 \cap T_2 \neq \phi$.

$F = F_1 \cup F_2 \cup \{(p_e, \varepsilon_1), (\varepsilon_1, p_{1e}), (\varepsilon_1, p_{2e}), (p_{1x}, \varepsilon_2), ((p_{2x}, \varepsilon_2), (\varepsilon_2, p_x)\}$

$M_e = (1, 0, M_{1c}, 0, 0, M_{2c}, 0, 0)$.

Graphically, the composite process (Figure 4.7) is obtained by:

1) adding a new entry place p_e and a dummy transition ε_1 for synchronizing the entry into B_1 and B_2;
2) $\forall t \in G$, merging transition t of B_1 and t of B_2; and
3) adding a new exit place p_x and a dummy exit transition ε_2 for synchronizing the exit from B_1 and the exit from B_2.

It is obvious that $(B_1|G|B_2, M_e)$ is also a marked process. The incidence matrix V of $B_1|G|B_2$ is formed from the matrices V_1 of B_1 and V_2 of B_2 as follows:

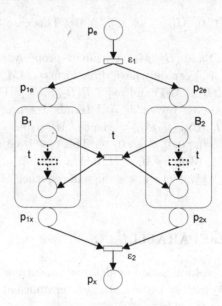

Fig. 4.7 $B_1|G|B_2$: Composition by the operator PARALLEL.

$$
\begin{array}{c}
\begin{array}{ccccc}
\varepsilon_1 & \varepsilon_2 & T_1 - G & G & T_2 - G
\end{array} \\
\begin{array}{c}
p_e \\ P_1 \\ P_2 \\ p_x
\end{array}
\left(
\begin{array}{ccccc}
-1 & 0 & 0 & 0 & 0 \\
p_{1e} & -p_{1x} & V_{11} & V_{12} & 0 \\
p_{2e} & -p_{2x} & 0 & V_{21} & V_{22} \\
0 & 1 & 0 & 0 & 0
\end{array}
\right)
\end{array}
$$

Explanation:

(a) $V_{11} = V_1[P_1, T_1 - G]$ and $V_{12} = V_1[P_1, G]$.
(b) $V_{21} = V_2[P_2, G]$ and $V_{22} = V_2[P_2, T_2 - G]$.

Synchronization has been one of the most complicated operations in the research of concurrent systems. Hence, as shown in Theorem 4.4 below, it is natural that extra conditions have to be added in order to preserve some of the properties.

Theorem 4.4. *(property preservation under PARALLEL $B_1|G|B_2$)*
Let (B_1, M_{1e}), (B_2, M_{2e}) and $(B_1|G| B_2, M_e)$ be the three marked PNPs involved in Definition 4.4. Then, the following propositions hold:

1) $B_1|G|B_2$ is not an SM, whether or not B_1 and B_2 are so.

2) *If both B_1 and B_2 are almost MGs, so is $B_1|G|\ B_2$.*

3) *Let both B_1 and B_2 be FC nets. There are three possible for $B_1|G|B_2$: (a) In general, $B_1|G|\ B_2$ may or may not be an FC net. (b) If $\forall t \in G$: $(^\bullet t)^\bullet = \{t\}$ in B_1 and B_2, then $B_1|G|B_2$ is an FC net. (c) If $\forall t \in G$: $(^\bullet t)^\bullet = \{t\}$ in B_1 or B_2, then $B_1|G|B_2$ is án AC net.*

4) *Let both B_1 and B_2 be AC nets. Two cases may hold for $B_1\ |G|B_2$: (a) In general, $B_1|G|B_2$ may or may not be an AC net. (b) $B_1|G|B_2$ is an AC net if $\forall t \in G$: $(^\bullet t)^\bullet = \{t\}$ in B_i and $(^\bullet t)^\bullet \supseteq \{t\}$ in B_j, where i, j = 1 or 2 and $i \neq j$.*

5) *If both B_1 and B_2 are conservative, so is $B_1|G|B_2$.*

6) *If both B_1 and B_2 are structurally bounded, so is $B_1|G|B_2$.*

7) *If $\exists \beta_1 = (x_1, \ldots, x_{|T_1|}, z_1) \geq 1$, $\beta_2 = (y_1, \ldots, y_{|T_2|}, z_2) \geq 1$: $((V_{1a}\beta_1 = 0$ and $V_{2a}\beta_2 = 0)$ and $(\forall t \in G$: $(z_1\beta_2[t] = z_2\beta_1[t]))$, then $B_1|G|B_2$ is almost consistent.*

8) *If $\exists \beta_1 = (x_1, \ldots, x_{|T_1|}, z_1) \geq 1$, $\beta_2 = (y_1, \ldots, y_{|T_2|}, z_2) \geq 1$: $((V_{1a}\beta_1 \geq 0$ and $V_{2a}\beta_2 \geq 0)$ and $(\forall t \in G$: $(z_1\beta_2[t] = z_2\beta_1[t]))$, then $B_1|G|B_2$ is almost repetitive.*

9) *$Rank(B_1|G|B_2) \geq Rank(B_1) + Rank(B_2) - |G| + 2$.*

10) *If, in B_1 (or B_2), $\forall t_1, t_2 \in G$: $(t_1 \neq t_2 \Rightarrow [t_1] \neq [t_2])$, then $|C(B_1|G|B_2) = |C(B_1)| + |C(B_2)| - |G| + 1$.*

11) *If both B_1 and B_2 satisfy the RC-property, then $B_1|G|B_2$ satisfies the RC-property provided that:*

    (a) $Rank(B_1|G|B_2) = Rank(B_1) + Rank(B_2) - |G| + 2$; and
    (b) In B_1 (or B_2), $\forall t_1, t_2 \in G$: $(t_1 \neq t_2 \Rightarrow [t_1] \neq [t_2])$.

12) *If both B_1 and B_2 almost have a minimal SM-cover, so does $B_1|G|B_2$.*

13) *Suppose D is a siphon of B_1 (resp., B_2). Then,*

    (a) D is a siphon of $B_1|G|B_2$ iff $p_{1e} \notin D$ (resp., $p_{2e} \notin D$);
    (b) $D \cup \{p_e\}$ is a siphon of $B_1|G|B_2$ provided that $p_{1e} \in D$ (resp., $p_{2e} \in D$).

14) *Suppose S is a trap of B_1 (resp., B_2). Then,*

    (a) S is a trap of $B_1|G|B_2$ iff $p_{1x} \notin S$ (resp., $p_{2x} \notin S$);
    (b) $S \cup \{p_x\}$ is a trap of $B_1|G|B_2$ provided that $p_{1x} \in S$ (resp., $p_{2x} \in S$).

15) *$Z(B_1|G|B_2) = Z(B_1) + Z(B_2) + |G|$.*

16) *$(B_1|G|B_2, M_e)$ may have some dead transitions even if both (B_1, M_{1e}) and (B_2, M_{2e}) have no dead transitions.*

17) *If both (B_1, M_{1e}) and (B_2, M_{2e}) are bounded, so is $(B_1|G|B_2, M_e)$.*

18) *In general $(B_1|G|B_2, M_e)$ may or may not terminate properly even if both (B_1, M_{1e}) and (B_2, M_{2e}) terminate properly.*

19) *If both (B_1, M_{1e}) and (B_2, M_{2e}) are almost live, then $(B_1|G|B_2, M_e)$ is almost live provided that:*

   (a) *B_1 and B_2 are both conservative;*

   (b) *$\exists \beta_1 = (x_1, \ldots, x_{|T_1|}, z_1) \geq 1$, $\beta_2 = (y_1, \ldots, y_{|T_2|}, z_2) \geq 1$: $((V_{1a}\beta_1 = 0$ and $V_{2a}\beta_2 = 0)$ and $(\forall t \in G: (z_1\beta_2[t] = z_2\beta_1[t])));$*

   (c) *B_1 and B_2 both satisfy the RC-property;*

   (d) *$Rank(B_1|G|B_2) = Rank(B_1) + Rank(B_2) - |G| + 2;$*

   (e) *In B_1 (or B_2), $\forall t_1, t_2 \in G: (t_1 \neq t_2 \Rightarrow [t_1] \neq [t_2]);$ and*

   (f) *Every P-invariant α of $B_1|G|B_2$ satisfies $\alpha M_e > 0$.*

20) *If both (B_1, M_{1e}) and (B_2, M_{2e}) are almost reversible and terminate properly, then $(B_1|G|B_2, M_e)$ is almost reversible provided that:*

   (a) *B_1 and B_2 are both conservative;*

   (b) *$\exists \beta_1 = (x_1, \ldots, x_{|T_1|}, z_1) \geq 1$, $\beta_2 = (y_1, \ldots, y_{|T_2|}, z_2) \geq 1$: $((V_{1a}\beta_1 = 0$ and $V_{2a}\beta_2 = 0)$ and $(\forall t \in G: (z_1\beta_2[t] = z_2\beta_1[t])));$*

   (c) *B_1 and B_2 both satisfy the RC-property;*

   (d) *$Rank(B_1|G|B_2) = Rank(B_1) + Rank(B_2) - |G| + 2;$*

   (e) *In B_1 (or B_2), $\forall t_1, t_2 \in G: (t_1 \neq t_2 \Rightarrow [t_1] \neq [t_2]);$ and*

   (f) *Every P-invariant α of $B_1|G|B_2$ satisfies $\alpha M_e > 0$.*

Proof.

1-2) $B_1|G|B_2$ is not an SM because of $|\varepsilon_1^{\bullet}| > 1$ and $|^{\bullet}\varepsilon_2| > 1$. However, it satisfies the definition of MG even after the fusion of transitions in G.

3) It is obvious that $B_1|G|B_2$ may or may not be an FC net. For any two places $p_1, p_2 \in P$ satisfying $p_1^{\bullet} \cap p_2^{\bullet} \neq \phi$, consider three cases. *Case 1* ($p_1^{\bullet} \cap G = \phi$ and $p_2^{\bullet} \cap G = \phi$): Then, $p_1, p_2 \in P_1$ or $p_1, p_2 \in P_2$. Since B_1 and B_2 are FC nets, $(p_1^{\bullet} = p_2^{\bullet})$ in $(B_1|G|B_2)$. *Case 2* ($p_1^{\bullet} \cap G = \phi$ and $p_2^{\bullet} \cap G \neq \phi$): Suppose p_1 in B_1. Since $p_1^{\bullet} \cap p_2^{\bullet} \neq \phi$ and $p_1^{\bullet} \cap G = \phi$, $p_2 \in B_1$. Since B_1 is an FC net, this case is impossible. *Case 3* ($p_1^{\bullet} \cap G \neq \phi$ and $p_2^{\bullet} \cap G \neq \phi$): If both p_1 and p_2 are in B_1 or B_2, then $(p_1^{\bullet} = p_2^{\bullet})$ in $B_1|G|B_2$ since both B_1 and B_2 are FC nets. Suppose $p_1 \in B_1$ and $p_2 \in P_2$ and $t \in p_1^{\bullet} \cap p_2^{\bullet} \cap G$. If $((^{\bullet}t)^{\bullet} = \{t\})$ in B_1 and B_2, then $(p_1^{\bullet} = p_2^{\bullet})$ in $B_1|G|B_2$ since both B_1 and B_2 are FC nets. If $((^{\bullet}t)^{\bullet} = \{t\})$ in B_1 or B_2, then $(p_1^{\bullet} \subseteq p_2^{\bullet}$ or $p_2^{\bullet} \subseteq p_1^{\bullet})$ in $B_1|G|B_2$ since both B_1 and B_2 are FC nets.

4) It is obvious that $B_1|G|B_2$ may or may not be an AC net. For any p_1, $p_2 \in P$ satisfying $p_1^\bullet \cap p_2^\bullet \neq \phi$, consider three cases. *Case 1 $(p_1^\bullet \cap G = \phi$ and $p_2^\bullet \cap G = \phi)$:* Then, p_1, $p_2 \in P_1$ or p_1, $p_2 \in P_2$. Since B_1 and B_2 are AC nets, $(p_1^\bullet \subseteq p_2^\bullet$ or $p_2^\bullet \subseteq p_1^\bullet)$ in $B_1|G|B_2$. *Case 2 $(p_1^\bullet \cap G = \phi$ and $p_2^\bullet \cap G \neq \phi)$:* Suppose p_1 in B_1. Since $p_1^\bullet \cap p_2^\bullet \neq \phi$ and $p_1^\bullet \cap G = \phi$, $p_2 \in B_1$. Since B_1 is an AC net, $p_2^\bullet \subseteq p_1^\bullet$. *Case 3 $(p_1^\bullet \cap G \neq \phi$ and $p_2^\bullet \cap G \neq \phi)$:* If both p_1 and p_2 are in B_1 or B_2, then $(p_1^\bullet \subseteq p_2^\bullet$ or $p_2^\bullet \subseteq p_1^\bullet)$ in $B_1|G|B_2$ since both B_1 and B_2 are AC nets. Suppose $p_1 \in B_1$ and $p_2 \in P_2$ and $t \in p_1^\bullet \cap p_2^\bullet \cap G$. Since $((^\bullet t)^\bullet = \{t\})$ in B_i and $((^\bullet t)^\bullet \supseteq \{t\})$ in B_j, where i, $j = 1$ or 2 and $i \neq j$. If $((^\bullet t)^\bullet = \{t\})$ in B_1, then, $(p_1^\bullet = \{t\})$ in B_1, this means that $(p_1^\bullet \subseteq p_2^\bullet)$ in $B_1|G|B_2$. If $((^\bullet t)^\bullet \supseteq \{t\})$ in B_1, then $((^\bullet t)^\bullet = \{t\})$ in B_2, then, $(p_2^\bullet = \{t\})$ in B_2, this means that $(p_2^\bullet \subseteq p_1^\bullet)$ in $B_1|G|B_2$. Hence, $B_1|G|B_2$ is also an AC net.

5) Since B_1 and B_2 are conservative, by Characterization 2.1, $\exists \alpha_1 = (x_1, \ldots, x_{|P_1|}) \geq 1$ and $\alpha_2 = (y_1, \ldots, y_{|P_2|}) \geq 1$ such that $\alpha_1 V_1 = 0$ and $\alpha_2 V_2 = 0$. Let $\alpha = (x_1 + y_1, \alpha_1, \alpha_2, x_{|P_1|} + y_{|P_2|})$ and V be shown in Figure 4.8. Then, $\alpha \geq 1$ and $\alpha V = (\alpha V[P, \varepsilon_1], \alpha V[P, \varepsilon_2], \alpha V[P, T_1 - G], \alpha V[P, G], \alpha V[P, T_2 - G]) = (-x_1 - y_1 + x_1 + y_1, -x_{|P_1|} - y_{|P_2|} + x_{|P_1|} + y_{|P_2|}, \alpha_1 V_1[P_1, T_1 - G], \alpha_1 V_1[P_1, G] + \alpha_2 V_2[P_2, G], \alpha_2 V_2[P_2, T_2 - G]) = 0$. It follows from Characterization 2.1 that $B_1|G|B_2$ is conservative.

6) Suppose B_1 and B_2 are structurally bounded. The same argument as in 5) shows that $\exists \alpha \geq 1$ such that $\alpha V \leq 0$. It follows from Characterization 2.2 that $B_1|G|B_2$ is structurally bounded.

7) Since B_1 and B_2 are almost consistent, by Characterization 2.3, If $\exists \beta_1 = (x_1, \ldots, x_{|T_1|}, z_1) \geq 1$, and $\beta_2 = (y_1, \ldots, y_{|T_2|}, z_2) \geq 1$ such that $V_{1a}\beta_1 = 0$ and $V_{2a}\beta_2 = 0$. Let $\beta = (z_1 z_2, z_1 z_2, z_2 \beta_1[T_1], z_1 \beta_2[T_2 - T_1], z_1 z_2)$ and V_a be shown in Figure 4.8. Then, $\beta \geq 1$. Since $V[P_i, \varepsilon_1] + V[P_i, \varepsilon_2] = V_{ia}[P_i, t_{ia}]$ for $i = 1$, 2, and $\forall t \in G$: $(z_1 \beta_2[t] = z_2 \beta_1[t])$, we have $V_a \beta = (V_a[p_e, T_a]\beta, V_a[P_1, T_a]\beta, V_a[P_2, T_a]\beta, V_a[p_x, T_a]\beta) = (-z_1 z_2 + z_1 z_2, z_2 V_{1a}\beta_1, z_1 V_{2a}\beta_2, z_1 z_2 - z_1 z_2) = 0$. It follows from Characterization 2.3 that $B_1|G|B_2$ is almost consistent.

8) Suppose B_1 and B_2 are almost repetitive. The same argument as in 7) shows that $\exists \beta \geq 1$ such that $V_a \beta \geq 0$. It follows from Characterization 2.4 that $B_1|G|B_2$ is almost repetitive.

9) The incidence matrix V of $B_1|G|B_2$ (Figure 4.8) can be obtained from the incidence matrix V' of $B_1|||B_2$ (Figure 4.6) in two steps:

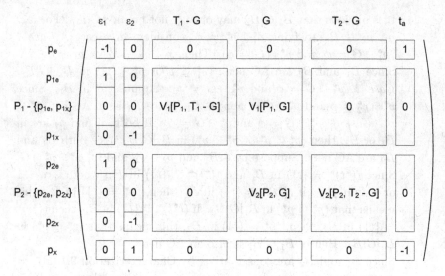

Fig. 4.8   Incidence matrix V_a of $(B_1|G|B_2)_a$ (V of $B_1|G|B_2$ is the part of V_a without column t_a).

Step 1) Each column t of G in T_2 is added to column t of G in T_1. Step 2) The columns of G are removed from T_2. Step 1, including just Gaussian eliminations, does not modify the rank of V'. Hence, $\text{Rank}(B_1|G|B_2) \geq \text{Rank}(B_1 |||B_2) - |G| = \text{Rank}(B_1) + \text{Rank}(B_2) - |G| + 2$. The lower bound is attained if every column t of G in V is a linearly independent column in V.

10) In constructing $B_1|G|B_2$, two clusters $[p_e] = \{p_e, \varepsilon_1\}$ and $[p_x] = \{p_x\}$ are created. Also, $[p_{1x}]$ of B_1 and $[p_{2x}]$ of B_2 are merged to form $[\varepsilon_2] = \{p_{1x}, p_{2x}, \varepsilon_2\}$. In addition, according to the given condition that $\forall t_1, t_2 \in G$: $(t_1 \neq t_2 \Rightarrow [t_1] \neq [t_2]$ in either B_1 or B_2 but not necessarily in both of them), fusing each pair of transitions t from B_1 and t' from B_2 results in merging two clusters $[t]$ of B_1 and $[t']$ of B_2 into one. Furthermore, any other clusters of B_1 or B_2 are also clusters of $B_1|G|B_2$. Hence, $|C(B_1|G|B_2)| = |C(B_1)| + |C(B_2)| - |G| + 1$.

11) It follows from 9) and 10) that $\text{Rank}(B_1|G|B_2) = |C(B_1|G|B_2)| - 1$.

12) $B_1|G|B_2$ and $B_1||| B_2$ have the same structure except the fusing of transitions of G. However, the fusion does not eliminate any original SM-components in B_1 and B_2. Hence, the result can be proved in the same way as in $B_1|||B_2$.

13-14) $B_1|G|B_2$ and $B_1||| B_2$ have the same structure except the fusing of transitions of G. However, the fusion does not affect the contents in $^\bullet D$ and $D^\bullet$. Hence, the result can be proved in the same way as in $B_1|||B_2$.

15) $Z(B_1|G|B_2) = |F| - |P \cup T| + 2 = |F_1| + |F_2| + 6 - (|P_1| + |P_2| + 2 + |T_1| + |T_2| + 2 - |G|) + 2 = |F_1| - (|P_1| + |T_1|) + 2 + |F_2| - (|P_2| + |T_2|) + 2 + |G| = Z(B_1) + Z(B_2) + |G|.$

16) It is very possible that some transition in G never be enabled in $(B_1|G|B_2, M_e)$, see the Example 4.1 below.

17) Note that fusing transitions of G may cause deadlocks but never change the token distributions of those reachable non-deadlocked markings of $(B_1|G|B_2, M_e)$. Deadlocked markings are obviously bounded. Proof of boundedness of non-deadlocked markings is similar to that for $B_1|||B_2$.

18) Proper termination requires that, $\forall M \in R((B_1|G|B_2)_a, M_e)$, $M(p_x) \geq 1$ implies $M = (0, M_{1c}, M_{2c}, 1)$. Since $(B_1|G|B_2, M_e)$ may occur the situation of deadlock, it is obvious that $(B_1|G| B_2, M_e)$ may or may not terminate properly even if both (B_1, M_{1e}) and (B_2, M_{2e}) terminate properly.

19) According to the given conditions a) to f), it follows from 5), 7), 11) and Characterization 2.11 that $((B_1|G|B_2)_a, M_e)$ is live.

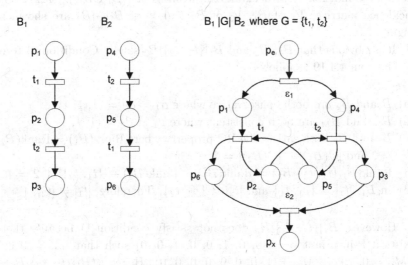

Fig. 4.9 An example for Theorem 4.4.19.

84 Property-Preserving Petri Net Process Algebra in Software Engineering

$$
B_1 = \begin{array}{c} \\ p_1 \\ p_2 \\ p_3 \end{array}
\begin{array}{cc} t_1 & t_2 \\ \left(\begin{array}{cc} -1 & 0 \\ 1 & -1 \\ 0 & 1 \end{array}\right) \end{array}
\qquad
B_2 = \begin{array}{c} \\ p_4 \\ p_5 \\ p_6 \end{array}
\begin{array}{cc} t_1 & t_2 \\ \left(\begin{array}{cc} 0 & -1 \\ -1 & 1 \\ 1 & 0 \end{array}\right) \end{array}
\qquad
V = \begin{array}{c} \\ p_e \\ p_1 \\ p_2 \\ p_3 \\ p_4 \\ p_5 \\ p_6 \\ p_x \end{array}
\begin{array}{cccc} \varepsilon_1 & \varepsilon_2 & t_1 & t_2 \\ \left(\begin{array}{cccc} -1 & 0 & 0 & 0 \\ 1 & 0 & -1 & 0 \\ 0 & 0 & 1 & -1 \\ 0 & -1 & 0 & 1 \\ 1 & 0 & 0 & -1 \\ 0 & 0 & -1 & 1 \\ 0 & -1 & 1 & 0 \\ 0 & 1 & 0 & 0 \end{array}\right) \end{array}
$$

Fig. 4.10 Incidence matrix of the Example 4.1.

20) According to the given conditions a) to f), the same argument as in 19) can prove that $(B_1|G|B_2)_a$ is live. That is, it has no deadlock and M_x can be reached. Since both (B_1, M_{1e}) and (B_2, M_{2e}) are almost reversible and terminate properly, similar argument as in the proof of Theorem 4.3.20 for $(B_1|||B_2, M_e)$ can conclude that $(B_1|G|B_2, M_e)$ is almost reversible.

□

Example 4.1. This example shown in Figure 4.9 illustrates the importance of Condition f) in deciding the liveness of $((B_1|G|B_2)_a, M_e)$. The incidence matrices V_1 of B_1, V_2 of B_2 and V of $B_1|G|B_2$ are shown in Figure 4.10.

It is obvious that B_1, B_2 and $B_1|\{t_1, t_2\}|B_2$ satisfy Condition a) to e) of Theorem 4.4.19 as follows.

(a) B_1 and B_2 are both conservative where $\alpha_1 = \alpha_2 = (1, 1, 1)$.
(b) B_{1a} and B_{2a} are both consistent where $\beta_1 = \beta_2 = (1, 1, 1)$.
(c) B_1 and B_2 both satisfy the RC-property where $\text{Rank}(B_1) = \text{Rank}(B_2) = 2$ and $|C(B_1)| = |C(B_2)| = 3$.
(d) $\text{Rank}(B_1|\{t_1, t_2\}|B_2) = \text{Rank}(B_1) + \text{Rank}(B_2) - |\{t_1, t_2\}| + 2 = 4$.
(e) In B_1, $[t_1] = \{p_1, t_1\}$ and $[t_2] = \{p_2, t_2\}$. Therefore, $[t_1] \neq [t_2]$.

However, $B_1|\{t_1, t_2\}|B_2$ does not satisfy condition f) because there exists a P-invariant $\alpha = (0, 0, 1, 0, 0, 1, 0, 0)$ such that $\alpha V = 0$ but $\alpha M_e = 0$, where $M_e = (1, 0, 0, 0, 0, 0, 0, 0)$. Hence, $((B_1|\{t_1, t_2\}|B_2)_a, M_e)$ is not live.

4.5 The Operator DISABLE

DISABLE models a potential interruption of process B_1 by another process B_2. According to LOTOS, the composite process behaves in three possible ways during an execution: (1) B_2 never starts and B_1 continues until termination. (2) If B_2 starts while B_1 is executing, B_1 will be stopped and B_2 continues until termination. (3) If B_2 starts first, B_1 will never start and B_2 will continue until termination. One application of Disable is to model task interruption in a system.

Definition 4.5. (composition by the operator DISABLE) (Figure 4.11) For two marked processes (B_i, M_{ie}),$i = 1$, 2, their composition by *DISABLE*, in notation $(B_1 \, [\rangle B_2, M_e)$ is defined as follows:

$P = (P_1 - \{p_{1x}\}) \cup (P_2 - \{p_{2x}\}) \cup \{p_e, p_x, p_d\}$, where $p_x = p_{1x} = p_{2x}$

$T = T_1 \cup T_2 \cup \{\varepsilon_1, t_d\}$

$F = F_1' \cup F_2' \cup \{(p_e, \varepsilon), (\varepsilon, p_{1e}), (\varepsilon, p_{2e}), (p_d, t_d), (t_d, p_{2e})\} \cup \{(p_{2e}, t) : t \in T_1\} \cup \{(t, p_d) : t \in T_1 - {}^\bullet p_{1x}\}$, where

$F_i' = F_i - \{(t, p_{ix}) : t \in {}^\bullet p_{ix}\} \cup \{(t, p_x) : t \in {}^\bullet p_{ix}\}$, $i = 1$, 2

$M_e = (1, 0, M_{1c}, 0, 0, M_{2c}, 0)$.

Graphically, the composite process (Figure 4.11) is obtained by:

1) adding a new entry place p_e and a dummy transition ε for allowing the selective execution of B_1 or B_2;

2) adding a set of *forward disabling arcs*, connecting place p_{2e} of B_2 to every transition t of B_1; (Note: By such an arrangement, whenever $B_2$is started, the token in p_{2e} will be taken away and all transitions in B_1will be immediately disabled, causing an interruption on B_1).

3) adding a set *backward disabling arcs*, a *disabling place* p_d and a *disabling transition* t_d , etc., connecting every transition of B_1 (except those in p_x) to p_{2e}; (Note: By such an arrangement, whenever any transition B_1 (except those in p_x) has been fired, a token is put back into p_{2e}, allowing B_2 to start again) and

4) merging exit places p_{1x} of B_1 and p_{2x} of B_2 to form a new exit place p_x.

It is obvious that $(B_1 \, [\rangle B_2, M_e)$ is also a marked process. The incidence matrix V of $B_1 \, [\rangle B_2$ is formed from the incidence matrices V_1 of B_1 and V_2

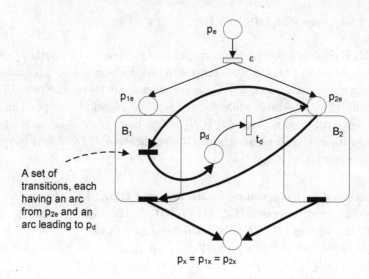

Fig. 4.11 $B_1 [\rangle B2$: Composition by the operator DISABLE.

of B_2 as follows:

$$
\begin{array}{c}
\\
p_e \\
P_1 - \{p_{1x}\} \\
P_2 - \{p_{2x}\} \\
p_x \\
p_d
\end{array}
\begin{array}{ccccc}
\varepsilon & T_{11} & T_{12} & T_2 & t_d \\
\left(\begin{array}{ccccc}
-1 & 0 & 0 & 0 & 0 \\
p_{1e} & V_{11} & V_{12} & 0 & 0 \\
p_{2e} & -p_{2e} & -p_{2e} & V_{21} & p_{2e} \\
0 & 0 & 1 & V_{22} & 0 \\
0 & 1 & 0 & 0 & -1
\end{array} \right)
\end{array}
$$

Explanation:

(a) $T_{11} = T_1 - {}^\bullet p_{1x}$ and $T_{12} = {}^\bullet p_{1x}$.
(b) $V_{11} = V_1[P_1 - \{p_{1x}\}, T_1 - {}^\bullet p_{1x}]$ and $V_{12} = V_1[P_1 - \{p_{1x}\}, {}^\bullet p_{1x}]$.
(c) $V_{21} = V_2[P_2 - \{p_{2x}\}, T_2]$ and $V_{22} = V_2[p_{2x}, T_2]$.

It can be seen that this composition fulfills the three requirements in the behavior of $B_1 [\rangle B_2$ as mentioned above. The definition here is an improvement over Cheung's version [CZ (1994)] of Disable. However, for simplicity, it will be used in the proofs of some of the theorems in the next chapters.

Theorem 4.5. *(property preservation under DISABLE $B_1 [\rangle B_2$)*
Let (B_1, M_{1e}), (B_2, M_{2e}) and $(B_1 [\rangle B_2, M_e)$ be the three marked PNPs described in Definition 4.5. Then, the following propositions hold:
(Note: For simplicity in the proofs of sub-theorems 4.5.17 to 4.5.20, M_e is constructed using the non-pure version of the operator DISABLE.)

1) $B_1 [\rangle B_2$ is not an SM, whether or not B_1 and B_2 are so.

2) Suppose both B_1 and B_2 are almost MGs. Two cases hold: (a) $B_1 [\rangle B_2$ is not almost an MG. (b) $B_1 [\rangle B_2$ is an AC net.

3) Suppose both B_1 and B_2 are FC nets. Two cases hold: (a) $B_1 [\rangle B_2$ is not an FC. (b) $B_1 [\rangle B_2$ is an AC net.

4) Suppose both B_1 and B_2 are AC nets. Two cases hold: (a) In general, $B_1 [\rangle B_2$ may or may not be an AC. (b) $B_1 [\rangle B_2$ is an AC net $\forall p \in P_2$ satisfying $((p^\bullet \cap p_{2e}^\bullet) \neq \phi \Rightarrow p^\bullet \subseteq p_{2e}^\bullet)$ in B_2.

5) (a) If both B_1 and B_2 are conservative, $B_1 [\rangle B_2$ is structurally bounded.

   (b) $B_1 [\rangle B_2$ is not almost conservative even if B_1 and B_2 are both almost conservative.

6) If both B_1 and B_2 are structurally bounded, so is $B_1 [\rangle B_2$.

7) If both B_1 and B_2 are almost consistent, then $B_1 [\rangle B_2$ is almost repetitive.

8) $B_1 [\rangle B_2$ is almost repetitive provided that:
   $\exists \beta_1 = (x_1, \ldots, x_{|T1|}, z_1) \geq 1, \beta_2 = (y_1, \ldots, y_{|T2|}, z_2) \geq 1: ((V_{1a}\beta_1 \geq 0 \text{ and } V_{2a}\beta_2 \geq 0) \text{ and } (V_{1a}[p_{1x}, T_{1a}]\beta_1 \leq V_{2a}[p_{2e}, T_{2a}]\beta_2))$.

9) $Rank(B_1 [\rangle B_2) = Rank(B_1) + Rank(B_2) + 2$.

10) $|C(B_1 [\rangle B_2)| = |C(B_2)| + 2$.

11) $B_1 [\rangle B_2$ does not satisfy the RC-property, whether or not B_1 and B_2 are so.

12) $B_1 [\rangle B_2$ does not almost have a minimal SM-cover, whether or not B_1 and B_2 have so.

13) Suppose D is a siphon of B_1 (resp., B_2). Then,

   (a) D is a siphon of $B_1 [\rangle B_2$ iff $p_{1e} \notin D$ and $p_{1x} \notin D$ (resp., $p_{2e} \notin D$ and $p_{2x} \notin D$);

   (b) $D \cup \{p_e\}$ (resp., $D \cup \{p_e, p_d\}$) is a siphon of $B_1 [\rangle B_2$ if $p_{1e} \in D$ and $p_{1x} \notin D$ (resp., $p_{2e} \in D$ and $p_{2x} \notin D$).

14) Suppose S is a trap of B_1. Then, S is a trap of $B_1 [\rangle B_2$. Suppose S is a trap of B_2. Then,

   (a) S is a trap of $B_1 [\rangle B_2$ if and only if $p_{2e} \notin S$;

   (b) $S \cup \{p_e, p_d, p_x\}$ is a trap of $B_1 [\rangle B_2$ if $p_{2e} \in S$.

15) $Z(B_1 [\rangle B_2) = Z(B_1) + Z(B_2) + 2|T_1| - |^\bullet p_{1x}|$.

16) If both (B_1, M_{1e}) and (B_2, M_{2e}) have no dead transitions, then $(B_1 [\rangle B_2, M_e)$ does not have either.

17) If both (B_1, M_{1e}) and (B_2, M_{2e}) are bounded, so is $(B_1 [\rangle B_2, M_e)$.

18) $(B_1 [\rangle B_2, M_e)$ does not terminate properly whether or not (B_1, M_{1e}) and (B_2, M_{2e}) do so.

19) $(B_1 [\rangle B_2, M_e)$ may or may not be almost live even if both (B_1, M_{1e}) and (B_2, M_{2e}) are almost live and terminate properly.

20) $(B_1 [\rangle B_2, M_e)$ may or may not be almost reversible even if both (B_1, M_{1e}) and (B_2, M_{2e}) are almost reversible.

Proof.

1-2) According to the structure of $B_1 [\rangle B_2$, adding p_e, ε and the Disable arcs violates the definitions of SM (because $|\varepsilon^\bullet| > 1$) and MG (because $|p_{2e}^\bullet| > 1$ and $|^\bullet p_x| > 1$). Suppose both B_1 and B_2 are MGs. Since $(\forall p \in P_1 \cup P_2\colon |p^\bullet| = |^\bullet p| = 1$ in B_1 or $B_2)$ and $(^\bullet(p_e^\bullet) = \{p_e\}$, $^\bullet(p_d^\bullet) = \{p_d\}$ and $p_{2x}^\bullet = \phi)$ in $(B_1 [\rangle B_2)$, by the structure of $B_1 [\rangle B_2$ and the definition of an AC net, we know that $B_1 [\rangle B_2$ is also an AC net.

3) It is obvious that $B_1 [\rangle B_2$ is not an FC net. For any p_1, $p_2 \in P_1 \cup P_2$ where $(p_1^\bullet \cap p_2^\bullet \neq \phi)$ in $(B_1 [\rangle B_2)$, consider three cases: (a) p_1, $p_2 \in P_1$. Since B_1 is an FC net, $(p_1^\bullet = p_2^\bullet)$ in B_1 and thus in $B_1 [\rangle B_2$. (b) p_1, $p_2 \in P_2$. Since B_2 is an FC net, $(p_1^\bullet = p_2^\bullet)$ in B_2. If p_1, $p_2 \neq p_{2e}$, then $(p_1^\bullet = p_2^\bullet)$ in $B_1 [\rangle B_2$. If p_1 or $p_2 = p_{2e}$, $(p_2^\bullet \subseteq p_1^\bullet)$ or $(p_1^\bullet \subseteq p_2^\bullet)$ in $B_1 [\rangle B_2$. (c) $p_1 \in P_1$ and $p_2 \in P_2$. Then $p_2 = p_{2e}$ because of $(p_1^\bullet \cap p_2^\bullet \neq \phi)$ in $B_1 [\rangle B_2$. By the structure of $B_1 [\rangle B_2$, we know that $(p_1^\bullet \subseteq p_2^\bullet)$ in $B_1 [\rangle B_2$. Note that $(^\bullet(p_e^\bullet) = \{p_e\}$, $^\bullet(p_d^\bullet) = \{p_d\}$ and $p_{2x}^\bullet = \phi)$ in $B_1 [\rangle B_2$, $B_1 [\rangle B_2$ is an AC net.

4) It is obvious that $B_1 [\rangle B_2$ may or may not be an AC net. For any p_1, $p_2 \in P_1 \cup P_2$ where $(p_1^\bullet \cap p_2^\bullet \neq \phi)$ in $B_1 [\rangle B_2$, consider three cases: (a) p_1, $p_2 \in P_1$. Since B_1 is an AC net, $(p_2^\bullet \subseteq p_1^\bullet)$ or $(p_1^\bullet \subseteq p_2^\bullet)$ in B_1 and thus in $B_1 [\rangle B_2$. (b) p_1, $p_2 \in P_2$. Since B_2 is an AC net, $(p_2^\bullet \subseteq p_1^\bullet)$ or $(p_1^\bullet \subseteq p_2^\bullet)$ in B_2. If p_1, $p_2 \neq p_{2e}$, then $(p_2^\bullet \subseteq p_1^\bullet)$ or $(p_1^\bullet \subseteq p_2^\bullet)$ in $B_1 [\rangle B_2$. If $p_1 = p_{2e}$, $(p_2^\bullet \subseteq p_1^\bullet)$ in $B_1 [\rangle B_2$ because of $(p^\bullet \cap p_{2e}^\bullet) \neq \phi \Rightarrow p^\bullet \subseteq p_{2e}^\bullet)$ in B_2. (c) $p_1 \in P_1$ and $p_2 \in P_2$. Then $p_2 = p_{2e}$ because of $(p_1^\bullet \cap p_2^\bullet \neq \phi)$ in $B_1 [\rangle B_2$. By the structure of $B_1 [\rangle B_2$, we know that $(p_1^\bullet \subseteq p_2^\bullet)$ in $B_1 [\rangle B_2$. Note that $(^\bullet(p_e^\bullet) = \{p_e\}$, $^\bullet(p_d^\bullet) = \{p_d\}$ and $p_{2x}^\bullet = \phi)$ in $B_1 [\rangle B_2$, $B_1 [\rangle B_2$ is an AC net.

5) (a) Since B_1 and B_2 are conservative, by Characterization 2.1, $\exists \alpha_1 = (x_1, \ldots, x_{|P_1|}) \geq 1$ and $\alpha_2 = (y_1, \ldots, y_{|P_2|}) \geq 1$ such that $\alpha_1 V_1 = 0$ and $\alpha_2 V_2 = 0$. Let $\alpha = (x_1 y_{|P_2|} + y_1 x_{|P_1|}, y_{|P_2|} \alpha_1 [P_1 - \{p_{1x}\}], x_{|P_1|} \alpha_2 [P_2 - \{p_{2x}\}], x_{|P_1|} y_{|P_2|}, x_{|P_1|} y_1)$ and V be shown in Figure 4.12. We have $\alpha \geq 1$ and $\alpha V = (\alpha V[P, \varepsilon], \alpha V[P, T_1 - {}^\bullet p_{1x}], \alpha V[P, {}^\bullet p_{1x}], \alpha V[P, T_2], \alpha V[P, t_d]) = (-x_1 y_{|P_2|} - y_1 x_{|P_1|} +$

$y_{|P_2|}x_1 + x_{|P_1|}y_1,\ y_{|P_2|}\alpha_1 V_1[P_1,T_1 - {}^{\bullet}p_{1x}]+(-x_{|P_1|}y_1 + x_{|P_1|}y_1)^u,$
$y_{|P_2|}\alpha_1 V_1[P_1,{}^{\bullet}p_{1x}]+(-x_{|P_1|}y_1)^v,\ x_{|P_1|}\alpha_2 V_2[P_2,T_2],\ x_{|P_1|}y_1 - x_{|P_1|}y_1)$
$= (0, 0, (-x_{|P_1|}y_1)^v, 0, 0) \le 0$, where $u = |T_1 - {}^{\bullet}p_{1x}|$, $v = |{}^{\bullet}p_{1x}|$ and $(w)^d$ represents a vector $(w\ \ w \ldots\ w)$ of dimension d. It follows from Characterization 2.1 that $B_1\ [\rangle B_2$ is structurally bounded.

(b) If there exists an marking M_e such that $((B_1\ [\rangle\ B_2)_a, M_e)$ has a firing transition sequence σ and $\sigma = \varepsilon\sigma' t'$, where $\sigma' \in T_2{}^*$ and $t' \in T_2 \cap {}^{\bullet}p_x$, then every firing of σ will increase a token to p1e since $\varepsilon^{\bullet} = \{p_{1e}, p_{2e}\}$. Hence, $((B_1\ [\rangle\ B_2)_a, M_e)$ will change unbounded. This implies that $(B_1\ [\rangle B_2)_a$ is not conservative, that is, $B_1\ [\rangle\ B_2$ is not almost conservative.

	ε	$T_1 - {}^{\bullet}p_{1x}$	${}^{\bullet}p_{1x}$	T_2	t_d	t_a
p_e	-1	0	0	0	0	1
p_{1e}	1					
$P_1 - \{p_{1e}, p_{1x}\}$	0	$V_1[P_1 - \{p_{1x}\},\ T_1 - {}^{\bullet}p_{1x}]$	$V_1[P_1 - \{p_{1x}\},\ {}^{\bullet}p_{1x}]$	0	0	0
p_{2e}	1	-1	-1		1	
$P_2 - \{p_{2e}, p_{2x}\}$	0	0	0	$V_2[P_2 - \{p_{2x}\},\ T_2]$	0	0
$p_x = p_{1x} = p_{2x}$	0	$V_1[p_{1x},\ T_1 - {}^{\bullet}p_{1x}]$	$V_1[p_{1x},\ {}^{\bullet}p_{1x}]$	$V_2[p_{2x},\ T_2]$	0	-1
p_d	0	1	0	0	-1	0

Fig. 4.12 Incidence matrix V_a of $(B_1[B_2\rangle_a$ (V of $B_1[\rangle B_2$ is the part of V_a without column t_a.

6) Suppose B_1 and B_2 are structurally bounded. The same argument as in 5) shows that $\alpha \ge 1$ such that $\alpha V \le 0$. It follows from Characterization 2.2 that $B_1\ [\rangle B_2$ is structurally bounded.

7) Since B_1 and B_2 are almost consistent, by Characterization 2.3, $\exists \beta_1 = (x_1, \ldots, x_{|T_1|}, z_1) \ge 1$ and $\beta_2 = (y_1, \ldots, y_{|T_2|}, z_2) \ge 1$ such that $V_{1a}\beta_1 = 0$ and $V_{2a}\beta_2 = 0$. Let $\beta = (z_1 + z_2, \beta_1, \beta_2, \sum_{t_1 \in T_1 - {}^{\bullet}p_{1x}}\beta_1[t_1], z_1 + z_2)$ and V_a be shown in Figure 4.12. Then, $\beta \ge 1$. Since B_1 is consistent, implying that $z_1 - \sum_{t \in {}^{\bullet}p_{1x}}\beta_1[t] = -V_{1a}[p_{1x}, T_{1a}]\beta_1 = 0$, we have $V_a\beta = (V_a[p_e, T_a]\beta, V_a[p_{1e}, T_a]\beta, V_a[P_1 - \{p_{1e}, p_{1x}\}, T_a]\beta,$ $V_a[p_{2e}, T_a]\beta, V_a[P_2 - \{p_{2e}, p_{2x}\}, T_a]\beta, V_a[p_x, T_a]\beta, V_a[p_d, T_a]\beta) =$

$(-z_1 - z_2 + z_1 + z_2, \ z_2 + V_{1a}[p_{1e}, T_{1a}]\beta_1, \ V_{1a}[P_1 - \{p_{1e}, p_{1x}\}, T_{1a}]\beta_1, z_1 - \sum_{t1 \in \bullet p1x} \beta_1[t_1] + V_{2a}[p_{2e}, T_{2a}]\beta_2, \ V_{2a}[P_2 - \{p_{2e}, p_{2x}\}, T_{2a}]\beta_2, \ V_{1a}[p_{1x}, T_{1a}]\beta_1 + V_{2a}[p_{2x}, T_{2a}]\beta_2, \ 0) \geq 0$. It follows from Characterization 2.3 that $B_1 \, [\rangle B_2$ is almost repetitive.

8) Suppose B_1 and B_2 are almost repetitive. Given that $V_{1a}[p_{1x}, T_{1a}]\beta_1 \leq V_{2a}[p_{2e}, T_{2a}]\beta_2)$, the same argument as in 7) shows that $\exists \beta \geq 1$ such that $V_a\beta \geq 0$. It follows from Characterization 2.4 that $B_1 \, [\rangle B_2$ is almost repetitive.

9) It is easy to see that the columns of $V[P, \varepsilon]$, $V[P, T_1]$, $V[P, T_2]$ and $V[P, t_d]$ are linearly independent. Hence, $\mathrm{Rank}(B_1 \, [\rangle \, B_2) = \mathrm{Rank}(B_1) + \mathrm{Rank}(B_2) + 2$.

10) $[p_e] = \{p_e, \varepsilon\}$ and $[p_d] = \{p_d, t_d\}$ are the only new clusters created. In $B_1 \, [\rangle B_2$, since all transitions of B_1 become output transitions of p_{2e}, all clusters of B_1 except $[p_{1x}]$ are merged into $[p_{2e}]$ of B_2. $[p_{1x}]$ of B_1 is merged with $[p_{2x}]$ of B_2 to form $[p_x] = \{p_x\}$ of $B_1 \, [\rangle B_2$. Any other clusters of B_2 are also clusters of $B_1 \, [\rangle B_2$. Hence, $|C(B_1 \, [\rangle B_2)| = |C(B_2)| + 2$.

11) This follows from 9) and 10).

12) To prove that $B_1 \, [\rangle B_2$ is not almost SM-coverable, we show that there does not exist any SM-component S' in $(B_1 \, [\rangle \, B_2)_a$ that embeds an SM-component of B_1, which includes p_{1e} and p_{1x}. Assume that there exists such an SM-component S' which contains p_e and p_x. Hence, S' must include $\bullet\{p_e, p_x\} \cup \{p_e, p_x\}\bullet$. Since S' is strongly connected, it must also include p_{2e}, and the arc (ε_1, p_{2e}). In order to embed any SM-component of B_1, S' must include the arc (ε_1, p_{1e}). It follows that S' is not an SM-component because $|\varepsilon_1^\bullet| > 1$. Hence, there does not exist any minimal SM-cover for $(B_1 \, [\rangle B_2)_a$, even if B_1 and B_2 have minimal SM-covers individually.

13) For $D \subseteq P_1$, the proof is the same as for the operator $[\,]$ except that $\{\varepsilon_1\}$ is replaced with $\{\varepsilon\}$ in Part a) and $\{\varepsilon_1, \varepsilon_2\}$ is replaced with $\{\varepsilon\}$ in Part b).

For $D \subseteq P_2$, consider two cases:

(a) $p_{2e} \notin D$. Then $\bullet D$ in $B_1 \, [\rangle B_2 = \bullet D$ in B_2 and $D\bullet$ in $B_1 \, [\rangle B_2 = D\bullet$ in B_2. Hence, $(\bullet D \subseteq D\bullet)$ in $B_1 \, [\rangle B_2$ if and only if $(\bullet D \subseteq D\bullet)$ in B_2.

(b) $p_{2e} \in D$ and $p_{2x} \notin D$. Then, $\bullet(D\cup \{p_e, p_d\})$ in $B_1 \, [\rangle B_2 = \{\varepsilon, t_d\} \cup (\bullet D$ in $B_2)$ and $(D\cup \{p_e, p_d\})\bullet$ in $B_1 \, [\rangle B_2 \supseteq \{\varepsilon, t_d\} \cup (D\bullet$ in $B_2)$. Hence, $(\bullet(D\cup \{p_e, p_d\}) \subseteq (D\cup \{p_e, p_d\})\bullet)$ in $B_1 \, [\rangle B_2$ provided that $(\bullet D \subseteq D\bullet)$ in B_2.

14) For $S \subseteq P_1$, $S^\bullet$ in $B_1 \,[\rangle B_2 = S^\bullet$ in B_1 but $S^\bullet$ in $B_1 \,[\rangle B_2 \supseteq S^\bullet$ in B_1. Since $(S^\bullet \subseteq {}^\bullet S)$ in B_1, it follows that $(S^\bullet \subseteq {}^\bullet S)$ in $B_1 \,[\rangle B_2$.
   For $S \subseteq P_2$, consider two cases:

   (a) If $p_{2e} \notin S$, $S^\bullet$ in $B_1 \,[\rangle B_2 = S^\bullet$ in B_2 and ${}^\bullet S$ in $B_1 \,[\rangle B_2 = {}^\bullet S$ in B_2. Hence, $(S^\bullet \subseteq {}^\bullet S)$ in $B_1 \,[\rangle B_2$ if $(S^\bullet \subseteq {}^\bullet S)$ in B_2. If $p_{2e} \in S$, then $(S^\bullet \not\subseteq {}^\bullet S)$ in $B_1 \,[\rangle B_2$ because $(\varepsilon \in {}^\bullet S$ but $\varepsilon \notin S^\bullet)$ in $B_1 \,[\rangle B_2$.

   (b) If $p_{2e} \in S$, $(S \cup \{p_e, p_d, p_x\})^\bullet$ in $B_1 \,[\rangle B_2 = (S^\bullet$ in $B_2) \cup T_1 \cup \{\varepsilon, t_d\}$ and ${}^\bullet(S \cup \{p_e, p_d, p_x\})$ in $B_1 \,[\rangle B_2 = ({}^\bullet S$ in $B_2) \cup T_1 \cup \{\varepsilon, t_d\}$. Hence, $(S \cup \{p_e, p_d, p_x\})^\bullet \subseteq {}^\bullet(S \cup \{p_e, p_d, p_x\})$ in $B_1 \,[\rangle B_2$ provided that $(S^\bullet \subseteq {}^\bullet S)$ in B_2.

15) $Z(B_1 \,[\rangle B_2) = |F| - |P \cup T| + 2 = |F_1| + |F_2| + 2|T_1| - |{}^\bullet p_{1x}| + 5 - (|P_1| + |P_2| + 1 + |T_1| + |T_2| + 2) + 2 = (|F_1| - |P_1 \cup T_1| + 2) + (|F_2| - |P_2 \cup T_2| + 2) + 2|T_1| - |{}^\bullet p_{1x}| = Z(B_1) + Z(B_2) + 2|T_1| - |{}^\bullet p_{1x}|$.

16) Similar to the proof of 16) of Theorem 4.3.

17) Any $M \in R(B_1 \,[\rangle B_2, M_e)$ can only have two forms: 1) $M = M_e$. M is obviously bounded. 2) $M = (0, m_1, 0, m_2, M(p_x))$. There are two sub-cases. (a) $(m_1, M(p_x)) \in R(B_1, M_{1e})$ and $m_2 = M_{2e}$ (i.e., B_1 executes to completion). (b) $(m_1, 0) \in R(B_1, M_{1e})$ and $(m_2, M(p_x)) \in R(B_2, M_{2e})$ (i.e., B_2 starts when B_1 is executing). For both cases, M is bounded. Therefore, $(B_1 \,[\rangle B_2, M_e)$ is bounded.

18) According to the structure of $(B_1 \,[\rangle B_2)_a$, when B_1 is disabled by B_2 before any exit transitions of B_1 is fired, a token is left in B_1. It follows from 17) that termination of $(B_1 \,[\rangle B_2, M_e)$ is not proper.

19) According to the structure of $(B_1 \,[\rangle B_2)_a$, when B_1 is disabled by B_2 before any exit transitions of B_1 is fired, a token is left in B_1. When B_1 is re-entered in the second run, this token will create markings that cover the original markings of B_{1a} and leads to the concurrent execution of the current B_{1a} and the previous instance of B_{1a}. Hence, we cannot decide whether $(B_1 \,[\rangle\, B_2, M_e)_a$ is live or not.

20) According to the structure of $(B_1 \,[\rangle B_2)_a$, when B_1 is disabled by B_2 before any exit transitions of B_1 is fired, a token is left in B_1. When B_1 is re-entered in the second run, a new token will also be deposited in B_1. Even though B_1 is not disabled by B_2 in this run, one of these tokens remains in $(B_1 \,[\rangle B_2, M_e)_a$. It follows that $(B_1 \,[\rangle\, B_2, M_e)_a$ is not reversible.

$\square$

4.6   The Operator DISABLE-RESUME

DISABLE-RESUME functions the same as the DISABLE operator $B_1 \;[\rangle\; B_2$ except that, if B_2 is ever executed and completed, then there will be three options: (1) The entire process terminates. (2) Execution on B_1 resumes from wherever it is interrupted. (3) B_2 is repeated. This operator is not part of LOTOS. It is added by the authors into the algebra.

Definition 4.6. (composition by the operator DISABLE-RESUME) (Figure 4.13)
For two marked processes (B_i, M_{ie}), $i = 1, 2$, their composition by the operator *DISABLE-RESUME*, in notation $(B_1 \;[r\rangle\; B_2, M_e)$, is defined as follows:

$$P = P_1 \cup P_2 \cup \{p_e, p_d\},\ p_x = p_{1x}$$
$$T = T_1 \cup T_2 \cup \{\varepsilon_1, \varepsilon_2, t_d\}$$
$$F = F_1 \cup F_2 \cup \{(p_e, \varepsilon_1), (\varepsilon_1, p_{1e}), (\varepsilon_1, p_{2e}), (p_{2x}, \varepsilon_2), (\varepsilon_2, p_{2e}), (p_d, t_d),$$
$$(t_d, p_{2e})\} \cup \{(p_{2e}, t) : t \in T_1\} \cup \{(t, p_d) : t \in T_1 - {}^\bullet p_{1x}\}$$
$$M_e = (1, 0, M_{1c}, 0, 0, 0, M_{2c}, 0)$$

Graphically, the composite process (Figure 4.13) is obtained by:

1) adding a new entry place p_e and a dummy transition ε_1 for synchronizing the entry into B_1 and B_2;
2) adding *forward Disable-resume arcs*, connecting place p_{2e} of B_2 to every transition of B_1;
3) adding *backward Disable-resume arcs*, connecting every transition of B_1, except for those exit transitions, back to place p_{2e} of B_2, via *Disable-resume place p_d* and *Disable-resume transition t_d*; and
4) adding a *loop back transition* ε_2, connecting p_{2x} of B_2 to p_{2e} of B_2.

It is obvious that $(B_1 \;[r\rangle\; B_2, M_e)$ is also a marked process. The incidence matrix V of $B_1 \;[r\rangle\; B_2$ is formed from the incidence matrices V_1 of B_1 and V_2 of B_2 as follows:

	ε_1	ε_2	T_{11}	T_{12}	T_2	t_d
p_e	-1	0	0	0	0	0
$P_1 - \{p_{1x}\}$	p_{1e}	0	V_{11}	V_{12}	0	0
p_{1x}	0	0	0	1	0	0
P_2	p_{2e}	$p_{2e} - p_{2x}$	$-p_{2e}$	$-p_{2e}$	V_2	p_{2e}
p_d	0	0	1	0	0	-1

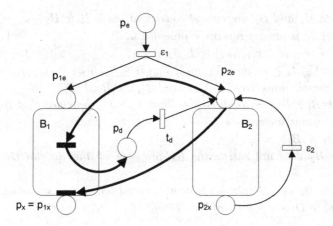

Fig. 4.13 $B_1[r > B_2$: Composition by the operator DISABLE-RESUME.

Explanation:

(a) $T_{11} = T_1 - {}^\bullet p_{1x}$ and $T_{12} = {}^\bullet p_{1x}$.
(b) $V_{11} = V_1[P_1 - \{p_{1x}\}, T_1 - {}^\bullet p_{1x}]$ and $V_{12} = V_1[P_1 - \{p_{1x}\}, {}^\bullet p_{1x}]$.

Theorem 4.6. *(property preservation under DISABLE-RESUME B_1 $[r\rangle B_2$)*
Let (B_1, M_{1e}), (B_2, M_{2e}) and $(B_1 [r\rangle B_2, M_e)$ be the three marked PNPs described in Definition 4.6. Then, the following propositions hold:
(Note: For simplicity in the proofs of sub-theorems 4.6.17 to 4.6.20, M_e is constructed using the non-pure version of the operator Disable-resume and p_x of $B_1 [r\rangle B_2$ occupies the column immediately after M_{1c} in M_c.)

1) *$B_1[r\rangle B_2$ is not an SM, whether or not B_1 and B_2 are so.*
2) *Suppose both B_1 and B_2 are almost MGs. Two cases hold: (a) $B_1[r\rangle B_2$ is not almost an MG. (b) $B_1[r\rangle$ B_2 is an AC net.*
3) *Suppose both B_1 and B_2 are FC nets. Two cases hold: (a) $B_1[r\rangle B_2$ is not an FC. (b) B_1 $[r\rangle B_2$ is an AC net.*
4) *Suppose both B_1 and B_2 are AC nets. Two cases hold: (a) In general, $B_1[r\rangle B_2$ may or may not be an AC. (b) B_1 $[r\rangle B_2$ is an AC net $\forall p \in P_2$ satisfying $(p^\bullet \cap p_{2e}^\bullet) \neq \phi \Rightarrow p^\bullet \subseteq p_{2e}^\bullet)$ in B_2.*
5) *If both B_1 and B_2 are conservative, then B_1 $[r\rangle$ B_2 is conservative provided that $M_{2e}[B_2, {}^*\rangle M_{2x}$.*
6) *If both B_1 and B_2 are structurally bounded, then B_1 $[r\rangle B_2$ is structurally bounded provided that $M_{2e}[B_2, {}^*\rangle M_{2x}$.*

7) *If both B_1 and B_2 are almost consistent, so is $B_1 [r\rangle B_2$.*

8) *$B_1 [r\rangle B_2$ is almost repetitive provided that:*
   $\exists \beta_1 = (x_1, \ldots, x_{|T_1|}, z_1) \geq 1, \beta_2 = (y_1, \ldots, y_{|T_2|}, z_2) \geq 1:\ ((V_{1a}\beta_1 \geq 0$ *and* $V_{2a}\beta_2 \geq 0)$ *and* $(V_{1a}[p_{1x}, T_{1a}]\beta_1 \leq V_{2a}[p_{2e}, T_{2a}]\beta_2))$.

9) *In general,* $Rank(B_1 [r\rangle B_2) = Rank(B_1) + Rank(B_2) + 3$.
   $Rank(B_1 [r\rangle B_2) = Rank(B_1) + Rank(B_2) + 2$ *provided that B_{2a} has a T-invariant including t_{2a}.*

10) *$|C(B_1 [r\rangle B_2)| = |C(B_2)| + 3$.*

11) *$B_1 [r\rangle B_2$ does not satisfy the RC-property whether or not B_1 and B_2 are so.*

12) *If both B_1 and B_2 almost have a minimal SM-cover, so does $B_1 [r\rangle B_2$.*

13) *Suppose D is a siphon of B_1. Then,*

   (a) *D is a siphon of $B_1 [r\rangle B_2$ if and only if $p_{1e} \notin D$;*
   (b) *$D \cup \{p_e\}$ is a siphon of $B_1 [r\rangle B_2$ if $p_{1e} \in D$.*

   Suppose D is a siphon of B_2. Then, $D \cup \{p_e\}$ is a siphon of $B_1 [r\rangle B_2$ if and only if $p_{2e} \notin D$.

14) *Suppose S is a trap of B_1. Then, S is a trap of $B_1 [r\rangle B_2$.*
   Suppose S is a trap of B_2. Then,

   (a) *S is a trap of $B_1 [r\rangle B_2$ if $p_{2e} \notin S$ and $p_{2x} \notin S$;*
   (b) *$S \cup \{p_e, p_d, p_{1x}, p_{2x}\}$ is a trap of $B_1 [r\rangle B_2$ provided that $p_{2e} \in S$.*

15) *$Z(B_1 [r\rangle B_2) = Z(B_1) + Z(B_2) + 2|T_1| - |{}^{\bullet}p_{1x}|$.*

16) *If both (B_1, M_{1e}) and (B_2, M_{2e}) have no dead transitions, then $(B_1 [r\rangle B_2, M_e)$ does not have either.*

17) *If both (B_1, M_{1e}) and (B_2, M_{2e}) are bounded and (B_2, M_{2e}) terminates properly, then $(B_1 [r\rangle B_2, M_e)$ is bounded.*

18) *If both (B_1, M_{1e}) and (B_2, M_{2e}) terminate properly, so does $(B_1 [r\rangle B_2, M_e)$.*

19) *If both (B_1, M_{1e}) and (B_2, M_{2e}) are almost live and terminate properly, then $(B_1 [r\rangle B_2, M_e)$ is almost live.*

20) *If both (B_1, M_{1e}) and (B_2, M_{2e}) are almost reversible and terminate properly, then $(B_1 [r\rangle B_2, M_e)$ is almost reversible.*

Proof.

1-2) According to the structure of $B_1 [r\rangle B_2$, adding p_e, ε_1 and the arcs violates the definitions of SM (because $|\varepsilon_1^{\bullet}| > 1$) and MG (because $|p_{2e}^{\bullet}| > 1$). Suppose both B_1 and B_2 are MGs. Since ($\forall p \in P_1 \cup P_2$: $|p^{\bullet}| = |{}^{\bullet}p| = 1$ in B_1 or B_2) and (${}^{\bullet}(p_e^{\bullet}) = \{p_e\}$, ${}^{\bullet}(p_d^{\bullet}) = \{p_d\}$, ${}^{\bullet}(p_{2x}^{\bullet})$

$= \{p_{2x}\}$ and $p_x^\bullet = \phi)$ in $B_1 \ [r\rangle B_2$, by the structure of $B_1 \ [r\rangle B_2$ and the definition of an AC net, we know that $B_1 \ [r\rangle B_2$ is also an AC net.

3) It is obvious that $B_1 \ [r\rangle B_2$ is not an FC net. For any $p_1, p_2 \in P_1 \cup P_2$ where $(p_1^\bullet \cap p_2^\bullet \neq \phi)$ in $B_1 \ [r\rangle B_2$, consider three cases: (a) $p_1, p_2 \in P_1$. Since B_1 is an FC net, $(p_1^\bullet = p_2^\bullet)$ in B_1 and thus in $B_1 \ [r\rangle B_2$. (b) p_1, $p_2 \in P_2$. Since B_2 is an FC net, $(p_1^\bullet = p_2^\bullet)$ in B_2. If $p_1, p_2 \neq p_{2e}$, then $(p_1^\bullet = p_2^\bullet)$ in $B_1 \ [r\rangle B_2$. If p_1 or $p_2 = p_{2e}$, $(p_2^\bullet \subseteq p_1^\bullet)$ or $(p_1^\bullet \subseteq p_2^\bullet)$ in $B_1 \ [r\rangle B_2$. (c) $p_1 \in P_1$ and $p_2 \in P_2$. Then $p_2 = p_{2e}$ because of $(p_1^\bullet \cap p_2^\bullet \neq \phi)$ in $B_1 \ [r\rangle B_2$. By the structure of $B_1 \ [r\rangle B_2$, we know that $(p_1^\bullet \subseteq p_2^\bullet)$ in $B_1 \ [r\rangle B_2$. Note that $(^\bullet(p_e^\bullet) = \{p_e\}$, $^\bullet(p_d^\bullet) = \{p_d\}$, $^\bullet(p_{2x}^\bullet) = \{p_{2x}\}$ and $p_x^\bullet = \phi)$ in $B_1 \ [r\rangle B_2$, $B_1 \ [r\rangle B_2$ is an AC net.

4) It is obvious that $B_1 \ [r\rangle B_2$ may or may not be an AC net. For any $p_1, p_2 \in P_1 \cup P_2$ where $(p_1^\bullet \cap p_2^\bullet \neq \phi)$ in $B_1 \ [r\rangle B_2$, consider three cases: (a) $p_1, p_2 \in P_1$. Since B_1 is an AC net, $(p_2^\bullet \subseteq p_1^\bullet)$ or $(p_1^\bullet \subseteq p_2^\bullet)$ in B_1 and thus in $B_1 \ [r\rangle B_2$. (b) $p_1, p_2 \in P_2$. Since B_2 is an AC net, $(p_2^\bullet \subseteq p_1^\bullet)$ or $(p_1^\bullet \subseteq p_2^\bullet)$ in B_2. If $p_1, p_2 \neq p_{2e}$, then $(p_2^\bullet \subseteq p_1^\bullet)$ or $(p_1^\bullet \subseteq p_2^\bullet)$ in $B_1 \ [r\rangle \ B_2$. If $p_1 = p_{2e}$, $(p_2^\bullet \subseteq p_1^\bullet)$ in $B_1 \ [r\rangle B_2$ because of $(p^\bullet \cap p_{2e}^\bullet) \neq \phi \Rightarrow p^\bullet \subseteq p_{2e}^\bullet)$ in B_2. (c) $p_1 \in P_1$ and $p_2 \in P_2$. Then $p_2 = p_{2e}$ because of $(p_1^\bullet \cap p_2^\bullet \neq \phi)$ in $B_1 \ [r\rangle B_2$. By the structure of $B_1 \ [r\rangle B_2$, we know that $(p_1^\bullet \subseteq p_2^\bullet)$ in $B_1 \ [r\rangle B_2$. Note that $(^\bullet(p_e^\bullet) = \{p_e\}$, $^\bullet(p_d^\bullet) = \{p_d\}$, $^\bullet(p_{2x}^\bullet) = \{p_{2x}\}$ and $p_x^\bullet = \phi)$ in $B_1 \ [r\rangle B_2$, $B_1 \ [r\rangle B_2$ is an AC net.

5) Since B_1 and B_2 are conservative, by Characterization 2.1, $\exists \alpha_1 = (x_1, \ldots, x_{|P_1|}) \geq 1$ and $\alpha_2 = (y_1, \ldots, y_{|P_2|}) \geq 1$ such that $\alpha_1 V_1 = 0$ and $\alpha_2 V_2 = 0$. Since $M_{2e}[B_2, {}^*\rangle M_{2x}$, $\alpha_2 M_{2e} = \alpha_2 M_{2x}$. This implies $y_1 = y_{|P_2|}$. Let $\alpha = (x_1 + y_1, \alpha_1[P_1 - \{p_{1x}\}], x_{|P_1|} + y_1, \alpha_2, y_1)$ and V be shown in Figure 4.14. Since for any column $V[P, t_1]$ such that $t_1 \in T_1 - {}^\bullet p_{1x}$, $\alpha V[P, t_1] = \alpha_1[P_1 - \{p_{1x}\}]V[P_1 - \{p_{1x}\}, t_1] + y_1(-1) + y_1(1) = 0$ and for any column $V[P, t_1]$ such that $t_1 \in {}^\bullet p_{1x}$, $\alpha V[P, t_1] = \alpha_1[P_1 - \{p_{1x}\}]V[P_1 - \{p_{1x}\}, t_1] + (x_{|P_1|} + y_1)(1) + y_1(-1) = \alpha_1[P_1, t_1] + y_1 - y_1 = 0$. Then, $\alpha \geq 1$ and $\alpha V = (\alpha V[P, \varepsilon_1], \alpha V[P, \varepsilon_2], \alpha V[P, T_1 - {}^\bullet p_{1x}], \alpha V[P, {}^\bullet p_{1x}], \alpha V[P, T_2], \alpha V[P, t_d]) = (-x_1 - y_1 + x_1 + y_1, y_1 - y_{|P_2|}, \alpha_1 V_1[P, T_1 - {}^\bullet p_{1x}] + (-y_1 + y_1)^u, \alpha_1 V_1[P, {}^\bullet p_{1x}] + (y_1 - y_1)^v, \alpha_2 V_2, y_1 - y_1) = 0$, where $u = |T_1 - {}^\bullet p_{1x}|$, $v = |{}^\bullet p_{1x}|$ and $(w)^d$ represents a vector $(w \ w \ \ldots \ w)$ of dimension d. It follows from Characterization 2.1 that $B_1 \ [r\rangle B_2$ is conservative.

6) Suppose B_1 and B_2 are structurally bounded. The same argument as in 5) shows that $\exists \alpha \geq 1$ such that $\alpha V \leq 0$. It follows from Characterization 2.2 that $B_1 \ [r\rangle B_2$ is structurally bounded.

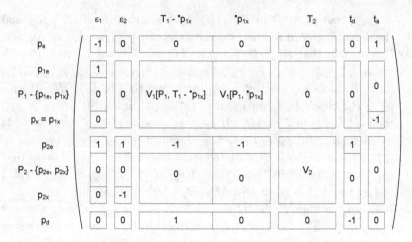

Fig. 4.14   Incidence matrix V_a of $(B_1[rB_2)_a$ (V of $B_1[rB_2$ is the part of V_a without column t_a).

7) Since B_1 and B_2 are almost consistent, by Characterization 2.3, $\exists \beta_1 = (x_1, \ldots, x_{|T_1|}, z_1) \geq 1$ and $\beta_2 = (y_1, \ldots, y_{|T_2|}, z_2) \geq 1$ such that $V_{1a}\beta_1 = 0$ and $V_{2a}\beta_2 = 0$. Let $\beta = (z_1, z_2, \beta_1, \beta_2, \sum_{t_1 \in T_1 - \bullet p_1 x}\beta_1[t_1], z_1)$ and V_a be shown in Figure 4.14. Then, $\beta \geq 1$. Since $z_1 - \sum_{t \in \bullet p_1 x}\beta_1[t] = -V_{1a}[p_{1x}, T_{1a}]\beta_1 = 0$, we have $V_a\beta = (V_a[p_e, T_a]\beta, V_a[P_1, T_a]\beta,$ $V_a[p_{2e}, T_a]\beta, V_a[P_2 - \{p_{2e}\}, T_a]\beta, V_a[p_d, T_a]\beta) = (-z_1 + z_1, V_{1a}\beta_1,$ $z_1 - \sum_{t \in \bullet p_1 x}\beta_1[t] + V_{2a}[p_{2e}, T_{2a}]\beta_2, V_{2a}[P_2 - \{p_{2e}\}, T_{2a}]\beta_2, 0) = 0$. It follows from Characterization 2.3 that $B_1 [r\rangle B_2$ is almost consistent.

8) Suppose B_1 and B_2 are almost repetitive. Given that $V_{1a}[p_{1x}, T_{1a}]\beta_1 \leq V_{2a}[p_{2e}, T_{2a}]\beta_2$, the same argument as in 7) shows that $\exists \beta \geq 1$ such that $V_a\beta \geq 0$. It follows from Characterization 2.4 that $B_1 [r\rangle B_2$ is almost repetitive.

9) It is easy to see that the columns of $V[P, \varepsilon_1]$, $V[P, T_1]$, $V[P, T_2]$ and $V[P, t_d]$ are linearly independent. If B_{2a} has a T-invariant including t_{2a}, the column $V[P, \varepsilon_2]$ is linearly dependent on the columns $V[P, T_2]$. Otherwise, they are linearly independent. Therefore, in, general, $\mathrm{Rank}(B_1 [r\rangle B_2) = \mathrm{Rank}(B_1) + \mathrm{Rank}(B_2) + 3$. If B_{2a} has a T-invariant including t_{2a}, $\mathrm{Rank}(B_1 [r\rangle B_2) = \mathrm{Rank}(B_1) + \mathrm{Rank}(B_2) + 2$.

10) $[p_e] = \{p_e, \varepsilon_1\}$ and $[p_d] = \{p_d, t_d\}$ are the only new clusters created. In $B_1 [r\rangle B_2$, since all transitions of B_1 become output transitions of p_{2e}, all clusters of B_1, except $[p_{1x}] = \{p_{1x}\}$, are merged into $[p_{2e}]$ of

B_2. Also, all clusters of B_2 are clusters of $B_1 \, [r\rangle B_2$. Hence, $|C(B_1 \, [r\rangle B_2)| = |C(B_2)| + 3$.

11) This follows from 9) and 10).

12) By definition, any SM-component of B_{ia} containing p_{ie} must also contain t_{ia} and p_{ix}, where $i = 1, 2$. Suppose K_1 and K_2 are minimal SM-covers of B_{1a} and B_{2a}, respectively. For any $S_i \in K_1$ which contains p_{1e} and p_{1x}, extend S_i by including p_e, ε_1, and the arcs $\langle p_e, \varepsilon_1 \rangle$ and $\langle \varepsilon_1, p_{1e} \rangle$ in it (with some trivial changes in labels). For any $S_j \in K_2$ which contains p_{2e} and p_{2x}, extend S_j by including p_e, p_x, p_d, ε_2, t_d, and the arcs $\{(p_e, \varepsilon_1), (\varepsilon_1, p_{2e}), (p_{2x}, \varepsilon_2), (\varepsilon_2, p_{2e}), (t_a, p_e), (p_x, t_a), (p_d, t_d), (t_d, p_{2e})\} \cup \{(p_{2e}, t_1): t_1 \in T_1\} \cup \{(t_1, p_d): t_1 \in T_1 - {}^\bullet p_{1x}\}) \}$ in it (with some trivial changes in labels). Obviously, these result in a minimal SM-cover of $(B_1 [r\rangle \, B_2)_a$.

13) Part 1:

   (a) "$\Leftarrow$" Suppose $p_{1e} \notin D$. Then, ${}^\bullet D$ in $B_1 \, [r\rangle B_2 = {}^\bullet D$ in B_1 and $D^\bullet$ in $B_1 \, [r\rangle B_2 = D^\bullet$ in B_1. Hence, $({}^\bullet D \subseteq D^\bullet)$ in $B_1 \, [r\rangle B_2$ if $({}^\bullet D \subseteq D^\bullet)$ in B_1. "$\Rightarrow$" Prove by contradiction. If $p_{1e} \in D$, then $(\varepsilon \in {}^\bullet D \text{but } \varepsilon \notin D^\bullet)$ in $B_1 \, [r\rangle B_2$. Hence, $({}^\bullet D \not\subseteq D^\bullet)$ in $B_1 \, [r\rangle B_2$ even if $({}^\bullet D \subseteq D^\bullet)$ in B_1.

   (b) If $p_{1e} \in D$, then, ${}^\bullet(D \cup \{p_e\})$ in $B_1 \, [r\rangle B_2 = \{\varepsilon_1\} \cup ({}^\bullet D$ in $B_1)$ and $(D \cup \{p_e\})^\bullet$ in $B_1 \, [r\rangle \, B_2 = (D^\bullet$ in $B_1) \cup \{\varepsilon_1\}$. Hence, $({}^\bullet(D \cup \{p_e\}) \subseteq (D \cup \{p_e\})^\bullet)$ in $B_1 \, [r\rangle B_2$ if $({}^\bullet D \subseteq D^\bullet)$ in B_1.

   For part 2, the proof is similar to the first section of part 1.

14) For $S \subseteq P_1$, it is obvious that S is a trap of $B_1 \, [r\rangle B_2$ if and only if S is a trap of B_1. For $S \subseteq P_2$, consider two cases:

   (a) $p_{2e} \notin S$ and $p_{2x} \notin S$. Then $S^\bullet$ in $B_1 \, [r\rangle B_2 = S^\bullet$ in B_2 and ${}^\bullet S$ in $B_1 \, [r\rangle B_2 = {}^\bullet S$ in B_2. Hence, $(S^\bullet \subseteq {}^\bullet S)$ in $B_1 \, [r\rangle B_2$ if and only if $(S^\bullet \subseteq {}^\bullet S)$ in B_2.

   (b) $p_{2e} \in S$. Then, $(S \cup \{p_e, p_d, p_{1x}, p_{2x}\})^\bullet$ in $B_1 \, [r\rangle B_2 = (S^\bullet$ in $B_2) \cup \{\varepsilon_1, \varepsilon_2, t_d\} \cup T_1$ and ${}^\bullet(S \cup \{p_e, p_d, p_{1x}, p_{2x}\})$ in $B_1 \, [r\rangle B_2 = ({}^\bullet S$ in $B_2) \cup \{\varepsilon_1, \varepsilon_2, t_d\} \cup T_1$. Hence, $((S \cup \{p_e, p_d, p_{1x}, p_{2x}\})^\bullet \subseteq {}^\bullet(S \cup \{p_e, p_d, p_{1x}, p_{2x}\}))$ in $B_1 \, [r\rangle B_2$ if $(S^\bullet \subseteq {}^\bullet S)$ in B_2.

15) $Z(B_1 \, [r\rangle B_2) = |F| - |P \cup T| + 2 = |F_1| + |F_2| + 2|T_1| - |{}^\bullet p_{1x}| + 7 - (|P_1| + |P_2| + 2 + |T_1| + |T_2| + 3) + 2 = (|F_1| - |P_1 \cup T_1| + 2) + (|F_2| - |P_2 \cup T_2| + 2) + 2|T_1| - |{}^\bullet p_{1x}| = Z(B_1) + Z(B_2) + 2|T_1| - |{}^\bullet p_{1x}|$.

16) It is obvious.

17) Any $M \in R(B_1 \ [r\rangle B_2, \ M_e)$ can only have three forms: 1)$M = M_e$. M is obviously bounded. 2)$M = (0, \ m_1, m_2)$, where $m_1 \in R(B_1, \ M_{1e})$ $- \{M_{1x}\}$ and $m_2 \in R(B_{2a}, \ M_{2e})$. Since termination of $(B_2, \ M_{2e})$ is proper, $(B_{2a}, \ M_{2e})$ is bounded. Hence, M is bounded. 3)$M = M_{1x}$. Since termination of $(B_2, \ M_{2e})$ is proper, M is bounded. Therefore, $(B_1 \ [r\rangle B_2, \ M_e)$ is bounded.

18) Consider any marking $M \in R(B_1 \ [r\rangle B_2, \ M_e)$ such that $M \geq p_x$. Since $M_e[\varepsilon_1\rangle(M_{1e} + M_{2e})$, $(B_{ia}, \ M_{ie})$ terminates properly, $i = 1, 2$, and the token in p_{2e} is removed when any exit transition of B_1 is fired, $M \geq p_x$ $\Rightarrow M = M_x$. (Note: termination of $(B_{2a}, \ M_{2e})$ is proper given that termination of $(B_2, \ M_{2e})$ is proper.)

19) Since $(B_1, \ M_{1e})$ and $(B_2, \ M_{2e})$ terminate properly, any marking $M \in R((B_1 \ [r\rangle B_2)_a, \ M_e)$ can only have two forms: (1)$M = (0 \ m_1 \ 0 \ m_2)$, where $m_1 \in R(B_{1a}, \ M_{1e})$. Consider two possible locations of an arbitrary transition t. (a) $t \in T_1 \cup \{\varepsilon_1, t_a\}$. There are three sub-cases: (i) $m_2 \in R(B_{2a}, \ M_{2e})$, where $m_2(p_{2e}) = 0$. Since $(B_1, \ M_{1e})$ is almost live and terminates properly, $\exists \sigma_1 \in T_{1a}^*$: $(m_1[B_{1a}, \sigma_1 t\rangle)$, where $\sigma_1 = \sigma_{11}$ or $\sigma_1 = \sigma_{11} t_{1a} \sigma_{12}$ and $\sigma_{11}, \ \sigma_{12} \in T_1^*$. Since $(B_{2a}, \ M_{2e})$ is live, $\exists \sigma_2 \in T_2^*$: $(m_2[B_{2a}, \ \sigma_2 t_{2a}\rangle)$. If $\sigma_1 = \sigma_{11}$, then $M[(B_1[r\rangle B_2)_a, \ \sigma t\rangle$, where $\sigma = \sigma_2 \varepsilon_2 \sigma_{11}$ for $t \in T_1 \cup \{t_a\}$ and $\sigma = \sigma_2 \varepsilon_2 \sigma_{11} t_a$ for $t = \varepsilon_1$. If $\sigma_1 = \sigma_{11} t_{1a} \sigma_{12}$, then $M[(B_1[r\rangle B_2)_a, \ \sigma t\rangle$, where $\sigma = \sigma_2 \varepsilon_2 \sigma_{11}$ for $t = t_a$, $\sigma = \sigma_2 \varepsilon_2 \sigma_{11} t_a$ for $t = \varepsilon_1$ and $\sigma = \sigma_2 \varepsilon_2 \sigma_{11} t_a \varepsilon_1 \sigma_{12}$ for $t \in T_1$. (ii) $m_2 = M_{2e}$. Similar argument as in (i) shows that t is live. (iii) $m_2 = M_{2c}$. That is, $m_1 = M_{1x}$. Hence, $M[(B_1 \ [r\rangle B_2)_a, \ t_a\rangle$. In addition, since $(B_{1a}, \ M_{1e})$ is live, $\exists \sigma_1 \in T_{1a}^*$: $(M_{1e}[\sigma_1 t\rangle)$. Then, $M[(B_1 \ [r\rangle B_2)_a, \ \sigma t\rangle$, where $\sigma = t_a$ for $t = \varepsilon_1$ and $\sigma = t_a \varepsilon_1 \sigma_1$ for $t \in T_1$. (b) $t \in T_2 \cup \{\varepsilon_1, t_a\}$. Similar argument as in (1) shows that t is live. (2) $M = (1, M_{1c}, M_{2c}, 0)$. This case becomes Case 1 after firing ε_1 at M.

20) Since $(B_1, \ M_{1e})$ and $(B_2, \ M_{2e})$ terminate properly, any $M \in R((B_1 \ [r\rangle B_2)_a, \ M_e)$ can only have two forms: (1) $M = (0 \ m_1 \ 0 \ m_2)$, where $m_1 \in R(B_{1a}, \ M_{1e})$. There are three sub-cases: (i) $m_2 \in R(B_{2a}, \ M_{2e})$, where $m_2(p_{2e}) = 0$. Since both $(B_1, \ M_{1e})$ and $(B_2, \ M_{2e})$ are almost reversible and terminate properly, $\exists \sigma_i \in T_i^*$: $(m_i[B_{ia}, \ \sigma_i t_{ia}\rangle M_{ie})$, $i = 1, 2$. Hence, $M[(B_1 \ [r\rangle B_2)_a, \ \sigma_2 \varepsilon_2 \sigma_1 t_a\rangle M_e$. (ii) $m_2 = M_{2e}$. Since $(B_1, \ M_{1e})$ is almost reversible and terminate properly, $\exists \sigma_1 \in T_1^*$: $(m_1[B_{1a}, \ \sigma_1 t_{1a}\rangle M_{1e})$. Hence, $M[(B_1 \ [r\rangle B_2)_a, \ \sigma_1 t_a\rangle M_e$. (iii) $m_2 = M_{2c}$. That is, $m_1 = M_{1x}$. Then, $M[(B_1 \ [r\rangle B_2)_a, \ t_a\rangle M_e$. (2) $M = (1, \ M_{1c}, M_{2c})$. Obviously, $M[(B_1 \ [r\rangle B_2)_a, \ \lambda\rangle M_e$, where λ is the null sequence of transitions. $\qquad \square$

Based on the results in this chapter, the following conclusions are correct:

Theorem 4.7.

1) *Suppose both (B_1, M_{1e}) and (B_2, M_{2e}) are bounded and terminate properly. Then $(B_1 \;[\;]\; B_2, M_e)$ is also bounded.*
2) *Even if B_1 and B_2 are both almost structurally bounded, $B_1 \;[\rangle\; B_2$ is not almost structurally bounded.*
3) *If (B_1, M_{1e}) and (B_2, M_{2e}) are both almost bounded, then $(B_1 \;[\rangle\; B_2, M_e)$ is almost bounded*
4) *Four primitive processes (B_1, M_{1e}), (B_2, M_{2e}), (B_3, M_{3e}) and (B_4, M_{4e}) listed in Figure 4.15 are selected as the constituent components. They include an MG, an AC net and a sequential SM. The composite process (B_0, M_0), where $B_0 \equiv ((t_5|B_1') \;[\;]\; B_2) >> (B_3|\{t_{18}\}|B_4)$ is shown in Figure 4.16 and $M_0 = p_1 + p_8 + p_{19}$. Then (B_0, M_0) satisfy the following properties by PPPA.*

   1) *B_0 is not an SM.*
   2) *B_0 is not almost an MG.*
   3) *B_0 is not an FC net.*
   4) *B_0 is an AC net.*
   5) *B_0 is conservative.*
   6) *B_0 is structurally bounded.*
   7) *B_0 is almost consistent.*
   8) *B_0 is almost repetitive.*
   9) *$Rank(B_0) = 19$.*
   10) *$|C(B_0)| = 19$.*
   11) *B_0 does not satisfy the RC-property.*
   12) *B_0 almost has a minimal SM-cover.*
   13) *$D_1 = \{p_{16}, p_{19}, p_{23}\}$ is a siphon of B_0, whereas $D_2 = \{p_{13}\}$ is not a siphon of B_0.*
   14) *$S_1 = \{p_7, p_8,\}$ a trap of B_0, whereas $S_2 = \{p_1, p_2, p_3, p_4, p_5, p_7, p_9, p_{11}, p_{12}\}$ is not a trap of B_0.*
   15) *$Z(B_0) = 12$.*
   16) *(B_0, M_0) does not have dead transitions.*
   17) *(B_0, M_0) is bounded.*
   18) *(B_0, M_0) terminates properly.*
   19) *(B_0, M_0) is almost live.*
   20) *(B_0, M_0) is almost reversible.*

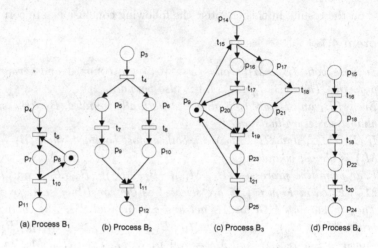

Fig. 4.15   Basic component processes B_1, B_2, B_3 and B_4.

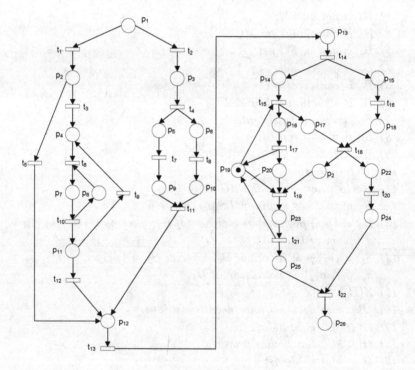

Fig. 4.16   A composite process $((t_5|B_1')[]B_2) >> (B_3|\{t_{18}\}|B_4)$.

Chapter 5

Property-Preserving Refinements

This chapter presents some operators for refinement in PPPA. Addition of operators for reduction and place merging will be presented in the subsequent two chapters.

In our Petri net based approach, a refinement replaces an individual transition or place of a Petri net with another Petri net process. From the viewpoint of system transformations, the refined place or transition represents a functional call and the refinement implies that the called process is embedded and integrated into the functioning of the calling process. As far as the design's specification is concerned, the refined Petri net has been transformed into another Petri net whose correctness has to be verified.

In the remaining parts of this chapter, we present the technical details for transition and place refinements and their preservation of properties. The refined net (B_1, M_r) is a general Petri net and the refinement net (B_2, M_s) is a Petri net process. For convenience, sometimes a marking uses the multi-set expression, e.g, $(M_e = p_e + M_c)$ is equivalent to $(M_e = (1 \ M_c \ 0))$ in a Petri net process.

5.1 A Property-Preserving Refinement for Transitions

A transition refinement expands a transition to a Petri net process. This section first formally describes the refinement technique and then derives some relationships among its various ingredients. Then, one of the two major results of this chapter, namely, preservation of properties under a transition refinement, is presented.

Definition 5.1. (refined net, refined transition) (B_1 in Figure 5.1).
A refined net B_1 is a 4-tuple (N_1, r_i, r_o, t_r), where

- $N_1 = (P_1, T_1, F_1)$ is an ordinary net
- $r_i \in P_1$ is the refinement inlet place, $r_o \in P_1$ is the refinement outlet place and $t_r \in T_1$ is the refined transition, where $r_i^\bullet = \{t_r\} = {}^\bullet r_o$ and $|{}^\bullet t_r| = |t_r^\bullet| = 1$
- t_r is not 2-enabled (or, r_i is 1-bounded)

A marked refined net (B_1, M_r) is a refined net with an initial marking M_r.

Definition 5.2. (transformation TR $B_1(t_r \to B_2)$) (Figure 5.1).
Let (B_1, M_r) be a marked refined net as defined in Definition 5.1, and (B_2, M_s) be a Petri net process (Definition 3.1). Replacing transition t_r of (B_1, M_r) with (B_2, M_s), where $B_2 = (N_2, p_e, p_x)$, results in a marked net (B, M_o), where

- $B = B_1(t_r \to B_2) = (N, p_i, p_o); N = (P, T, F)$
- $P = P_1 \bigcup P_2 \bigcup \{p_i, p_o\} - \{r_i, r_o, p_e, p_x\}; T = T_1 \bigcup T_2 - \{t_r\};$
  $F = F_1 \bigcup F_2 \bigcup (\{(p_i, x) | x \in p_e^\bullet\} \bigcup \{(x, p_o) | x \in {}^\bullet p_x\} \bigcup \{(x, p_i) | x \in {}^\bullet r_i\} \bigcup \{(p_o, x) | x \in r_o^\bullet\}) - (\{(r_i, t_r), (t_r, r_o)\} \bigcup \{(x, r_i) | x \in {}^\bullet r_i\} \bigcup \{(r_o, x) | x \in r_o^\bullet\} \bigcup \{(p_e, x) | x \in p_e^\bullet\} \bigcup \{(x, p_x) | x \in {}^\bullet p_x\})$
- p_i, is the inlet place and p_o is the outlet place
- The initial marking M_o of B is derived from M_r of B_1 and M_s of B_2 as follows:
  $M_o(p_i) = M_r(r_i); M_o(p_o) = M_r(r_o); M_o(p) = M_r(p)$ for $p \in P_1 - \{r_i, r_o\}$ and $M_o(p) = M_s(p)$ for $p \in P_2 - \{p_e, p_x\}$.

For the rest of this section, B_1 denotes the refined net, B_2 the refinement net process and $B = B_1(t_r \to B_2)$ the resulting net as described in Definition 5.2.

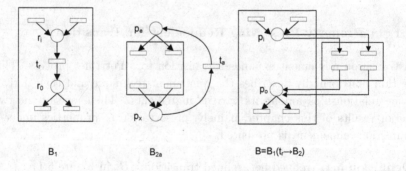

B$_1$ B$_{2a}$ B=B$_1$(t$_r \to$B$_2$)

Fig. 5.1   Transformation TR (Transition Refinement).

Definition 5.3. (mappings arising from Transformation TR).

Let $\sigma \in L(B, M_o)$ and $M_o[B, \sigma\rangle M$. The mappings of σ and M from B to B_1 and B_2 are defined below, where λ is a null sequence and γ is an e-sequence of B_2.

$f_1: T^* \to T_1^*$ is defined as follows: $f_1(\lambda) = \lambda$. If $f_1(\sigma)$ has been defined for σ, then

$$\begin{aligned}
f_1(\sigma t) &= f_1(\sigma) && if\ t \in T_2 - {}^\bullet p_o \\
&= f_1(\sigma)t && if\ t \in T_1 - \{t_r\} \\
&= f_1(\sigma)t_r && if\ t \in {}^\bullet p_o
\end{aligned}$$

M_1 is a restriction of M from P to P_1, where

$$\begin{aligned}
M_1(r_i) &= M(p_i) + (\ \#(\sigma, p_i^\bullet) - (\#(\sigma, {}^\bullet p_o)) \\
M_1(r_o) &= M(p_o) \\
M_1(p) &= M(p) && if\ p \in P_1 - \{r_i, r_o\}
\end{aligned}$$

$f_2: T^* \to T_2^*$ is defined as follows: $f_2(\lambda) = \lambda$. If $f_2(\sigma)$ has been defined for σ, then

$$\begin{aligned}
f_2(\sigma t) &= f_2(\sigma) && if\ t \in T_1 - \{t_r\} \\
&= f_2(\sigma)t && if\ t \in T_2 - {}^\bullet p_o \\
&= \gamma && if\ t \in {}^\bullet p_o
\end{aligned}$$

M_2 is a restriction of M from P to P_2, where

$$\begin{aligned}
M_2(p_e) &= M(p_i) \\
M_2(p_x) &= 0 \\
M_2(p) &= M(p) && if\ p \in P_2 - \{p_e, p_x\}
\end{aligned}$$

Lemma 5.1. *For any firable e-sequence γ of a properly terminated Petri net process (B_2, M_e), $M_e[B_2, \gamma\rangle M_x$.*

Proof. This follows from the definition of proper termination on (B_2, M_s).

The following results describe the relationships among the transition sequences σ, $f_1(\sigma)$, $f_2(\sigma)$ and the markings M, M_1, M_2 of B, B_1 and B_2, respectively. $\qquad\square$

Lemma 5.2. *Suppose (B_2, M_s) terminates properly in Definition 5.2. Then $\forall \sigma \in L(B, M_o)$, $\#(\sigma, {}^\bullet p_o) + 1 \geq \#(\sigma, p_i^\bullet) \geq \#(\sigma, {}^\bullet p_o)$.*

Proof. (by mathematical induction on σ) For $\sigma = \lambda$, Lemma 5.2 obviously holds because $\#(\lambda, {}^\bullet p_o) = \#(\lambda, p_i^\bullet) = 0$. Suppose $\forall \sigma' \in L(B, M_0)$ where $|\sigma'| \leq n$, $\#(\sigma', {}^\bullet p_o) + 1 \geq \#(\sigma', p_i^\bullet) \geq \#(\sigma', {}^\bullet p_o)$. Let $\sigma = \sigma' t$. For $t \notin p_i^\bullet \cup {}^\bullet p_o$, $\#(\sigma, p_i^\bullet) = \#(\sigma', p_i^\bullet)$ and $\#(\sigma, {}^\bullet p_o) = \#(\sigma', {}^\bullet p_o)$. Hence, $\#(\sigma, {}^\bullet p_o) + 1 \geq \#(\sigma, p_i^\bullet) \geq \#(\sigma, {}^\bullet p_o)$. For $t \in p_i^\bullet$, since B_2 terminates properly, we have $\#(\sigma, p_i^\bullet) = \#(\sigma', p_i^\bullet) + 1$ and $\#(\sigma, {}^\bullet p_o) = \#(\sigma', {}^\bullet p_o)$.

Hence, $\#(\sigma, {}^\bullet p_o) + 1 \geq \#(\sigma, p_i^\bullet) \geq \#(\sigma, {}^\bullet p_o)$. For $t \in {}^\bullet p_o$, the proof is similar. $\qquad\qquad\qquad\qquad\qquad\qquad\qquad\qquad\qquad\qquad\qquad\qquad\quad \square$

Theorem 5.1. *Suppose B_2 terminates properly and a transition sequence σ and a marking M satisfy $M_0[B, \sigma\rangle M$ in Definition 5.2. Let f_1, f_2, M_1 and M_2 be defined as in Definition 5.3. Then, the following two propositions hold: (1) $f_1(\sigma) \in L(B_1, M_r)$ and $f_2(\sigma) \in L(B_2, M_e)$. (2) $M_r[B_1, f_1(\sigma)\rangle M_1$ in B_1 and $M_e[B_2, f_2(\sigma)\rangle M_2$ in B_2.*

Proof. (by mathematical induction on the length of σ) For $\sigma = \lambda$, obviously $M = M_0$. By Definition 5.3, $f_1(\sigma) = f_2(\sigma) = \lambda$. Hence $f_1(\sigma) \in L(B_1, M_r)$ and $f_2(\sigma) \in L(B_2, M_e)$. Since $M_0 = M_r + M_s$ (with multi set expression) as defined in Definition 5.2, $M_r = M_1$ and $M_e = M_2$. Since obviously $M_r[B_1, \lambda\rangle M_r$ and $M_e[B_2, \lambda\rangle M_e$, it follows that $M_r[B_1, f_1(\sigma)\rangle M_1$ and $M_e[B_2, f_2(\sigma)\rangle M_2$. Next, suppose Propositions 1 and 2 hold for all μ where $|\mu| \leq n$. That is, $M_0[B, \mu\rangle M'$ implies:

1) $f_1(\mu) \in L(B_1, M_r)$; $f_2(\mu) \in L(B_2, M_e)$.
2) $M_r[B_1, f_1(\mu)\rangle M_1'$, where $M_1'(r_i) = M'(p_i) + (\#(\mu, p_i^\bullet) - \#(\mu, {}^\bullet p_o))$, $M_1'(r_o) = M'(p_o)$ and $M_1'(p) = M'(p)$ for $p \in P_1 - \{r_i, r_o\}$; $M_e[B_2, f_2(\mu)\rangle M_2'$, where $M_2'(p_e) = M'(p_i)$, $M_2'(p_x) = 0$, $M_2'(p) = M'(p)$ for $p \in P_2 - \{p_e, p_x\}$.

For $\sigma = \mu t \in L(B, M_0)$, i.e., $M_0[B, \mu\rangle M'[B, t\rangle M$, proof of Propositions 1 and 2 for σ proceeds as follows:
1) For Proposition 1, three cases should be considered:

(a) If $t \in T_2 - {}^\bullet p_o$, $f_1(\sigma) = f_1(\mu) \in L(B_1, M_r)$, $f_2(\sigma) = f_2(\mu)t$. By Definition 5.3, $M_2'(p_e) = M'(p_i)$, $M_2'(p_x) = 0$ and $M_2'(p) = M'(p)$ for $p \in P_2 - \{p_e, p_x\}$. Since p_x is not in ${}^\bullet t$ and t is enabled at M' in B, t is also enabled at M_2' in B_2. Hence, $f_2(\sigma) \in L(B_2, M_e)$.

(b) If $t \in T_1 - \{t_r\}$, $f_1(\sigma) = f_1(\mu)t$ and $f_2(\sigma) = f_2(\mu) \in L(B_2, M_e)$. Since $M_1' \geq M'$ by Definition 5.3 and the fact that t is enabled at M' in B, t is also enabled at M_1' in B_1. Hence, $f_1(\sigma) \in L(B_1, M_r)$.

(c) If $t \in {}^\bullet p_o$, $f_1(\sigma) = f_1(\mu)t_r$ and $f_2(\sigma) = \gamma \in L(B_2, M_e)$. Since μ is followed by $t \in {}^\bullet p_o$, Lemma 5.2 implies that $\#(\mu, p_i^\bullet) = \#(\mu, {}^\bullet p_o) + 1$. Hence, $M_1'(r_i) = M(p_i) + \#(\mu, p_i^\bullet) - \#(\mu, {}^\bullet p_o) \geq 1$ and t_r is enabled at M_1' in B_1. Hence, $f_1(\sigma) \in L(B_1, M_r)$.

2) By Proposition 1, we can assume that $M'[B, t\rangle M$, $M_1'[B_1, f_1(t)\rangle M_1$ and $M_2'[B_2, f_2(t)\rangle M_2$. Proposition 2 will follow if M_1 and M_2 can be derived

from M according to Definition 5.3. To show this, we consider the following five cases:

(a) If $t \in T_2 - p_e^\bullet - {}^\bullet p_x = T_2 - p_i^\bullet - {}^\bullet p_o$, then $\#(\sigma, p_i^\bullet) = \#(\mu, p_i^\bullet)$ and $\#(\sigma, {}^\bullet p_o) = \#(\mu, {}^\bullet p_o)$. Also, the token distribution within B_1 is not affected, i.e., $M_1(p) = M_1'(p) = M(p) \forall p \in P_1$. Within B_2, firing t does not affect the tokens in p_e and p_x. Hence, $M_2(p_e) = M_2'(p_e) = M'(p_i) = M(p_i)$ and $M_2(p_x) = M_2'(p_x) = 0$, $M_2(p) = M_2'(p) + W(t,p) - W(p,t) = M'(p) + W(t,p) - W(p,t) = M(p)$ for $p \in P_2 - \{p_e, p_x\}$.

(b) If $t \in T_1 - \{t_r\}$, then $\#(\sigma, p_i^\bullet) = \#(\mu, p_i^\bullet), \#(\sigma, {}^\bullet p_o) = \#(\mu, {}^\bullet p_o)$. Within B_1, we have: $M_1(r_i) = M_1'(r_i) + W(t, r_i) = M'(p_i) + (\#(\mu, p_i^\bullet) - \#(\mu, {}^\bullet p_o)) + W(t, r_i) = M(p_i) + \#(\sigma, p_i^\bullet) - \#(\sigma, {}^\bullet p_o)$, $M_1(r_o) = M_1'(r_o) - W(r_o, t) = M'(p_o) - W(r_o, t) = M(p_o)$ and $M_1(p) = M_1'(p) + W(t,p) - W(p,t) = M'(p) + W(t,p) - W(p,t) = M(p)$ for $p \in P_1 - \{r_i, r_o\}$. Within B_2, firing t does not affect the tokens distribution in B_2. Hence, $M_2(p_e) = M_2'(p_e) + W(t, p_e) = M'(p_i) + W(t, p_i) = M(p_i)$ and $M_2(p_x) = M_2'(p_x) = 0$, $M_2(p) = M_2'(p) = M(p)$ for $p \in P_2 - \{p_e, p_x\}$.

(c) If $t \in p_i^\bullet - {}^\bullet p_o$, then $\#(\sigma, p_i^\bullet) = \#(\mu, p_i^\bullet) + 1$, $\#(\sigma, {}^\bullet p_o) = \#(\mu, {}^\bullet p_o)$ and $f_1(t) = \lambda$. Within B_1, firing t does not affect the token distribution in B_1. $M(p) = M'(p)$ for $p \in P_1$ and $M(p_i) = M'(p_i) - 1$. Hence, $M_1(r_i) = M_1'(r_i) = M'(p_i) + \#(\mu, p_i^\bullet) - \#(\mu, {}^\bullet(p_o) = M(p_i) + \#(\sigma, p_i^\bullet) - \#(\sigma, {}^\bullet p_o)$, $M_1(r_o) = M_1'(r_o) = M_1'(p_o) = M(p_o)$ and $M_1(p) = M_1'(p) = M'(p) = M(p)$ for $p \in P_1 - \{r_i, r_o\}$. Within B_2, it follows from Definitions 3.1 and 5.3 that $f_2(\mu) = \lambda$ or γ, and $f_2(\sigma) = t$. Hence, $M_2(p_e) = M_2'(p_e) - 1 = M'(p_e) - 1 = M(p_e)$, $M_2(p) = M_2'(p) + W(t,p) - W(p,t) = M'(p) + W(t,p) - W(p,t) = M(p)$ for $p \in P_2 - \{p_e, p_x\}$ and $M_2(p_x) = M_2'(p_x) = 0$.

(d) If $t \in {}^\bullet p_o - p_i^\bullet$, then $\#(\sigma, p_i^\bullet) = \#(\mu, p_i^\bullet)$, $\#(\sigma, {}^\bullet p_o) = \#(\mu, {}^\bullet p_o) + 1$ and $f_1(t) = t_r$. Within B_1, after firing t in B, $M(p) = M'(p)$ for $p \in P_1 - \{r_o\}$, $M(p_i) = M'(p_i)$, $M(p_o) = M'(p_o) + 1$ and $M_1(r_i) = M_1'(r_i) - 1$. Hence, $M_1(r_i) = M_1'(r_i) - 1 = (M'(p_i) + \#(\mu, p_i^\bullet) - \#(\mu, {}^\bullet p_o)) - 1 = M'(p_i) + \#(\sigma, p_i^\bullet) - \#(\sigma, {}^\bullet p_o) = M(p_i) + \#(\sigma, p_i^\bullet) - \#(\sigma, {}^\bullet p_o)$, $M_1(r_o) = M_1'(r_o) + 1 = M'(p_o) + 1 = M(p_o)$ and $M_1(p) = M_1'(p) = M'(p) = M(p)$ for $p \in P_1 - \{r_i, r_o\}$. Within B_2, $f_2(\sigma) = \gamma$. $M_2(p_e) = M_2'(p_e) = M'(p_i) = M(p_i)$ and $M_2(p_x) = 0$. By Definitions 5.3, $M_2(p) = M_2'(p) + W(t,p) - W(p,t) = M'(p) + W(t,p) - W(p,t) = M(p)$ for $p \in P_2 - \{p_e, p_x\}$.

(e) If $t \in p_i^\bullet \cap {}^\bullet p_o$, then $\#(\sigma, p_i^\bullet) = \#(\mu, p_i^\bullet) + 1$, $\#(\sigma, {}^\bullet p_o) = \#(\mu, {}^\bullet p_o) + 1$

and $f_1(t) = t_r$. Within B_1, after firing t in B, $M(p) = M'(p)$ for $p \in P_1 - \{r_i, r_o\}$, $M(p_i) = M'(p_i) - 1$, $M(p_o) = M'(p_o) + 1$, $M_1(r_i) = M'_1(r_i) - 1$ and $M_1(r_o) = M'_1(r_o) + 1$. The rest of the proof is similar to Case d above. Within B_2, by Definitions 3.1 and 5.3, $f_2(\mu) = f_2(\sigma) = \gamma$. Hence, $M_2(p_e) = M'_2(p_e) - 1 = M'(p_i) - 1 = M(p_i)$, $M_2(p_x) = 0$ and $M_2(p) = M'_2(p) + W(t, p) - W(p, t) = M'(p) + W(t, p) - W(p, t) = M(p)$ for $p \in P_2 - \{p_e, p_x\}$.

$\square$

Lemma 5.3. *(1) Any $\mu \in L(B_2, M_e)$ is a firable subsequence of some $\sigma \in (B, M_0)$ at a marking M where $M(p_e) \geq 1$. (2) Suppose (B_2, M_e) has at least one firable e-sequence. Then, for any $\mu \in L(B_1, M_r)$, there exists $\sigma \in L(B, M_0)$ such that $\mu = f_1(\sigma)$.*

Proof.

1) It is obvious that if $M_e[B_2, \mu\rangle M_2$, then $\exists \sigma \in L(B, M_0)$ such that $(M_1 + M_e)[B, \sigma\rangle(M_1 + M_2)$, where μ is a subsequence of σ and M_1 is any marking of B_1.

2) Let $\gamma = t_i \ldots t_o$ be a firable e-sequence of B_2, where $t_i \in p_e^{\bullet}$ and $t_o \in {}^{\bullet}p_x$. Any $\mu \in L(B_1, M_r)$ can be expressed as $\mu_1 \ldots t_r \ldots \mu_2 \ldots t_r \ldots \mu_3 \ldots$. It follows from Definition 5.3 that $\mu = f_1(\sigma)$. Next, we shall show that $\sigma \in L(B, M_0)$ by induction on μ. For $\mu = \lambda$, it is trivial that $\sigma = \lambda \in L(B, M_0)$. Suppose that, $\forall \mu' \in L(B_1, M_r)$ where $|\mu'| \leq n$, $\exists \sigma' \in L(B, M_0)$ such that $\mu' = f_1(\sigma')$. Let $M'_1 = M_r[B_1, \mu'\rangle$ and $M' = M_0[B, \sigma'\rangle$. By Definition 5.3, $M'(p_i) = M'_1(r_i) - (\#(\sigma', p_i^{\bullet}) - \#(\sigma', {}^{\bullet}p_o))$ and $M'(p) = M'_1(p)$ for $p \in P_1 - \{r_i\}$. By Lemma 5.2, $M'(p) \geq M_1(p) \forall p \in P_1$. For $\mu = \mu' t \in L(B_1, M_r)$, consider two cases: (a) $t \in T_1 - \{t_r\}$. Let $\sigma = \sigma' t$. Since t is firable at M'_1 in B_1, it is firable at M' in B. Hence, $\sigma \in L(B, M_0)$. (b) $t = t_r$. Let $\sigma = \sigma' \gamma$. By the result of Part(1), $\sigma \in L(B, M_0)$.

$\square$

Theorem 5.2. *For the composite net obtained by Definition 5.2, the following two propositions hold: (1) B_2 can be initiated only from B_1. (2) As far as token distribution in B_1 is concerned, executing one cycle of B_2 is equivalent to firing t_r once in B_1.*

Proof. Proposition 1 follows from the fact that B_2 initiates and terminates properly and that B_2 itself is initially marked with M_s which cannot initiate any firing in B_2. Proposition 2 follows from two facts: (a) B_2 is not re-enterable (i.e., B_2 cannot be initiated again unless its current execution cycle has terminated at M_x and the token deposited at p_x has been removed. This is guaranteed by the assumptions of proper initiation and proper termination on B_2. (b) The only initial marking that B_1 can create for B_2 is $M_e = p_e + M_c$. This is guaranteed by the assumptions that B_1 is not 2-enabled at t_r and that B_2 is non-reenterable. From the viewpoint of B_1, a token deposited into r_i is held there during execution of B_2 and is removed only after B_2 has terminated. Hence, since B_1 is not 2-enabled at t_r, no more token can be put into r_i during the execution of B_2.

Let us consider the implication of two cases of Theorem 5.1: Case 1. $\#(\sigma, p_i^\bullet) - \#(\sigma, {}^\bullet p_o) > 0$. In this case, execution of B_2 has initiated but has not terminated yet. Furthermore, $M(p_i) = 0$ but $M_1(r_i) = 1$. This apparent inconsistence can be explained as follows. From the viewpoint of B, the token at p_i that initiates B_2 has been removed; whereas, from the viewpoint of B_1, this token till stays in r_i. Then, since t_r is not 2-enabled, B_2 cannot be re-entered. Case 2. $\#(\sigma, p_i^\bullet) - \#(\sigma, {}^\bullet p_o) = 0$. In this case, execution of B_2 either has never occurred or has terminated. If the last transition t of σ is in ${}^\bullet p_o$, then t is projected onto the single transition t_r within B_1 and the entire σ is the same as a firable e-sequence γ when observed within B_2. Furthermore, we have $M_1(r_o) = M(p_o) = 1$ but $M_2(p_x) = 0$. This implies that any token in p_o is considered as belonging to B_1 rather than to B_2. That is, we have $M_e[B_2, \gamma\rangle M_c$ but $M_0[B, \gamma\rangle M$, where $M(p_i) = 0$ and $M(p_o) = 1$.

In general, it is difficult to relate the ranks of the incidence matrices of the nets involved in Definition 5.2. Theorem 5.3 provides such a relation for several important special cases. □

Theorem 5.3. *(relation of ranks of Petri nets in Transformation TR) (Figure 5.2).*
(Note: In the statements below, S, S', S_1 and S_2 are P-invariants of $B_1 \backslash \{t_r\}$ or B_2. Accordingly, p_i should be replaced by r_i or p_e and p_o by r_o or p_x.) $Rank(B_1(t_r \to B_2)) = Rank(B_1) + Rank(B_2) - h + d - 2$, where

$h = 0$ *if t_r appears in at least one T-invariant of B_1*
$ = 1$ *otherwise*

$d = 2$ *if $\exists S_1, S_2$ such that one of the following conditions holds: (1) $p_i \in S_1$, $p_o \notin S_1$, $p_o \in S_2$. (2) $p_o \in S_2$, $p_i \notin S_2$, $p_i \in S_1$.*
$ = 1$ *if one of the conditions holds: (1) One and only one of $\{p_i, p_o\}$ appears in some P-invariant. (2) $\exists S_1$ such that $\{p_i, p_o\} \subseteq S_1$, and*

$\forall S' \neq S_1$, $S' \cap \{p_i, p_o\} = \phi$. *(3)* $B_1 \backslash \{t_r\}$ *has an P-invariant* S_1 *of the form:* $a_1 p_i + b_1 p_o + c_1 = 0$ *(where* $a_1 \neq 0$ *and* $b_1 \neq 0$*) and* B_2 *has an P-invariant* S_2 *of the form:* $a_2 p_i + b_2 p_o + c_2 = 0$ *(where* $a_2 \neq 0$ *and* $b_2 \neq 0$*) such that* $b_2/a_2 = b_1/a_1$*. Furthermore,* $B_1 \backslash \{t_r\}$ *and* B_2 *have no other P-invariants containing* p_i *or* p_o.
$= 0$ *if neither* p_i *nor* p_o *appears in any P-invariant of* $B_1 \backslash \{t_r\}$ *or* B_2.

Proof.   (Figure 5.2) (Note: The proof presented below is based on two facts: 1) Eliminations are always achieved by applying Gaussian row operations over V though their effects may be considered within individual submatrices. 2) If a non-isolated place (transition) is contained in a P-invariant (T-invariant), its corresponding row (column) in the incidence matrix can be expressed as a linear combination of the other rows (columns) of the invariant.) The conclusion of this theorem follows from two results: (a) $Rank(B_1) = Rank(B_1 \backslash \{t_r\}) + h$. (b) $Rank(B_1(t_r \to B_2)) = Rank(B_1 \backslash \{t_r\}) + Rank(B_2) + d - 2$.

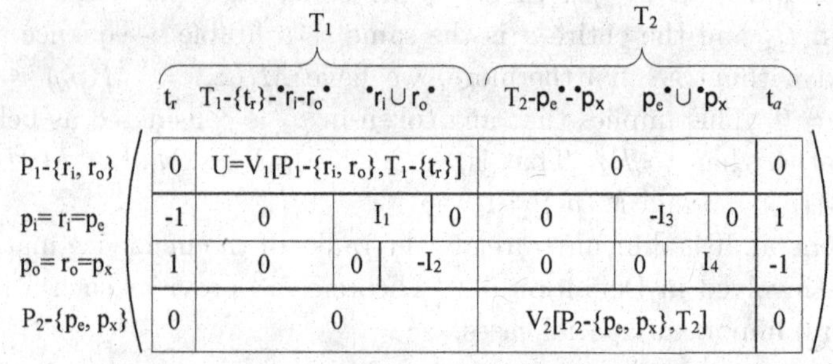

	t_r	T_1-$\{t_r\}$-${}^\bullet r_i$-$r_o{}^\bullet$	${}^\bullet r_i \cup r_o{}^\bullet$	T_2-$p_e{}^\bullet$-${}^\bullet p_x$	$p_e{}^\bullet \cup {}^\bullet p_x$	t_a
P_1-$\{r_i, r_o\}$	0	$U = V_1[P_1$-$\{r_i, r_o\}, T_1$-$\{t_r\}]$	0		0	
$p_i = r_i = p_e$	-1	0	I_1	0	0	-I_3 0	1
$p_o = r_o = p_x$	1	0	0 -I_2	0	0 I_4	-1
P_2-$\{p_e, p_x\}$	0	0	$V_2[P_2$-$\{p_e, p_x\}, T_2]$		0

(with brace labels: T_1 spanning columns 2–4, T_2 spanning columns 5–6)

Fig. 5.2   Compositions of V_1, V_2, V_{2a} and V (See explanation below).

(a) It is obvious that $Rank(B_1) = Rank(B_1 \backslash \{t_r\}) + h$, where $h = 0$ or 1. If t_r occurs in any T-invariant of B_1, column t_r is linearly dependent on the other columns and hence $h = 0$. Otherwise, column t_r is linearly independent and $h = 1$.

(b) Let V be divided into three horizontal blocks: BK_1 contains its top $|P_1| - 2$ rows, BK_2 its middle 2 rows, and BK_3 its bottom $|P_2| - 2$ rows. Consider the following three cases:

Case 1. $d = 2$. We shall prove for Condition (1) only. Proof for Condition (2) is similar. Consider four cases: (i) Both S_1 and S_2 belong to $B_1 \backslash \{t_r\}$. Within $B_1 \backslash \{t_r\}$, since $p_o \in S_2$, row p_o can be expressed as a linear combination of rows of BK_1 and row p_i. It can thus be eliminated to zero. Also, since $p_i \in S_1$ and $p_o \notin S_1$, row p_i can be expressed as a linear combination of the rows of BK_1 alone and can be eliminated to zero. After these two eliminations, V becomes diagonal with diagonal blocks $BK_1[T_1 - \{t_r\}]$ and B_2. Also, $Rank(BK_1[T_1 - \{t_r\}]) = Rank(B_1\backslash\{t_r\})$ and the rank of B_2 remains unchanged. Hence, $Rank(B) = Rank(BK_1[T_1 - \{t_r\}]) + Rank(B_2) = Rank(B_1\backslash\{t_r\}) + Rank(B_2)$. (ii) S_1 and S_2 both belong to B_2. Proof is similar as (i). (iii) S_2 belongs to $B_1\backslash\{t_r\}$ whereas S_1 belongs to B_2. Without affecting the ranks of $B_1\backslash\{t_r\}$, B_2 and V, row p_o can be eliminated to zero within $B_1\backslash\{t_r\}$ and row p_i to zero within B_2 (by means of the rows of BK_3 alone). Again, V becomes diagonal and hence $Rank(B) = Rank(B_1\backslash\{t_r\}) + Rank(B_2)$. (iv) S_1 belongs to $B_1\backslash\{t_r\}$ whereas S_2 belongs to B_2. Proof is similar as (iii).

Case 2. $d = 1$. Without loss of generality, suppose that S_1 belongs to $B_1\backslash\{t_r\}$ and that $p_i \in S_1$. Then, within $B_1\backslash\{t_r\}$, row p_i is linearly dependent on the other rows and can thus be eliminated to zero without altering the ranks of $B_1\backslash\{t_r\}$, B_2 as well as V. On the other hand, for the reasons stated in Remark below, row p_o is linearly independent of the other rows within $B_1\backslash\{t_r\}$ and B_2 (and thus within V as well). Hence, by ignoring row p_o, V becomes diagonal and $Rank(B) = 1 + (Rank(B_1\backslash\{t_r\})-1)+(Rank(B_2)-1) = Rank(B_1\backslash\{t_r\})+Rank(B_2)-1$. (Remark: For Condition (1), since p_i already appears in S_1, p_o cannot appear in any P-invariant and hence cannot be dependent on the other rows within both $B_1\backslash\{t_r\}$ and B_2. For Condition (2), after p_i has been eliminated to zero within $B_1\backslash\{t_r\}$, row p_o cannot be linearly dependent on the other rows within $B_1\backslash\{t_r\}$, because, otherwise, $\exists S'$ of $B_1\backslash\{t_r\}$ such that $p_o \in S'$ but $p_i \notin S'$— contrary to Condition (2). Row p_o cannot be linearly dependent on the other rows within B_2 either because p_o does not belong to any P-invariant of B_2. Under Condition (3), the fact that row p_o cannot be linearly dependent on the other rows within $B_1\backslash\{t_r\}$ follows from the same argument as for Condition (2). Within B_2 it is due to the following argument: Within $B_1\backslash\{t_r\}$, in order to eliminate row p_i to zero by means of the P-invariant $a_1p_i+b_1p_o+c_1 = 0$, a multiple of (b_1/a_1) of row p_o (and some combination of other rows in BK_1) has to be added to row p_i. Since this is done over V, row p_i within B_2 becomes $p_i + (b_1/a_1)p_o$. Now, suppose row p_o is linearly dependent on the other rows within B_2, that is, $\exists a' \neq 0$ and $b' \neq 0$: $a'(p_i + (b_1/a_1)p_o) + b'p_o + c' = a'p_i + (a'b_1/a_1 + b')p_o + c' = 0$, where c' is a linear combination of the rows of BK_3. Then, since $a_2 p_i + b_2 p_o + c_2 = 0$ is the unique P-invariant of B_2 containing p_o, the coefficients of these two equations must be proportional. That is, $a'/a_2 = (a'b_1/a_1 + b')/b_2$ or $b'/a' = b_2/a_2 - b_1/a_1$. Since $b_2/a_2 = b_1/a_1$, $b' = 0$ — a contradiction.)

Case 3. $d = 0$. Rows p_i and p_o are each linearly independent of the other rows within both $B_1 \backslash \{t_r\}$ and B_2 (and hence also within V) because, otherwise, p_i or p_o will appear in at least one P-invariant. Then, after ignoring rows p_i and p_o, the ranks of $B_1 \backslash \{t_r\}$, B_2 and V will each be reduced by 2 and V becomes diagonal. Hence, $Rank(B) = 2 + (Rank(B_1 \backslash \{t_r\}) - 2) + (Rank(B_2) - 2) = Rank(B_1 \backslash \{t_r\}) + Rank(B_2) - 2$.

$\square$

Explanations for Figure 5.2:

- B_1 denotes a refined net, B_2 a Petri net process and $B = B_1(t_r \to B_2)$ the result of applying Transformation TR to replace transition t_r of B_1 with B_2.
- $V(P, T) = V[P_1 \cup P_2, T_1 \cup T_2 - \{t_r\}]$ is the incidence matrix of B.
- $V_1(P_1, T_1)$ of B_1 occupies the first $|P_1|$ rows and first $|T_1|$ columns.
- $V_2(P_2, T_2)$ of B_2 occupies the bottom $|P_2|$ rows and the columns under T_2.
- $V_{2a}(P_2, T_2 \cup \{t_a\})$ occupies the bottom $|P_2|$ rows and the rightmost $|T_2| + 1$ columns.
- Row p_i has values: $p_i(t) = 1$ if $t \in {}^\bullet r_i$, $p_i(t) = -1$ if $t \in \{t_r\} \cup p_e^\bullet$ and $p_i(t) = 0$ if $t \in T_1 \cup T_2 - \{t_r\} - {}^\bullet r_i - p_e^\bullet$; and the row p_o has values: $p_o(t) = -1$ if $t \in r_o^\bullet$, $p_o(t) = 1$ if $t \in \{t_r\} \cup {}^\bullet p_x$, and $p_i(t) = 0$ if $t \in T_1 \cup T_2 - \{t_r\} - {}^\bullet p_x - r_o^\bullet$.

Theorem 5.4. *(property preservation under Transformation TR $B_1(t_r \to B_2)$).*
The following propositions are valid under Transformation TR.

1) *If B_1 is an SM (resp., MG, FC net, AC net) and B_2 is an SM (resp., almost MG, FC net, AC net), then $B_1(t_r \to B_2)$ is also an SM (resp., MG, FC net, AC net).*

2) *If B_1 is conservative (resp., structurally bounded) and B_2 is almost conservative (resp., almost structurally bounded), then $B_1(t_r \to B_2)$ is conservative (resp., structurally bounded).*

3) *If B_1 is consistent (resp., repetitive) and B_2 is almost consistent (resp., almost repetitive), then $B_1(t_r \to B_2)$ is consistent (resp., repetitive).*

4) $Rank(B_1(t_r \to B_2)) = Rank(B_1) + Rank(B_2) - h + d - 2$, *where h and d are computed according to Theorem 5.3.*

5) $|C(N(t_r \to B))| = |C(N)| + |C(B)| - 2.$

6) *If both* B_1 *and* B_2 *satisfy the RC-property and* $Rank(B_1(t_r \to B_2)) = Rank(B_1) + Rank(B_2) - 1$, *then* $B_1(t_r \to B_2)$ *also satisfy the RC-property.*

7) *If* B_1 *has a minimal SM-cover and* B_2 *almost has a minimal SM-cover, then* $B_1(t_r \to B_2)$ *has a minimal SM-cover.*

8) *Suppose D is a siphon of* B_1. *Then, three cases hold for* $B_1(t_r \to B_2)$: *(a) If* r_i, $r_o \notin D$, *then D is a siphon of* $B_1(t_r \to B_2)$. *(b) If* $r_i \in D$ *but* $r_o \notin D$, *then* $D - \{r_i\} \cup \{p_i\}$ *is a siphon of* $B_1(t_r \to B_2)$. *(c) If* $r_o \in D$ *and* $({}^\bullet p_x \subseteq p_e^\bullet)$ *in* B_2, *then* $D - \{r_i, r_o\} \cup \{p_i, p_o\}$ *is a siphon of* $B_1(t_r \to B_2)$.
Suppose D is a siphon of B_2. *Then, three cases hold for* $B_1(t_r \to B_2)$: *(a) If* p_e, $p_x \notin D$, *then D is a siphon of* $B_1(t_r \to B_2)$. *(b) If* $p_e \notin D$ *but* $p_x \in D$, *then* $D - \{p_x\} \cup \{p_o\}$ *is a siphon of* $B_1(t_r \to B_2)$. *(c) If* $p_e \in D$ *and* $({}^\bullet r_i \subseteq r_o^\bullet)$ *in* B_1, *then* $D - \{p_e, p_x\} \cup \{p_i, p_o\}$ *is a siphon of* $B_1(t_r \to B_2)$.

9) *Suppose S is a trap of* B_1. *Then, three cases hold for* $B_1(t_r \to B_2)$: *(a) If* r_i, $r_o \notin S$, *then S is a trap of* $B_1(t_r \to B_2)$. *(b) If* $r_i \notin S$ *but* $r_o \in S$, *then* $S - \{r_o\} \cup \{p_o\}$ *is a trap of* $B_1(t_r \to B_2)$. *(c) If* $r_i \in S$ *and* $(p_e^\bullet \subseteq p_x^\bullet)$ *in* B_2, *then* $S - \{r_i, r_o\} \cup \{p_i, p_o\}$ *is a trap of* $B_1(t_r \to B_2)$.
Suppose S is a trap of B_2. *Then, three cases hold for* $B_1(t_r \to B_2)$: *(a) If* p_e, $p_x \notin S$, *then S is a trap of* $B_1(t_r \to B_2)$. *(b) If* $p_e \notin S$ *but* $p_x \in S$, *then* $S - \{p_x\} \cup \{p_o\}$ *is a trap of* $B_1(t_r \to B_2)$. *(c) If* $p_x \in S$ *and* $(r_o^\bullet \subseteq r_i^\bullet)$ *in* B_1, *then* $S - \{p_e, p_x\} \cup \{p_i, p_o\}$ *is a trap of* $B_1(t_r \to B_2)$.

10) *If* B_1 *is a Petri net process and* $M_r(r_i) = 0$, *then* $B_1(t_r \to B_2)$ *is also a Petri net process. Besides, two proposition hold: (a)* $Z(B_1(t_r \to B_2)) = Z(B_1) + Z(B_2) - 1$. *(b)* $(B_1(t_r \to B_2), M_0)$ *terminates properly if both* (B_1, M_r) *and* (B_2, M_e) *terminate properly.*

11) *Let both* (B_1, M_r) *and* (B_2, M_e) *be bounded. Three cases hold for* $(B_1(t_r \to B_2), M_0)$: *(a)* $(B_1(t_r \to B_2), M_0)$ *may or may not be bounded. (b) If* (B_1, M_r) *is* k_1-*bounded and* (B_2, M_e) *is* k_2-*bounded and terminates properly, then* $(B_1(t_r \to B_2), M_0)$ *is k-bounded, where* $k = max\{k_1, k_2\}$. *(c) If* (B_1, M_r) *is* k_1-*bounded and* (B_2, M_e) *is almost 1-bounded, then* $(B_1(t_r \to B_2), M_0)$ *is also* k_1- *bounded.*

12) *Let* (B_1, M_r) *be live and* (B_2, M_e) *be almost live and terminates properly, then* $(B_1(t_r \to B_2), M_0)$ *is live.*

13) *Let (B_1, M_r) be reversible and (B_2, M_e) be almost reversible and terminates properly, then $(B_1(t_r \to B_2), M_0)$ is reversible.*

Proof.

1) According to its construction, $B_1(t_r \to B_2)$ satisfies the definition of SM (resp., MG, FC and AC) if B_1 is an SM (resp., MG, FC net, AC net) and B_2 is an SM (resp., almost MG, FC net and AC net).

2) (See Figure 5.2) Since B_1 is conservative and B_2 is almost conservative, by Characterization 2.1, $\exists \alpha_1 = (x_{|p_1|}, \ldots, x_2, x_1) \geq 1$ and $\alpha_{2a} = (y_1, y_2, \ldots, y_{|P_2|}) \geq 1$ such that $\alpha_1 V_1 = 0$ and $\alpha_{2a} V_{2a} = 0$. Since $\alpha_1 V_1 = 0$, the first column of $\alpha_1 V_1$ is 0, i.e., $\alpha_1 V_1[P_1\{t_r\}] = 0$, leads to $x_1 = x_2$. Since $\alpha_2 V_{2a} = 0$, the last column of $\alpha_2 V_{2a}$ is 0, i.e., $y_1 = y_2$. Next, let $\alpha = (y_1 \alpha_1[P_1 - \{r_i, r_o\}], x_2 y_1, x_1 y_2, x_1 \alpha_2[P_2 - \{p_e, p_x\}]) = (x_{|p_1|} y_1, \ldots, x_3 y_1, x_2 y_1, x_1 y_2, x_1 y_3, \ldots, x_1 y_{|P_2|})$. Then, $\alpha \geq 1$ and $\alpha V = (-x_2 y_1 + x_1 y_2, y_1 \alpha_1[P_1 - \{r_i, r_o\}] V_1[P_1 - \{r_i, r_o\}, T_1 - \{t_r\}] + x_2 y_1 V_1[r_i, T_1 - \{t_r\}] + x_1 y_2 V_1[r_o, T_1 - \{t_r\}], x_2 y_1 V_2[p_e, T_2] + x_1 y_2 V_2[p_x, T_2] + x_1 \alpha_2[P_2 - \{p_e, p_x\}] V_2[P_2 - \{p_e, p_x\}, T_2]) = (0, y_1 \alpha_1 V_1[P_1, T_1 - \{t_r\}], x_1 \alpha_2 V_2) = 0$. By Characterization 2.1, $B_1(t_r \to B_2)$ is conservative. Similar to the argument as shown above, $\alpha_1 V_1 \leq 0$ and $\alpha_2 V_{2a} \leq 0$ will lead to $\alpha V \leq 0$. It follows from Characterization 2.2 that $B_1(t_r \to B_2)$ is structurally bounded.

3) Since B_1 is consistent and B_2 is almost consistent, by Characterization 2.3, $\exists \beta_1 = (x_1, \ldots, x_{|T_1|}) \geq 1$ such that $V_1 \beta_1 = 0$. and $\exists \beta_2 = (y_1, \ldots, y_{|T_2|}, z) \geq 1$ such that $V_{2a} \beta_2 = 0$. Let $\beta = (z \beta_1[T_1 - \{t_r\}], x_1 \beta_2[T_2]) = (z x_2, \ldots, z x_{|T_1|}, x_1 y_1, \ldots, x_1 y_{|T_2|})$. Then, $\beta \geq 1$. Since $V[P, t_r] + V[P, t_a] = 0$, we have $V\beta = (z V_1[P_1, T_1 - \{t_r\}] \beta_1[T_1 - \{t_r\}], 0) + (0, x_1 V_2 \beta_2[T_2]) = ((z V_1 \beta_1, 0) - z x_1 V[P, t_r]) + ((0, x_1 V_{2a} \beta_2) - x_1 z V[P, t_a]) = 0$. By Characterization 2.3, $B_1(t_r \to B_2)$ is consistent. Similar to the argument as shown above, $V_1 \beta_1 \geq 0$ and $V_{2a} \beta_2 \geq 0$ lead to $V\beta \geq 0$. It follows from Characterization 2.4 that $B_1(t_r \to B_2)$ is repetitive.

4) Refer to Theorem 5.3.

5) In forming $B_1(t_r \to B_2)$, since t_r is deleted, cluster $\{r_i, t_r\}$ of B_1 is destroyed and r_i is absorbed into the cluster $[p_e]$ of B_2 to form the new cluster $[p_i]$ of $B_1(t_r \to B_2)$. Also, the clusters $[r_o]$ of B_1 and $[p_x]$ of B_2 are merged to form the new cluster $[p_o]$ of $B_1(t_r \to B_2)$. Also, no other clusters are created or destroyed. Hence, $|C(B_1(t_r \to B_2))| = |C(B_1)| + |C(B_2)| - 2$.

6) If both B_1 and B_2 satisfy the RC-property, then $Rank(B_1) = C(B_1) - 1$

and $Rank(B_2) = C(B_2) - 1$. By 5), we know that $|C(B_1(t_r \to B_2))| = |C(B_1)| + |C(B_2)| - 2$. Hence, if $Rank(B_1(t_r \to B_2)) = Rank(B_1) + Rank(B_2) - 1$, then $Rank(B_1(t_r \to B_2)) = Rank(B_1) + Rank(B_2) - 1 = (|C(B_1)| - 1) + (|C(B_2)| - 1) - 1 = |C(B_1)| + |C(B_2)| - 3 = |C(B_1(t_r \to B_2))| - 1$.

7) Let K_1 and K_2 be minimal SM-covers of B_1 and B_{2a}, respectively. Any SM-component in K_1 containing r_o must also contain r_i, t_r, and a directed path from r_o to r_i. Similarly, any SM-component in K_2 containing p_x must also contain p_e, t_a and a directed path from p_x to p_e. A minimal SM-cover for $B_1(t_r \to B_2)$ can be created as follows. Let $H_1 = \{S \in K_1 | \{r_i, t_r, r_o\} \subseteq S\}$ and $H_2 = \{S \in K_2 | \{p_e, t_a, p_x\} \subseteq S\}$. Create S' by merging all SM-components of H_1 with all SM-components of H_2 at their common places p_i and p_o (i.e., fuse r_i with p_e and r_o with p_x and delete all t_r and t_a). Obviously, S' is an SM-component. Then, $K_1 \cup K_2 \cup \{S'\} - H_1 - H_2$ forms a minimal SM-cover of $B_1(t_r \to B_2)$.

8) For $D \subseteq P_1$, consider three cases: Case 1. r_i, $r_o \notin D$. This implies that $(t_r \notin {}^\bullet D \cup D^\bullet)$ in B_1 and ${}^\bullet D$ in $B = {}^\bullet D$ in B_1 and $D^\bullet$ in $B = D^\bullet$ in B_1. Hence, $({}^\bullet D \subseteq D^\bullet)$ in B provided that $({}^\bullet D \subseteq D^\bullet)$ in B_1. Case 2. $r_i \in D$ but $r_o \notin D$. This implies that $(t_r \in D^\bullet - {}^\bullet D)$ in B_1. Let $D' = D - \{r_i\} \cup \{p_i\}$. Since $(D')^\bullet$ in $B = D^\bullet$ in $B_1 - \{t_r\} \cup p_e^\bullet$ in B_2 and ${}^\bullet('D)$ in $B_1(t_r \to B_2) = {}^\bullet D$ in B_1, $({}^\bullet('D) \subseteq (D')^\bullet)$ in $B_1(t_r \to B_2)$ provided that $({}^\bullet D \subseteq D^\bullet)$ in B_1. Case 3. $r_o \in D$. Since $({}^\bullet D \subseteq D^\bullet)$ in B_1, $(t_r \in D^\bullet)$ in B_1. This means $(r_i \in D)$ in B_1. Let $D' = D - \{r_i, r_o\} \cup \{p_i, p_o\}$. Then, ${}^\bullet D'$ in $B_1(t_r \to B_2) = {}^\bullet D$ in $B_1 - \{t_r\} \cup ({}^\bullet p_i \cup {}^\bullet p_o)$ in $B_1(t_r \to B_2)$ and $(D')^\bullet$ in $B_1(t_r \to B_2) = D^\bullet$ in $B_1 - \{t_r\} \cup ({}^\bullet p_i \cup {}^\bullet p_o)$ in $B_1(t_r \to B_2)$. Since ${}^\bullet p_i$ in $B_1(t_r \to B_2) = r_i^\bullet$ in B_1, $({}^\bullet D' \subseteq (D')^\bullet)$ in $B_1(t_r \to B_2)$ provided that $({}^\bullet p_o \subseteq p_i^\bullet)$ in $B_1(t_r \to B_2)$, or equivalently, $({}^\bullet p_x \subseteq p_e^\bullet)$ inB_2.

For $D \subseteq P_2$, consider three cases: Case 1. p_e, $p_x \notin D$. This implies ${}^\bullet D$ in $B_1(t_r \to B_2) = {}^\bullet D$ in B_2 and $D^\bullet$ in $B_2 = D^\bullet$ in $B_1(t_r \to B_2)$. Hence, ${}^\bullet D$ in $B_1(t_r \to B_2) = {}^\bullet D$ in $B_2 \subseteq D^\bullet$ in $B_2 = D^\bullet$ in $B_1(t_r \to B_2)$. Case 2. $p_e \notin D$ but $p_x \in D$. Let $D' = D - \{p_x\} \cup \{p_o\}$. Then, $({}^\bullet('D) \subseteq (D')^\bullet)$ in $B_1(t_r \to B_2)$ provided that $({}^\bullet D \subseteq D^\bullet)$ in B_2. Case 3. $p_e \in D$. Then $p_x \in D$. Let $D' = D - \{p_e, p_x\} \cup \{p_i, p_o\}$. Since $(D')^\bullet$ in $B_1(t_r \to B_2) = D^\bullet$ in $B_2 \cup (p_i^\bullet \cup p_o^\bullet)$ in $B_1(t_r \to B_2)$ and ${}^\bullet D'$ in $B_1(t_r \to B_2) = {}^\bullet D$ in $B_2 \cup ({}^\bullet p_i \cup {}^\bullet p_o)$ in B_2, we have $({}^\bullet D' \subseteq (D')^\bullet)$ in $B_1(t_r \to B_2)$ provided that $({}^\bullet p_i \subseteq p_o^\bullet)$ in $B_1(t_r \to B_2)$, or equivalently, $({}^\bullet r_i \subseteq r_o^\bullet)$ in B_1.

9) This proof is similar to that of 8).

10) $B_1(t_r \to B_2)$ is obtained by first deleting t_r, (r_i, t_r) and (t_r, r_o) from B_1 and then fusing r_i with p_e and r_o with p_x. This results in a total loss of 3 nodes and 2 arcs. If B_1 and B_2 are both Petri net processes and $M_r(r_i) = 0$, then, by the definition of a Petri net process, $(B_1(t_r \to B_2), M_0)$ is a Petri net process. Hence, $Z(B_1(t_r \to B_2)) = |F| - |P \cup T| + 2 = (|F_1| + |F_2| - 2) - (|P_1 \cup T_2| + |P_1 \cup T_2| - 3) + 2 = |F_1| - |P_1 \cup T_1| + 2 + |F_2| - |P_2 \cup T_2| + 2 - 1 = Z(B_1) + Z(B_2) - 1$. It is obvious that $(B_1(t_r \to B_2), M_0)$ terminates properly if both (B_1, M_r) and (B_2, M_e) terminate properly.

11) Suppose both (B_1, M_r) and (B_2, M_e) are bounded. Then, it is obvious that $(B_1(t_r \to B_2), M_0)$ may or may not be bounded. Suppose (B_1, M_r) is k_1-bounded and (B_2, M_e) is k_2-bounded. If (B_2, M_e) terminates properly or is almost 1-bounded, then p_o is 1-bounded in $(B_1(t_r \to B_2), M_0)$. Let $M \in R(B, M_0)$. By Definition 5.3, $M(p) = M_1(p)$ if $p \in P_1$ and $M(p) = M_2(p)$ if $p \in P_2 - \{p_e, p_x\}$. By Theorem 5.1, $M_1 \in [B_1, M_r\rangle$ and $M_2 \in [B_2, M_e\rangle$. Since M_1 is bounded by k_1 and M_2 by k_2, $M(p) \leq max\{k_1, k_2\} \forall p \in P$.

12) Suppose (B_1, M_r) is live and (B_2, M_e) is almost live. Then, it is obvious that $(B_1(t_r \to B_2), M_0)$ may or may not be live. If (B_2, M_e) terminates properly, then $\forall M : M_0[B, \sigma\rangle M$, where $\sigma \in L(B, M_0)$, let $M = (m_1, m_2)$, where m_1 is the component of M over P_1 and m_2 over $P_2 - \{p_e, p_x\}$. $\forall t \in T = T_1 \cup T_2 - \{t_r\}$, consider four cases: Case 1. $t \in T_1 - \{t_r\}$ and $m_2 = M_c$. This implies that $M_r[B_1, f_1(\sigma)\rangle m_1$. Since t is live in B_1, $\exists \mu : m_1[B_1, \mu t\rangle$. Let $\sigma_1 = \mu$ if μ does not contain t_r. Otherwise, let σ_1 be the result of replacing each t_r within μ with a firable e-sequence of B_2. Then, in B, we have $M_i[B, \sigma\sigma_1 t\rangle$. That is, t is live in B. Case 2. $t \in T_1 - \{t_r\}$ and $m_2 \neq M_c$. Since B_2 is almost live and terminates properly, $\exists \sigma_2 : m_2[B_2, \sigma_2\rangle M_x$. Then, in B, we have $M_i[B, \sigma\sigma_2\rangle (m_1', M_c)$, where $m_1' = m_1 + p_o$. This becomes Case 1. Case 3. $t \in T_2$ and $m_2 = M_c$. This implies that $M_r[B_1, f_1(\sigma)\rangle m_1$. Since t_r is live in B_1, $\exists \sigma_1 : t_r \notin \sigma_1$ and $m_1[B_1, \sigma_1\rangle m_1'[t_r\rangle$, where $m_1'(p_e) = 1$. Since t is live in B_2, $\exists \mu : M_e[B_2, \mu t\rangle$. Then, in B, we have $M_i[B, \sigma\sigma_1\mu t\rangle$. That is, t is live in B. Case 4. $t \in T_2$ and $M_2 \neq M_c$. The proof follows from a combination of Case 2 and Case 3.

13) Suppose (B_1, M_r) is reversible and (B_2, M_e) is almost reversible. Then, it is obvious that $(B_1(t_r \to B_2), M_0)$ may or may not be reversible. If (B_2, M_e) terminates properly $\forall M : M_0[B, \sigma\rangle M$, where $\sigma \in L(B, M_0)$, let $M = (m_1, m_2)$, where m_1 is the component of M over P_1 and m_2 over $P_2 - \{p_e, p_x\}$. Consider two cases: Case 1. $m_2 = M_c$. Then, $m_1 =$

$M_r[B_1, f_1(\sigma)\rangle$. Since B_1 is reversible, $\exists \sigma_1$, such that $M_1[B_1, \sigma_1\rangle M_r$. By Lemma 5.3, $\exists \sigma_1'$, such that $f_1(\sigma\sigma_1') = f_1(\sigma)\sigma_1 \in L(B_1, M_r)$. Hence, $M[B, \sigma_1'\rangle M_0$. That is, B is reversible. Case 2. $m_2 \neq M_c$. Since B_2 terminates properly, $\exists \sigma_2 : m_2[B_2, \sigma_2\rangle M_x$. Then, in B, we have $M[B, \sigma_2\rangle(m_1', M_c)$, where $m_1' = m_1 + p_o$. This becomes Case 1. Hence, B is reversible.

$\square$

5.2 A Property-Preserving Refinement for Places

To refine a place, our approach is to first convert the place to a transition and then refine the transition by Transformation TR (Definition 5.2). The conversion includes splitting the place into two places connected by a transition. Therefore, in order to obtain similar results as for Transformation TR, we have to show that this conversion also preserves the same properties.

Definition 5.4. (transformation PS $N \to N'$) (Figure 5.3).
Let (N, M_r) be a marked net, where $N = (P, T, F)$. Splitting place $p_r \in P$ of (N, M_r) results in a marked net (N', M_r'), where

- $N' = (P', T', F')$
- $P' = P \cup \{r_i, r_o\} - \{p_r\}$; $T' = T \cup \{t_r\}$; $F' = F \cup \{(r_i, t_r), (t_r, r_o)\} \cup \{(x, r_i)|x \in {}^\bullet p_r\} \cup \{(r_o, x)|x \in p_r^\bullet\} - \{(x, p_r)|x \in {}^\bullet p_r\} - \{(p_r, x)|x \in p_r^\bullet\}$
- $M_r'(r_i) = M_r(p_r)$, $M_r'(r_o) = 0$ and $M_r'(p) = M_r(p)$ for $p \in P - \{p_r\}$

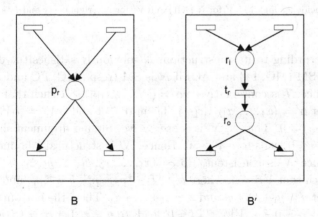

B   B'

Fig. 5.3 Transformation PS.

Theorem 5.5. *(property preservation under Transformation PS). Suppose (N', M_r') is obtained from (N, M_r) by Transformation PS. Then, the following propositions are valid.*

1) *If N is an SM (MG, FC net, AC net), so is N'.*
2) *If N is conservative (structurally bounded), so is N'.*
3) *If N is consistent (repetitive), so is N'.*
4) *$Rank(N') = Rank(N) + 1$.*
5) *$|C(N')| = |C(N)| + 1$.*
6) *If N satisfies the RC-property, so does N'.*
7) *If N has a minimal SM-cover, so does N'.*
8) *Suppose D is a siphon of N. If $p_r \notin D$, then D is a siphon of N'. If $p_r \in D$, then $D - \{p_r\} \cup \{r_i, r_o\}$ is a siphon of N'.*
9) *Suppose D is a trap of N. If $p_r \notin D$, then D is a trap of N'. If $p_r \in D$, then $D - \{p_r\} \cup \{r_i, r_o\}$ is a trap of N'.*
10) *Suppose N is a Petri net process. Then, $Z(N') = Z(N)$.*

11-13) *If (N, M_r) is bounded (live, reversible), so is (N', M_r').*

Proof. (Figure 5.4)

$$\begin{array}{c|ccc} & T-{}^\bullet p_r - p_r^\bullet & {}^\bullet p_r & p_r^\bullet \\ \hline P-\{p_r\} & \multicolumn{3}{c}{V(P-\{p_r\},T)} \\ p_r & 0 & I_1 & -I_2 \end{array} \qquad \begin{array}{c|cccc} & T-{}^\bullet r_i - r_o^\bullet & {}^\bullet r_i & r_o^\bullet & t_r \\ \hline P'-\{r_i,r_o\} & \multicolumn{3}{c}{V(P-\{p_r\},T)} & 0 \\ r_i & 0 & I_1 & 0 & -1 \\ r_o & 0 & 0 & -I_2 & 1 \end{array}$$

Fig. 5.4 Incidence matrices V for N (left) and V' for N' (right). (I_1 and I_2 are identity vectors).

1) According to its construction, N' obviously satisfies the definition of SM (MG, FC and AC) if N is SM (resp., MG, FC and AC).
2) Since N is conservative, $\exists \alpha = (x_1, \ldots, x_{|P|}) \geq 1$ such that $\alpha V = 0$. Let $\alpha' = (x_1, \ldots, x_{|P|}, x_{|P|})$, Then, $\alpha' \geq 1$ and $\alpha' V' = (\alpha V, -x_{|P|} + x_{|P|}) = 0$. That is, N' is conservative. Similar argument shows that $\alpha V \leq 0$ leads to $\alpha' V' \leq 0$. Hence, N' is structurally bounded.
3) Since N is consistent, $\exists \beta = (x_1, \ldots, x_a, y_1, \ldots, y_b, z_1, \ldots, z_c) \geq 1$ such that $V\beta = 0$, where $a = |T - {}^\bullet p_r - p_r^\bullet|$, $b = |{}^\bullet p_r|$ and $c = |p_r^\bullet|$. Let $y = y_1 + \ldots + y_b$ and $z = z_1 + \ldots + z_c$. Then, the bottom equation of $V\beta = 0$, i.e., $V[p_r, T]\beta = 0$, leads to $y = z$. Let $\beta' = (\beta, y)$. Then, $\beta' \geq 1$ and $V'\beta' = 0$. The latter follows from the three horizontal

components of V' : $V'[P' - \{r_i, r_o\}, T \cup \{t_r\}]\beta' = V[P - \{p_r\}, T]\beta + y \cdot 0 = 0$, $V'[r_i, T \cup \{t_r\}]\beta' = y_1 + y_2 + \ldots + y_a - y = 0$, and $V'[r_o, T \cup \{t_r\}]\beta' = -z_1 - z_2 - \ldots - z_c + y = -y + y = 0$. Hence, N' is consistent. Similar argument shows that $V'\beta \geq 0$ leads to $V'\beta' \geq 0$ and that N' is repetitive.

4-6) In V', after adding row r_o to row r_i, the top $|P' - \{r_o\}|$ rows become $(V, 0)$ and the bottom row $(0, 0, -I_2, 1)$ is linearly independent of the other rows. The rank of this new matrix is obviously $Rank(V) + 1$. Hence, $Rank(N') = Rank(N) + 1$. In N', cluster $[p_r]$ of N is replaced by cluster $[r_o]$ of N' and a new cluster $[r_i] = \{r_i, t_r\}$ is created. Hence, $|C(N')| = |C(N)| + 1$. Lastly, $Rank(N') = Rank(N) + 1 = |C(N)| - 1 + 1 = |C(N')| - 1$.

7) Let K be a minimal SM-cover of N. $\forall S \in K$: $(p_r \in S)$, create S' by replacing p_r with $\{r_i, r_o\} \cup \{t_r\} \cup \{(r_i, t_r)(t_r, r_o)\}$(with some obvious relabeling of connections) and replace S with S' in K. Since S' is a SM, the resulting K is a minimal SM-cover of N'.

8-9) If $p_r \notin D$, then $r_i \notin D'$, $r_o \notin D'$ and $t_r \notin {}^\bullet D' \cup (D')^\bullet$. This implies that ${}^\bullet D' = {}^\bullet D$ and $(D')^\bullet = D^\bullet$. Hence, ${}^\bullet D' \subseteq (D')^\bullet$ provided that ${}^\bullet D \subseteq D^\bullet$ and $(D')^\bullet \subseteq {}^\bullet D'$ provided that $D^\bullet \subseteq {}^\bullet D$. If $p_r \in D$, let $D' = D - \{p_r\} \cup \{r_i, r_o\}$. Then, $(D')^\bullet = D^\bullet \cup \{t_r\}$ and ${}^\bullet D' = {}^\bullet D \cup \{t_r\}$. Hence, ${}^\bullet D' \subseteq (D')^\bullet$ provided that ${}^\bullet D \subseteq D^\bullet$ and $(D')^\bullet \subseteq {}^\bullet D'$ provided that $D^\bullet \subseteq {}^\bullet D$.

10) $Z(N') = |F'| - |P' \cup T'| + 2 = |F| + 2 - (|P| + 1) - (|T| + 1) + 2 = Z(N)$.

11-13) Since, in N', r_i and r_o together play the same role as p_r in N, the initial marking, token distribution and firability of the transitions all remain unchanged. Hence, boundedness, liveness and reversibility of N are preserved in N'.

$\square$

Definition 5.5. (transformation PR $B_1(p_r \to B_2)$).
Let (B_1, M_r) be a marked net and (B_2, M_s) be a Petri net process. Place p_r of (B_1, M_r) can be replaced with (B_2, M_c) by the following two steps, resulting in a marked net (B, M_i)

1) $B_1 \to B_1'$ (by Transformation PS), creating a new transition t_r.
2) $B = B_1(p_r \to B_2)) = B_1'(t_r \to B_2)$ (by Definition 5.2).

Theorem 5.6. *(property preservation under Place Refinement $B_1(p_r \to B_2)$).*
Let B_1 *be a pure ordinary Petri net*, B_2 *be a Petri net process and*

$B_1(p_r \to B_2)$ *be obtained from* B_1 *and* B_2 *by Transformation PR. Then, the following propositions are valid.*

1) *If both* B_1 *and* B_2 *are SM (almost MG, FC nets, AC nets), so is* $B_1(p_r \to B_2)$.

2) *If both* B_1 *and* B_2 *are conservative, (structurally bounded), so is* $B_1(p_r \to B_2)$.

3) *If* B_1 *is consistent (repetitive) and* B_2 *is almost consistent (almost repetitive), then* $B_1(p_r \to B_2)$ *is consistent (repetitive).*

4) $Rank(B_1(p_r \to B_2)) = Rank(B_1) + Rank(B_2) - h + d - 1$, *where* h *and* d *are computed according to Theorem 5.3 (with* B_1 *replaced by* B', *the result of Transformation PS).*

5) $|C(B_1(p_r \to B_2))| = |C(B_1)| + |C(B_2)| - 1$.

6) *If both* B_1 *and* B_2 *satisfy the RC-property, so does* $B_1(p_r \to B_2)$ *provided that* $d - h = 1$ *in Proposition 4.*

7) *If* B_1 *has a minimal SM-cover and* B_2 *almost has a minimal SM-cover, then* $B_1(p_r \to B_2)$ *has a minimal SM-cover.*

8) *Suppose* D *is a siphon of* B_1. *If* $p_r \notin D$, *then* D *is a siphon of* $B_1(p_r \to B_2)$. *If* $p_r \in D$ *and* ${}^\bullet p_x \subseteq p_e^\bullet$ *in* B_2, *then* $D - \{p_r\} \cup \{p_i, p_o\}$ *is a siphon of* $B_1(p_r \to B_2)$.
   Suppose D *is a siphon of* B_2. *If* $p_e \notin D$, *then* D *is a siphon of* $B_1(p_r \to B_2)$. *If* $\{p_e, p_x\} \subseteq D$ *and* ${}^\bullet p_r \subseteq p_r^\bullet$ *in* B_1, *then* $D - \{p_e, p_x\} \cup \{p_i, p_o\}$ *is a siphon of* $B_1(p_r \to B_2)$.

9) *Suppose* D *is a trap of* B_1. *If* $p_r \notin D$, *then* D *is a trap of* $B_1(p_r \to B_2)$. *If* $p_r \in D$ *and* $p_e^\bullet \subseteq {}^\bullet p_x$ *in* B_2, *then* $D - \{p_r\} \cup \{p_i, p_o\}$ *is a trap of* $B_1(p_r \to B_2)$.
   Suppose D *is a trap of* B_2. *If* $p_x \notin D$, *then* D *is a trap of* $B_1(p_r \to B_2)$. *If* $p_e, p_x \in D$ *and* $p_r^\bullet \subseteq {}^\bullet p_r$ *in* B_1, *then,* $D - \{p_e, p_x\} \cup \{p_i, p_o\}$ *is a trap of* $B_1(p_r \to B_2)$.

10) *Suppose* B_1 *is also a Petri net process with or without the re-initiation path. Then,* $Z(B_1(p_r \to B_2)) = Z(B_1) + Z(B_2) - 1$.

11) *Suppose* B_1 *is also a Petri net process. Let* $LP(B_1, p_r)$ *be the longest path within* B_1 *containing* p_r. *Then,* $LP(B_1(p_r \to B_2)) = max\{LP(B_1), LP(B_1, p_r) + LP(B_2)\}$.

12) *If* (B_1, M_r) *is bounded by* k_1 *and* (B_2, M_e) *is bounded by* k_2, *then* $(B_1(p_r \to B_2), M_i)$ *is bounded by* $max(k_1, k_2)$.

13) *If* (B_1, M_r) *is live,* p_r *has at least one input transition and* (B_2, M_e) *is almost live, then* $(B_1(p_r \to B_2), M_i)$ *is live.*

14) If (B_1, M_r) is reversible and (B_2, M_e) is almost reversible, $(B_1(p_r \to B_2), M_i)$ is reversible.

Proof. By combining the corresponding properties of Theorems 5.4 and Theorem 5.5. □

5.3 An Example for Transformation TR

Description of a complaint-processing workflow system (Figure 5.5 and Table 5.1): The workflow system (B, p_1) operates as follows: When a complaint is launched, it will be registered. Then, a questionnaire is sent to the complainant while initial evaluation of the complaint is commenced. The response will be processed if returned within two weeks. Otherwise, it is discarded. Based on the result of the initial evaluation, the complaint is either formally processed or ignored. The actual processing of the complaint is delayed until the questionnaire is processed or a time-out has occurred. Processing of the complaint is monitored until all issues have been resolved and reprocessing may be warranted if there are any unsolved issues. Finally, the complaint is archived together with the questionnaire.

The Petri net model and the specification are shown in Figure 5.5 and Table 5.1.

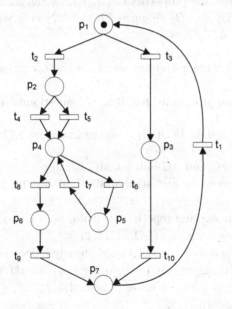

Fig. 5.5   Petri net representation of a workflow system B for processing complaints.

Table 5.1 Interpretation of places and transitions of the workflow system (Figure 5.5).

Place/Transition	Interpretation
p_1	Register a complaint
p_2	Waiting for the return of a questionnaire
p_3	The activity of normal processing of a complaint
p_4	Ready to process or finish the whole processing of a questionnaire
p_5	Ready to hold a questionnaire for further processing or reprocessing of a complaint
p_6	Finish the whole processing of a questionnaire
p_7	Archive a complaint together with a questionnaire
t_1	Register a new complaint
t_2	Send a questionnaire to a complainant
t_3	Evaluate a complaint
t_4	Discard a questionnaire not returned within two weeks
t_5	Process a questionnaire
t_6	Process a complaint
t_7	Hold the questionnaire for further processing or reprocessing of a complaint
t_8	Quit the activity of the questionnaire
t_9 and t_{10}	Finish the whole processing of the complaint

In order to verify the properties of (B, p_1), we can consider two smaller Petri net models B_1 and B_2 (Figures 5.6 and 5.7) satisfying $B = B_1(t_r \to B_2)$ by Definition 5.2, where

a) B_1 specifies the main system, and t_r is the operation which triggers Process B_2.
b) B_2 specifies the process for handling the questionnaire.

Properties of B_1 (matrix V_1 of B_1 is shown in Figure 5.7):

1) B_1 is SM, FC and AC but not MG because $|p_1^{\bullet}| > 1$ and $|{}^{\bullet}p_7| > 1$.
2) B_1 is conservative and structurally bounded because $\alpha_1 \geq 1$ and $\alpha_1 V_1 = 0$, where $\alpha_1 = (x_5, x_4, \ldots, x_1) = (1, 1, 1, 1, 1)$.
3) B_1 is consistent and repetitive because $\beta_1 \geq 1$ and $V_1 \beta_1 = 0$, where $\beta_1 = (x_1, x_2, \ldots, x_6) = (1, 2, 1, 1, 1, 1)$.
4) Since the first four rows of V_1 are linearly independent and the last row is the negative sum of the other rows, $Rank(B_1) = 4$.
5) $|C(B_1)| = 5$. The 5 clusters are: $\{p_1, t_2, t_3\}$, $\{p_2, t_r\}$, $\{p_6, t_9\}$, $\{p_3, t_{10}\}$ and $\{p_7, t_1\}$.
6) B_1 satisfies the RC-property because $Rank(B_1) = |C(B_1)| - 1$.

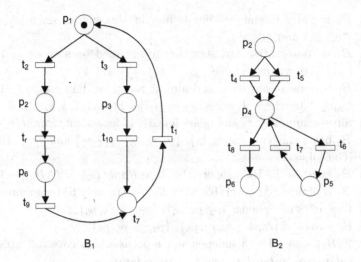

Fig. 5.6 Basic component processes B_1 and B_2 for B of Figure 5.5. (B_1 has a re-initiation path and B_2 does not.)

7) B_1 is SM-coverable with two SM-components: $p_1 t_2 p_2 t_r p_6 t_9 p_7 t_1 p_1$ and $p_1 t_3 p_3 t_{10} p_7 t_1 p_1$.

8-9) $D = \{p_1, p_2, p_3, p_6, p_7\}$ is both a siphon and a trap of B_1 because $^\bullet D = D^\bullet$.

10) Since B_1 has a reinitiation path $p_7 t_1 p_1$, $z(B_1) = 2 - 1 + 1 = 2$. The 2 independent e-paths which cover all arcs of B_1 are: $p_1 t_2 p_2 t_r p_6 t_9 p_7$ and $p_1 t_3 p_3 t_{10} p_7$.

11-13) (B_1, p_1) is safe, live and reversible.

Properties of B_2 (matrices V_2 and V_{2a} of B_2 are shown in Figure 5.7):

	t_r	t_1	t_3	t_{10}	t_2	t_9
p_1	0	1	-1	0	-1	0
p_3	0	0	1	-1	0	0
p_7	0	-1	0	1	0	1
p_2	-1	0	0	0	1	0
p_6	1	0	0	0	0	-1

	t_6	t_7	t_4	t_5	t_8	t_{2a}
p_2	0	0	-1	-1	0	1
p_6	0	0	0	0	1	-1
p_4	-1	1	1	1	-1	0
p_5	1	-1	0	0	0	0

Fig. 5.7 Incidence matrices V_1 for B_1 and V_{2a} for B_{2a}. (V_2 for B_2 is the part of V_{2a} without column t_{2a}.)

1) B_2 is SM, FC and AC but is not an almost MG since $|p_2^{\bullet}| > 1$, $|^{\bullet}p_4| > 1$ and $|p_4^{\bullet}| > 1$.

2) B_2 is conservative and structurally bounded because $\alpha_2 \geq 1$ and $\alpha_2 V_2 = 0$, where $\alpha_2 = (y_1, y_2, y_3, y_4) = (1, 1, 1, 1)$.

3) B_2 is almost consistent and almost repetitive because $\beta_2 \geq 1$ and $V_{2a}\beta_2 = 0$, where $\beta_2 = (y_1, y_2, y_3, y_4, y_5, z) = (1, 1, 1, 1, 2, 2)$.

4) Since columns t_4, t_6 and t_8 are linearly independent, $Rank(B_2) = 3$.

5) B_2 has 4 clusters: $\{p_2, t_4, t_5\}$, $\{p_4, t_6, t_8\}$, $\{p_5, t_7\}$ and $\{p_6\}$. Hence, $|C(B_2)| = 4$.

6) B_2 satisfies the RC-property because $Rank(B_2) = |C(B_2)| - 1$.

7) B_2 is almost SM-coverable with B_{2a} as the only SM-component.

8) Examples of siphons: $\{p_2, p_4, p_5\}$, $\{p_2, p_4, p_5, p_6\}$.

9) Examples of traps: $\{p_4, p_5, p_6\}$, $\{p_2, p_4, p_5, p_6\}$.

10) $Z(B_2) = 3$. The 3 independent e-paths which cover all arcs are: $p_2 t_4 p_4 t_8 p_6$, $p_2 t_5 p_4 t_8 p_6$ and $p_2 t_5 p_4 t_6 p_5 t_7 p_4 t_8 p_6$.

11-13) (B_2, p_2) is bounded, almost live and reversible.

Properties of B: The results listed below follow from Theorem 5.4 and V (Figure 5.8).

	t_1	t_3	t_{10}	t_2	t_9	t_6	t_7	t_4	t_5	t_8
p_1	1	-1	0	-1	0	0	0	0	0	0
p_3	0	1	-1	0	0	0	0	0	0	0
p_7	-1	0	1	0	1	0	0	0	0	0
p_2	0	0	0	1	0	0	0	-1	-1	0
p_6	0	0	0	0	-1	0	0	0	0	1
p_4	0	0	0	0	0	-1	1	1	1	-1
p_5	0	0	0	0	0	1	-1	0	0	0

Fig. 5.8   Incidence matrix V for B (column t_r is not included).

1) B is an SM, not an almost MG, an FC net and an AC net.

2) B is conservative and structurally bounded because $\alpha \geq 1$ and $\alpha V = 0$, where $\alpha = (y_1\alpha_1[P_1 - \{p_2, p_4\}], x_2 y_1, x_1 y_2, x_1\alpha_2[P_2 - \{p_2, p_4\}]) = (1, 1, 1, 1, 1, 1, 1)$. (Note: α_1 and α_2 are taken from the descriptions of B_1 and B_2 given above, respectively.)

3) B is almost consistent and repetitive because $\exists \beta = (z\beta_1[T_1 - \{t_r\}], x_1\beta_2[T_2]) = (zx_2, \ldots, zx_6, x_1 y_1, \ldots, x_1 y_5) = (4, 2, 2, 2, 2, 1, 1, 1, 1, 2) \geq$

1, such that $V\beta = 0$. (Note: β_1 and β_2 are taken from the descriptions of B_1 and B_2 given above, respectively.)

4) In B_1, $t_r = (0,0,0,-1,1)$. t_r occurs in the T-invariant: $t_r + t_1 + t_2 + t_9 = 0$. Hence, $h = 0$. Also, $B_1 \backslash \{t_r\}$ has the unique P-invariant: $p_1 + p_3 + p_7 + p_2 + p_6 = 0$ and B_2 has the unique P-invariant: $p_2 + p_6 + p_4 + p_5 = 0$. This satisfies Condition (3) in Case 2 of Theorem 5.3. Hence, $d = 1$. It follows from Theorem 5.3 that $Rank(B) = Rank(B_1) + Rank(B_2) - h + d - 2 = 4 + 3 - 1 = 6$.

5) $|C(B)| = |C(B_1)| + |C(B_2)| - 2 = 5 + 4 - 2 = 7$. The 7 clusters are: $\{p_1, t_2, t_3\}$, $\{p_2, t_4, t_5\}$, $\{p_4, t_6, t_8\}$, $\{p_5, t_7\}$, $\{p_6, t_9\}$, $\{p_3, t_{10}\}$ and $\{p_7, t_1\}$.

6) Since both B_1 and B_2 satisfy the RC-property and $d - h = 1$, B satisfies the RC-property. (This result is consistent with propositions 4) and 5))

7) B is SM-coverable. The two SM-components of B are shown in Figure 5.9.

8-9) Siphon (trap) $D_1 = \{p_1, p_2, p_6, p_3, p_7\}$ of B_1 is not a siphon (trap) of B because it does not satisfy the condition of Theorem 5.4. Similarly, siphons $D_2 = \{p_2, p_4, p_5\}$ and $D_3 = \{p_2, p_4, p_5, p_6\}$ of B_2 are not siphons of B. D_3 is not a trap of B either.

10) $Z(B) = Z(B_1) + Z(B_2) - 1 = 2 + 3 - 1 = 4$. By applying the cyclomatic complexity formula directly, $Z(B) = 20 - (7 + 10) + 1 = 4$. The 4 linearly independent e-paths which cover all arcs except (p_7, t_1) and (t_1, p_1) are: $p_1 t_2 p_2 t_4 p_4 t_8 p_6 t_9 p_7$, $p_1 t_2 p_2 t_5 p_4 t_8 p_6 t_9 p_7$, $p_1 t_2 p_2 t_5 p_4 t_6 p_5 t_7 p_4 t_8 p_6 t_9 p_7$ and $p_1 t_3 p_3 t_{10} p_7$.

11-13) (B, p_1) is bounded, live and reversible.

In dealing with complex systems that have many desirable properties, current research in the top-down approach for system design aims at finding transformations that can preserve as many properties as possible. For Petri nets, transition and place refinements are important transformations for building complex systems from simple components. Previous study on refinement techniques [SM (1983); VAL (1979)] focused mainly on their preservation of liveness and boundedness. Based on a more general version of transition and place refinements, where the refinement net simulates a software process and has multiple entry transitions and exit transitions, this chapter proposes various conditions under which many common properties will be preserved. These properties cover most of the important properties of Petri nets. The results may serve at least two purposes: (1) To guide the

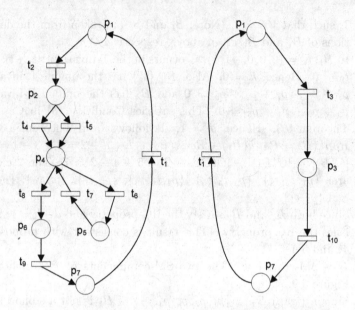

Fig. 5.9   The two SM-components of B.

design of a refinement net so that, when substituted into the refined net, it will not destroy the latter's desirable properties. (2) To determine the properties of a Petri net after some of its places or transitions have been refined.

Chapter 6

Property-Preserving Reductions

Chapter 5 introduces some operators for refinement in PPPA. This chapter presents some operators for reduction in PPPA.

Technically, from the viewpoint of Petri net transformations, a reduction is the reverse of a refinement. It abstracts the functional details of some part of a Petri net into a single place or a single transition. In component-based system design, for example, a component is abstracted into a single function. In software programs, a complex subprogram is abstracted into a single data type. In manufacturing engineering, a set of resources is represented by a single 'super' resource. For verification purposes, a reduction may be used to simplify the processing of checking whether a system is valid when every replaced part is assumed to be correct but operates as a single function in the 'super' program. This is not a simple task, especially if the parts under abstraction form subsystems with multiple entries and exits. A logical mishandling will make the reduction erroneous.

This chapter presents three operators for reduction. The first is to reduce a place-bordered path to a single place, the second is to reduce a transition-bordered path to a single transition and the third is to reduce a place-bordered subnet to a single place. The first two operators have major applications in abstracting programs having a single entry and a single exit. They had been studied in the literature [LF (1987); ESP (1994); MAK (2001)] but under much more restrictive conditions on the start and end transitions or places. The third operator is an extension of the above two path reductions wherein the place-bordered path or transition-bordered path is changed to a place-bordered subnet. This operator has major application in abstracting subprograms with multiple entries and/or multiple exits into single functions.

In conformance with our approach for component-based system design, similar to refinements, our reduction techniques also preserve two features of the original Petri net:

(a) It is still an element of PPPA so that it can be applied again within the algebra.
(b) It preserves the correctness of the original Petri net process.

To preserve these two features, conditions are proposed for both forwardly and backwardly preserving properties such as liveness, boundedness, reversibility, etc. on each of the three operators.

The rest of this chapter presents the technical details of our results. Path and subnet reductions and their preservation of properties are presented in Sections 6.1, 6.2 and 6.3, respectively. Section 6.4 illustrates how to apply the three reductions to the verification of two systems design and some remarks are presented in at last.

6.1   Reducing a Transition-Bordered Path to a Transition

This section studies an operator that reduces an elementary path to a single transition. The path both starts and ends at a transition. This operator is formally stated below, where the place p may represent an elementary directed path starting and ending at a place. The entire path may also represent a subsystem that has a single entry and a single exit.

Definition 6.1. (Reduce-T-Path) (Figure 6.1)
Let (N, M_0), where $N = (P, T, F)$, be an ordinary Petri net. Suppose there exist $\varepsilon_i,\ \varepsilon_o \in T$ and $p \in P$ such that $\varepsilon_i \neq \varepsilon_o$, $^\bullet p = \{\varepsilon_i\}$, $p^\bullet = \{\varepsilon_o\}$ and $^\bullet\varepsilon_i \cap {}^\bullet\varepsilon_o = \varepsilon_i^\bullet \cap \varepsilon_o^\bullet = \phi$. Reduce-T-Path transforms (N, M_0) to (N', M_0') as follows:

$$P' = P - \{p\}$$
$$T' = (T - \{\varepsilon_i, \varepsilon_o\}) \cup \{\varepsilon'\}$$
$$F' = F - (\{(x, \varepsilon_i)|x \in {}^\bullet\varepsilon_i\} \cup \{(\varepsilon_i, x)|x \in \varepsilon_i^\bullet\} \cup \{(x, \varepsilon_o)|x \in {}^\bullet\varepsilon_o\} \cup \{(\varepsilon_o, x)|x \in \varepsilon_o^\bullet\}) \cup (\{(x, \varepsilon')\ |x \in {}^\bullet\varepsilon_i \cup {}^\bullet\varepsilon_o - \{p\}\} \cup \{(\varepsilon', x)|x \in \varepsilon_o^\bullet \cup \varepsilon_i^\bullet\ \{p\}\})$$ and
$$M_0'(p) = M_0(p)\forall p \in P'$$

The reduction rules studied in [LF (1987); MAK (2001)] are special cases of Reduce-T-Path in the sense that they satisfy an additional set of conditions: a) $\varepsilon_o^\bullet \neq \phi$, $^\bullet\varepsilon_o = \{p\}$ and $M_0(p) = 0$ or b) $^\bullet\varepsilon_i \neq \phi$, $\varepsilon_i^\bullet = \{p\}$ and $M_0(p) = 0$.

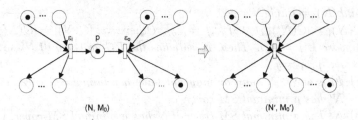

(N, M₀)   (N', M₀')

Fig. 6.1   Petri nets before and after applying Reduce-T-Path.

Theorem 6.1. *(property preservation under Reduce-T-Path)*
Let (N, M) and (N', M') be the two Petri nets involved in Definition 6.1.
Then, the following propositions hold:

1) *N is an SM iff N' is an SM.*
2) *N is an MG iff N' is an MG.*
3) *(a) In general, N (resp., N') is not an FC net even if N' (resp., N)
   is an FC net.*
   (b) N is an FC net if N' is an FC net and $((^\bullet\varepsilon')^\bullet = \{\varepsilon'\})$ in N'.
   *(c) Suppose N is an FC net. N' is an FC net if $(^\bullet\varepsilon_i)^\bullet = \{\varepsilon_i\}$ and
   $(^\bullet\varepsilon_o)^\bullet = \{\varepsilon_o\}$ in N. N' is an AC net if either $((^\bullet\varepsilon_i)^\bullet = \{\varepsilon_i\}$ or
   $(^\bullet\varepsilon_o)^\bullet = \{\varepsilon_o\})$ in N.*
4) *(a) In general, N (resp., N') is not an AC net even if N' (resp., N)
   is an AC net.*
   (b) N is an AC net if N' is an AC net and $(^\bullet\varepsilon')^\bullet = \{\varepsilon'\}$ in N'.
   *(c) N' is an AC net if N is an AC net and either $(^\bullet\varepsilon_i)^\bullet = \{\varepsilon_i\}$ or
   $(^\bullet\varepsilon_o)^\bullet = \{\varepsilon_o\}$ holds.*
5) *(a) In general, N may or may not be conservative even if N' is conservative.*
   *(b) N is conservative if $\exists\alpha' > 0$ such that $\alpha'V' = 0$ (i.e., N' is con-
   servative) and $\alpha'V[P - \{p\}, \varepsilon_o] > 0$.*
   (c) N' is conservative if N is conservative.
6) *(a) In general, N may or may not be structurally bounded even if N'
   is structurally bounded.*
   *(b) N is structurally bounded if $\exists\alpha' > 0$ such that $\alpha'V' \leq 0$ (i.e., N'
   is structurally bounded) and $\alpha'[P - \{p\}, \varepsilon_o] > 0$.*
   (c) N' is structurally bounded if N is structurally bounded.
7) *N is consistent iff N' is consistent.*
8) *N is repetitive iff N' is repetitive.*

9) $Rank(N) = Rank(N') + 1$.

10) $|C(N)| = |C(N')| + 1$ if $[\varepsilon_i] \neq [\varepsilon_o]$. Otherwise, $|C(N)| = |C(N')|$.

11) Suppose $[\varepsilon_i] \neq [\varepsilon_o]$. Then, N satisfies the RC-property iff N' satisfies the RC-property.

12) (a) In general, N may or may not have a minimal SM-cover even if N' has a minimal SM-cover.

    (b) N' has a minimal SM-cover if N has a minimal SM-cover.

13) (a) Suppose D' is a siphon of N'. Then, $D' \cup \{p\}$ is a siphon of N if $(D' \cap {}^\bullet\varepsilon_i \neq \phi)$ in N', otherwise, D' is a siphon of N.

    (b) Suppose D is a siphon of N. Then, $D - \{p\}$ is a siphon of N'.

14) (a) Suppose S' is a trap of N'. Then, $S' \cup \{p\}$ is a trap of N.

    (b) Suppose S is a trap of N. Then, $S - \{p\}$ is a trap of N'.

15) Suppose $M_0(p) = 0$. Then, the following propositions hold:

    (a) N is a process and the entry place $p_e \neq p$ and the exit place $p_x \neq p$ iff N' is a process.

    (b) Suppose both N and N' are processes. Then, (b.1) $Z(N') = Z(N)$. (b.2) In general, (N, M_0) (resp., (N', M_0')) may or may not terminate properly even if (N', M_0') (resp., (N, M_0)) terminates properly. (b.3) Suppose $\varepsilon_o^\bullet \neq \phi$ and ${}^\bullet\varepsilon_o = \{p\}$ in N. Then (N, M_0) terminates properly iff (N', M_0') terminates properly. (b.4) Suppose ${}^\bullet\varepsilon_i \neq \phi$ and $\varepsilon_i^\bullet = \{p\}$ in N. Then (N', M_0') terminates properly if (N, M_0) terminates properly.

16) (a) In general, (N, M_0) may or may not be bounded even if (N', M_0') is bounded.

    (b) If $(\varepsilon_o^\bullet \neq \phi$ and ${}^\bullet\varepsilon_o = \{p\})$ in N and $M_0(p) = 0$, then (N, M_0) is bounded iff (N', M_0') is bounded.

    (c) (N', M_0') is bounded if (N, M_0) is bounded.

17) (a) In general, (N, M_0) may or may not be live even if (N', M_0') is live and (N', M_0') may not be live even if (N, M_0) is live.

    (b) If $(\varepsilon_o^\bullet \neq \phi$ and ${}^\bullet\varepsilon_o = \{p\})$ in N and $M_0(p) = 0$, then (N, M_0) is live iff (N', M_0') is live.

    (c) (N', M_0') is live if (N, M_0) is live, $({}^\bullet\varepsilon_i \neq \phi$ and $\varepsilon_i^\bullet = \{p\})$ in N and $M_0(p) = 0$.

18) (a) In general, (N, M_0) may or may not be reversible even if (N', M_0') is reversible and (N', M_0') may not be reversible even if (N, M_0) is reversible.

    (b) If $(\varepsilon_o^\bullet \neq \phi$ and ${}^\bullet\varepsilon_o = \{p\})$ in N and $M_0(p) = 0$, then (N, M_0) is reversible iff (N', M_0') is reversible.

(c) (N', M_0') *is reversible if* (N, M_0) *is reversible,* $(\bullet \varepsilon_i \neq \phi$ *and* $\varepsilon_i^\bullet = \{p\})$ *in* N *and* $M_0(p) = 0$.

Proof.

1-2) Since $|\bullet \varepsilon_i| = |\varepsilon_i^\bullet| = |\bullet \varepsilon_o| = |\varepsilon_o^\bullet| = 1$ in N iff $|\bullet \varepsilon'| = |(\varepsilon')^\bullet| = 1$ in N' and no other transitions are modified, N is an SM iff N' is an SM. Since Reduce-T-Path does not alter the connectivity degree of any place, N is an MG iff N' is an MG.

3) (a) It is obvious N (resp., N') may not be an FC net even if N' (resp., N) is an FC net.

(b) Since the only change is due to the fusing of ε_i and ε_o, just those place q in $\bullet \varepsilon_i$ or $\bullet \varepsilon_o$ may be affected and thus have to be considered. If $((\bullet \varepsilon')^\bullet = \{\varepsilon'\})$ in N', then $((\bullet \varepsilon_i)^\bullet = \{\varepsilon_i\}$ and $(\bullet \varepsilon_o)^\bullet = \{\varepsilon_o\})$ in N. Hence, N is an FC net if N' is an FC net and $((\bullet \varepsilon')^\bullet = \{\varepsilon'\})$ in N'.

(c) For any $p, q \in P'$ satisfying $(p^\bullet \cap q^\bullet \neq \phi)$ in N', consider the following cases: (1) $(p, q \notin \bullet \varepsilon')$ in N'. Then, $(p, q \notin \bullet \varepsilon_i \cup \bullet \varepsilon_o)$ in N. Since N is an FC net, $(p^\bullet = q^\bullet)$ in N and thus in N'. (2) $(p \in \bullet \varepsilon'$ and $q \notin \bullet \varepsilon')$ in N'. Then, $(p \in \bullet \varepsilon_i \cup \bullet \varepsilon_o$ and $q \notin \bullet \varepsilon_i \cup \bullet \varepsilon_o)$ in N, i.e., $(\varepsilon_i, \varepsilon_o \notin q^\bullet)$ in N. Since N is an FC net, Case (2) is impossible. (3) $(p, q \in \bullet \varepsilon')$ in N'. Then, $(p, q \in \bullet \varepsilon_i \cup \bullet \varepsilon_o)$ in N. Hence, if $((\bullet \varepsilon_i)^\bullet = \{\varepsilon_i\}$ and $(\bullet \varepsilon_o)^\bullet = \{\varepsilon_o\})$ in N, then $(p^\bullet = q^\bullet)$ in N'. If $((\bullet \varepsilon_i)^\bullet = \{\varepsilon_i\}$ or $(\bullet \varepsilon_o)^\bullet = \{\varepsilon_o\})$ in N, then $(p^\bullet \subseteq q^\bullet$ or $q^\bullet \subseteq p^\bullet)$ in N'. This means that N' is an FC net if $((\bullet \varepsilon_i)^\bullet = \{\varepsilon_i\}$ and $(\bullet \varepsilon_o)^\bullet = \{\varepsilon_o\})$ in N and that N' is an AC net if $((\bullet \varepsilon_i)^\bullet = \{\varepsilon_i\}$ or $(\bullet \varepsilon_o)^\bullet = \{\varepsilon_o\})$ in N.

4) (a) It is obvious that N (resp., N') may or may not be an AC net even if N' (resp., N) is an AC net.

(b) If N' is an AC net and $((\bullet \varepsilon')^\bullet = \{\varepsilon'\})$ in N', then it is obvious that N is also an AC net.

(c) For any $p, q \in P'$ satisfying $(p^\bullet \cap q^\bullet \neq \phi)$ in N', consider the following cases: (1) $(p, q \notin \bullet \varepsilon')$ in N'. Then, $(p, q \notin \bullet \varepsilon_i \cup \bullet \varepsilon_o)$ in N. Since N is an AC net, $(p^\bullet \subseteq q^\bullet$ or $q^\bullet \subseteq p^\bullet)$ in N and thus in N'. (2) $(p \in \bullet \varepsilon'$ and $q \notin \bullet \varepsilon')$ in N'. Then, $(p \in \bullet \varepsilon_i \cup \bullet \varepsilon_o$ and $q \notin \bullet \varepsilon_i \cup \bullet \varepsilon_o)$ in N, i.e., $(\varepsilon_i, \varepsilon_o \notin q^\bullet)$ in N. Since N is an AC net, $q^\bullet \subseteq p^\bullet$ in N and thus in N'. (3) $(p, q \in \bullet \varepsilon')$ in N'. Then, $(p, q \in \bullet \varepsilon_i \cup \bullet \varepsilon_o)$ in N. If $((\bullet \varepsilon_i)^\bullet = \{\varepsilon_i\}$ or $(\bullet \varepsilon_o)^\bullet = \{\varepsilon_o\})$ in N, then $(p^\bullet \subseteq q^\bullet$ or $q^\bullet \subseteq p^\bullet)$ in N'. This means that N' is an AC net. (Note that $p^\bullet \cap q^\bullet = \phi$ in N is possible for this case.)

5) (a) (Figure 6.2) It is obvious that N may or may not be conservative even if N' is conservative.

(b) Suppose $\exists \alpha' > 0$ such that $\alpha' V' = 0$ (i.e., N' is conservative) and $\alpha' V[P - \{p\}, \varepsilon_o] > 0$. Let $\alpha = (\alpha' V[P - \{p\}, \varepsilon_o], \alpha')$. Then $\alpha > 0$. Since $\alpha' V[P - \{p\}, \varepsilon_i] = -\alpha' V[P - \{p\}, \varepsilon_o]$, $\alpha V = 0$, by Characterization 2.1, N is conservative.

(c) If N is conservative, then by Characterization 2.1, $\exists \alpha = (x_1, x_2, \ldots, x_{|P|}) > 0$ such that $\alpha V = 0$. Let $\alpha' = (x_2, \ldots, x_{|P|})$. Then, $\alpha = (x_1, \alpha')$ and $\alpha' \geq 1$. By splitting V into three vertical components as in Figure 6.2, $\alpha V = 0$ is the same as $x_1 + \alpha' V[P - \{p\}, \varepsilon_i] = 0$, $-x_1 + \alpha' V[P - \{p\}, \varepsilon_o] = 0$ and $\alpha' V[P - \{p\}, T - \{\varepsilon_i, \varepsilon_o\}] = 0$. By adding up the first two components, $\alpha' V' = 0$. That is, N' is conservative according to Characterization 2.1.

$$
\begin{array}{c}
\quad\quad \varepsilon_i \quad\quad\quad\quad \varepsilon_0 \quad\quad\quad\quad T - \{\varepsilon_i, \varepsilon_o\} \\
\begin{array}{c} a \\ b \end{array}
\left(
\begin{array}{ccc}
1 & -1 & 0 \\
V[P - p, \varepsilon_i] & V[P - \{p\}, \varepsilon_0] & V[P - \{p\}, T - \{\varepsilon_i, \varepsilon_o\}]
\end{array}
\right)
\end{array}
$$

$$
\begin{array}{c}
\quad\quad\quad \varepsilon' \quad\quad\quad\quad\quad\quad\quad T - \{\varepsilon'\} = T - \{\varepsilon_i, \varepsilon_o\} \\
P' \left(
\begin{array}{cc}
V[P - p, \varepsilon_i] + V[P - \{p\}, \varepsilon_o] & V[P - \{p\}, T - \{\varepsilon_i, \varepsilon_o\}]
\end{array}
\right)
\end{array}
$$

Fig. 6.2   Matrix V of N (top) and matrix V' of N' (bottom).

6) Similar to the proof of 5).

7) "$\Leftarrow$" (Figure 6.2) If N' is consistent, then by Characterization 2.3, $\exists \beta' = (y_1, y_2, \ldots, y_{|T|-1}) > 0$ such that $V' \beta' = 0$. Let $\beta = (y_1, \beta')$. Then, it is obvious that $\beta > 0$ and $V \beta = 0$, i.e., N is consistent.

"$\Rightarrow$"If N is consistent, then $\exists \beta = (z_1, z_2, \ldots, z_{|T|}) \geq 1$ such that $V \beta = 0$. The top row of this system of equations implies $z_1 = z_2$. Let $\beta' = (z_2, \ldots, z_{|T|})$. Then, $(0, V' \beta') = (z_1 - z_2, V' \beta') = V \beta = 0$. Hence, $\beta' > 0$ and $V' \beta' = 0$. By Characterization 2.3, N' is consistent.

8) Similar to the proof of 7).

9) As shown in Figure 6.2, adding column 1 to column 2 in V changes its top row to $(1, 0, \ldots, 0)$. Then, all the elements in column 1 except the topmost one can be reduced to zero by using Gaussian row operations. This results in a matrix with diagonal elements 1 and V' and remaining elements 0. Hence, $\text{Rank}(N) = 1 + \text{Rank}(N')$.

10) After applying Reduce-T-Path, clusters $[\varepsilon_o]$ and $[\varepsilon_i]$ are merged. Hence, $|C(N)| = |C(N')| + 1$ if $[\varepsilon_i] \neq [\varepsilon_o]$, otherwise $|C(N)| = |C(N')|$.

11) Together with 9) and 10), we know that, if $[\varepsilon_i] \neq [\varepsilon_o]$, then N satisfies the RC-property iff N' satisfies the RC-property.

12) (a) It is obvious that N may or may not have a minimal SM-cover even if N' has a minimal SM-cover.

    (b) Let K be a minimal SM-cover of N. Replace every SM-component $S \in K$ containing $\{\varepsilon_i, p, \varepsilon_o\}$ with $S' = S \cup \{\varepsilon'\} \cup \{(x, \varepsilon') \mid (x, \varepsilon_i) \in S\} \cup \{(\varepsilon', x) \mid (\varepsilon_o, x) \in S\} - \{\varepsilon_i, p, \varepsilon_o\} - \{(x, \varepsilon_i) \mid (x, \varepsilon_i) \in S\} - \{(\varepsilon_o, x) \mid (\varepsilon_o, x) \in S\} - \{(\varepsilon_i, p), (p, \varepsilon_o)\}$. (Note that, being strongly connected, every SM-component containing p must also contain both ε_i and ε_o.) Obviously, $|{}^\bullet \varepsilon'$ in $S'| = |{}^\bullet \varepsilon_i$ in $S| = 1$, $|(\varepsilon')^\bullet$ in $S'| = |\varepsilon_o^\bullet$ in $S| = 1$ and the connectivity of other transitions in S' is the same as in S. Hence, S' is an SM-component and the resulting K forms a minimal SM-cover of N'.

13) (a) Suppose D' is a siphon of N'. If $(\varepsilon_i \notin {}^\bullet (D'))$ in N, it is obvious that D' is also a siphon of N. Hence, $D' \cup \{p\}$ is a siphon of N. If $(\varepsilon_i \in {}^\bullet (D'))$ in N', then $(\varepsilon_i \in (D')^\bullet)$ in N' because D' is a siphon. Since ${}^\bullet(D' \cup \{p\})$ in $N = ({}^\bullet D'$ in $N') \cup \{\varepsilon_i\} \subseteq (D')^\bullet \cup \{\varepsilon_i, \varepsilon_o\} = (D' \cup \{p\})^\bullet$ in N, $D' \cup \{p\}$ is a siphon of N.

    (b) Suppose D is a siphon of N. Let $D' = D - \{p\}$. For any $t \in {}^\bullet D'$ in N', $(t \in {}^\bullet D \subseteq D^\bullet)$ in N, implying that there exists $q \in D$ such that $(t \in q^\bullet)$ in N. Consider two cases: *Case 1.* $q \neq p$. Since only p has been eliminated, q remains in D' and $t \in q^\bullet$ in $N \subseteq (D')^\bullet$ in N'. *Case 2.* $q = p$. Since $({}^\bullet p = \{\varepsilon_i\}$ and $p^\bullet = \{\varepsilon_o\})$ in N, $t = \varepsilon_o$ in N. Hence, $t = \varepsilon'$ in N' and $(\varepsilon_i \in {}^\bullet p \subseteq {}^\bullet D \subseteq D^\bullet)$ in N. This implies that there exists $s \in D$ such that $(\varepsilon_i \in s^\bullet)$ in N. Since $(\varepsilon_i \notin p^\bullet)$ in N, $s \neq p$. By similar argument as in Case 1, $(\varepsilon_i \in s^\bullet)$ in N implies that $\varepsilon' \in s^\bullet$ in $N \subseteq (D')^\bullet$ in N'. Hence, in both cases, $(t \in {}^\bullet D')$ in N' implies that $(t \in (D')^\bullet)$ in N'.

14) The proof for a trap is similar to the proof for a siphon in 13).

15) Suppose $M_0(p) = 0$. (a) By the definition of a process, (N, M_0) is a process iff (N', M_0') is a process. (b) Suppose both N' and N are processes. Then, (b.1) $Z(N') = |F'| - |P' \cup T'| + 2 = |F| - 2 - (|P| - 1) - (|T| - 1) + 2 = Z(N)$.

(b.2) It is obvious that, in general, proper termination may not be preserved. The example in Figure 6.3 shows that (N, M_0) terminates properly but (N', M_0') does not. The special cases are also obvious.

16) (a) It is obvious that (N, M_0) may or may not be bounded even if (N', M_0') is bounded.

    (b) see [MAK (2001)] and [LF (1987)].

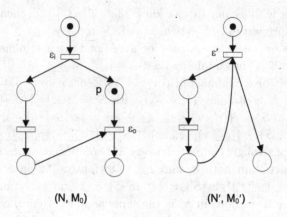

(N, M₀)   (N', M₀')

Fig. 6.3   It is obvious that (N, M_0) may or may not be bounded even if (N', M_0') is bounded.

(c) Suppose (N, M_0) is bounded. If (N', M_0') is unbounded, then there exists an infinite firing sequence $\sigma' = \sigma_1\varepsilon'\sigma_2\varepsilon'.\ldots$ (or $\sigma' = \sigma_1\sigma_2\ldots$) (where every σ_i does not contain ε') such that $M_0'[\sigma'\rangle M'$ and M' becomes unbounded. Let $\sigma = \sigma_1\varepsilon_i\varepsilon_o\sigma_2\varepsilon_i\varepsilon_o \ldots$ be obtained from σ' by replacing each ε' with $\varepsilon_i\varepsilon_o$ (or let $\sigma = \sigma_1\sigma_2\ldots$). Obviously, σ is firable in N and $M_0[\sigma\rangle M$, where $M(q) = M'(q)$ for every $q \in P - \{p\}$ and $M(p) = M_0(p)$. Hence, (N, M_0) is unbounded – a contradiction.

17) (a) It is obvious that (N, M_0) may or may not be live even if (N', M_0') is live. It is also obvious that (N', M_0') may not be live even if (N, M_0) is live. For example, in Figure 6.4, (N, M_0) is live but (N', M_0') is not. For the particular cases (b) and (c), refer to [MAK (2001)] and [LF (1987)], respectively.

18) (a) It is obvious that (N, M_0) may or may not be reversible even if (N', M_0') is reversible. It is also obvious that (N', M_0') may not be reversible even if (N, M_0) is reversible. For example, in Figure 6.2, (N, M_0) is reversible but (N', M_0') is not.

(b) "$\Leftarrow$" $\forall M = M_0[\sigma\rangle$, where $\sigma = \sigma_1 \varepsilon_i\varepsilon_o \sigma_2 \varepsilon_i\varepsilon_o \ldots$, by the assumption, $\exists M' = M_0' [\sigma'\rangle$, where $\sigma' = \sigma_1 \varepsilon' \sigma_2 \varepsilon' \ldots$, every σ_i does not contain ε', such that $M'(s) = M(s)$ for every $s \in P'$. Since N' is reversible, $\exists \sigma_r' = \sigma_1 {}'\varepsilon'\sigma_2'\varepsilon'\sigma_3 {}'\varepsilon'\sigma_4'.\ldots$ such that $M' [\sigma_r'\rangle M_0'$. Let $\sigma_r = \sigma_1{}'\varepsilon_i\varepsilon_o \sigma_2'\varepsilon_i\varepsilon_o\sigma_3'\varepsilon_i\varepsilon_o\sigma_4'.\ldots$. Then, obviously we have $M[\sigma_r\rangle M_0$. "$\Rightarrow$" $\forall M' = M_0' [\sigma'\rangle$, where $\sigma' = \sigma_1 \varepsilon' \sigma_2 \varepsilon' \ldots$, every

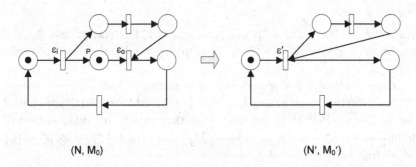

(N, M₀)   (N', M₀')

Fig. 6.4   An example showing that the properties liveness and reversibility are not preserved under Reduce-T-Path.

σ_i does not contain ε', $\exists M = M_0[\sigma\rangle$, where $\sigma = \sigma_1 \; \varepsilon_i\varepsilon_o \; \sigma_2 \; \varepsilon_i\varepsilon_o$ ... such that $M(s) = M'(s)$ for every $s \in P - \{p\}$ and $M(p) = 0$. Since N is reversible, $\exists \sigma_r = \sigma_1'\varepsilon_i\sigma_2'\varepsilon_o\sigma_3'\varepsilon_i\sigma_4'\varepsilon_o...$ such that $M[\sigma_r\rangle M_0$. By the condition, firability of ε_i in N implies firability of ε' in N' and firing ε' in N' has the same effect as firing both ε_i and ε_o. Hence, in σ_r, we can replace ε_i with ε' and ignore ε_o and the resulting $\sigma_r' = \sigma_1'\varepsilon'\sigma_2'\sigma_3'\varepsilon'\sigma_4'...$ is firable in N'. Also, M' $[\sigma_r'\rangle M_0'$.

(c) Similar to the proof above by letting $\sigma_r' = \sigma_1'\sigma_2'\varepsilon'\sigma_3'\sigma_4'\varepsilon'...$, we have M' $[\sigma_r'\rangle M_0'$. □

6.2   Reducing a Place-Bordered Path to a Place

This section studies an operator that reduces an elementary path to a single place. The path both starts and ends at a place. This operator is formally · stated below, where the transition ε may represent an elementary directed path starting and ending at a transition. The entire path may also represent a subsystem that has a single entry and a single exit.

Definition 6.2. (Reduce-P-Path) (Figure 6.5)
Let (N, M_0), where $N = (P, T, F)$, be an ordinary Petri net. Suppose there exist $p_i, p_o \in P$ and $\varepsilon \in T$ such that $p_i \neq p_o$, $^\bullet\varepsilon = \{p_i\}$, $\varepsilon^\bullet = \{p_o\}$ and $^\bullet p_i \cap {}^\bullet p_o = p_i^\bullet \cap p_o^\bullet = \phi$. Reduce-P-Path transforms (N, M_0) to (N', M_0') as follows:

$$P' = (P - \{p_i, p_o\}) \cup \{p'\}$$

$T' = T - \{\varepsilon\}$

$F' = F - (\{(x, p_i)|x \in {}^{\bullet}p_i\} \cup \{(p_i, x)|x \in p_i^{\bullet}\} \cup \{(x, p_o)|x \in {}^{\bullet}p_o\} \cup \{(p_o, x) | x \in p_o^{\bullet}\}) \cup (\{(x, p')|x \in {}^{\bullet}p_i \cup {}^{\bullet}p_o - \{\varepsilon\}\} \cup \{(p', x)|x \in p_o^{\bullet} \cup p_i^{\bullet} - \{\varepsilon\}\})$

$M_0'(p) = M_0(p)$ if $p \neq p'$ and $M_0'(p') = M_0(p_i) + M_0(p_o)$.

The reduction rules studied in [LF (1987); MAK (2001)] are special cases of Reduce-P-Path in the sense that they satisfy an additional set of conditions: (a) $p_o^{\bullet} \neq \phi$, ${}^{\bullet}p_o = \{\varepsilon\}$ and $M_0(p_o) = 0$ or (b) ${}^{\bullet}p_i \neq \phi$, $p_i^{\bullet} = \{\varepsilon\}$ and $M_0(p_o) = 0$.

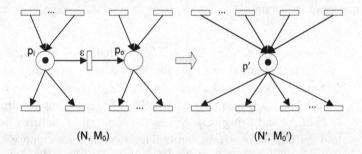

(N, M_0) $\qquad\qquad\qquad\qquad\qquad$ (N', M_0')

Fig. 6.5   Petri nets before and after applying Reduce-P-Path.

Lemma 6.1. *Let (N, M_0) and (N', M_0') be the two Petri nets defined in Definition 6.2. Then, the following propositions hold:*

1) For every $M \in R(N, M_0)$, there exists $M' \in R(N', M_0')$ such that $M'(p') = M(p_i) + M(p_o)$ and $M'(p) = M(p)$ for $p \in P' - \{p'\}$.

2) If, for each $t_i \in p_i^{\bullet} - \{\varepsilon\}$, there exists $t_o \in p_o^{\bullet}$ such that $t_i^{\bullet} = t_o^{\bullet}$ and ${}^{\bullet}t_i - \{p_i\} = {}^{\bullet}t_o - \{p_o\}$, then, for every $M' \in R(N', M_0')$, there exists $M \in R(N, M_0)$ such that $M(p_i) + M(p_o) = M'(p')$ and $M(p) = M'(p)$ for $p \in P - \{p_i, p_o\}$.

Proof.

1) Since $M \in R(N, M_0)$, $\exists \sigma \in L(N, M_0)$ such that $M_0[N, \sigma\rangle M$. $\forall \varepsilon \in \sigma$, let M_1 and M_2 be the two markings just before and just after firing ε in N. Then, according to the way ε is eliminated in Reduce-P-Path, $M_i'(p') = M_i(p_i) + M_i(p_o)$ and $M_i'(p) = M_i(p)$ for $p \in P' - \{p'\}$, $i = 1, 2$ in N'. Suppose that σ' is the transition sequence obtained from σ

by deleting all such ε, the above argument shows that $\sigma' \in L(N', M_0')$ and $M_0'[N', \sigma'\rangle M'$.

2) Since $M' \in R(N', M_0')$, $\exists \sigma' \in L(N', M_0')$ such that $M_0'[N', \sigma'\rangle M'$. Let us try to fire σ' in N. Consider any $t \in \sigma'$. If $t \notin p_i^\bullet - \{\varepsilon\}$, then t is always firable in N. If $t = t_i \in p_i^\bullet - \{\varepsilon\}$, then t_i may or may not be firable in N. Each t_i that is firable in N is kept in σ', possibly having to insert a ε into σ' if necessary. For each t_i that is not firable in N, since there exists $t_o \in p_o^\bullet$ such that $t_i^\bullet = t_o^\bullet$ and $^\bullet t_i - \{p_i\} = {}^\bullet t_o - \{p_o\}$, t_i is replaced with this t_o. Since t_i is firable in N', t_o is also firable in N', resulting in the same marking as firing t_i. This replacement results in a transition sequence σ such that $M_0[N, \sigma\rangle M$. $\qquad\Box$

Theorem 6.2. *(preservation of liveness, boundedness, and reversibility under Reduce-P-Path)*
Let (N, M_0) and (N', M_0') be the two Petri nets defined in Reduce-P-Path.

1) *Suppose $^\bullet p_i \neq \phi$, $p_i^\bullet = \{\varepsilon\}$ in N and $M_0(p_o) = 0$. Then (N', M_0') is live (resp., bounded, reversible) iff (N, M_0) is live (resp., bounded, reversible).*

2) *Suppose at least one of the following conditions is valid: (a) $p_o^\bullet \neq \phi$, $^\bullet p_o = \{\varepsilon\}$ and $M_0(p_o) = 0$. (b) For each $t_i \in p_i^\bullet - \{\varepsilon\}$, there exists $t_o \in p_o^\bullet$ such that $t_i^\bullet = t_o^\bullet$ and $^\bullet t_i - \{p_i\} = {}^\bullet t_o - \{p_o\}$. Then, if (N, M_0) is live (resp., bounded, reversible), (N', M_0') is live (resp., bounded, reversible).*

Proof.

1) See [MAK (2001)] and [LF (1987)].

2) For Condition (a), proof can be found in [LF (1987)] and [MAK (2001)], respectively. For Condition (b), the proof proceeds as follows: For any reachable marking M' of (N', M_0') and any transition t in N', under the assumption of Condition (c), Lemma 6.1 implies that $\exists M \in R(N, M_0)$ such that $M(p_i) + M(p_o) = M'(p')$ and $M(p) = M'(p)$ for $p \in P - \{p_i, p_o\}$. Since (N, M_0) is live, $\exists M_1 \in R(N, M)$, such that t is firable at M_1. By Lemma 6.1, $\exists M_1' \in R(N', M')$ such that $M_1'(p') = M_1(p_i) + M_1(p_o)$ and $M_1'(p) = M_1(p)$ for $p \in P' - \{p'\}$. This implies that $\forall p \in {}^\bullet t$, $M_1'(p) \geq M_1(p)$. The fact that t is firable at M_1 implies that t is firable at M_1'. Hence, (N', M_0') is live. The proofs for boundedness and reversibility are similar to the above proof for liveness. They are omitted here. $\qquad\Box$

Example 6.1. In all cases considered below, the path $p_i \varepsilon p_o$ in N is reduced to p' in N'. In Figure 6.6 , (N, M_0) is live, bounded and reversible. Since for $t_i \in p_i^\bullet - \{\varepsilon\}$, there exists $t_o \in p_o^\bullet$ such that $t_i^\bullet = t_o^\bullet = \{p\}$ and $^\bullet t_i - \{p_i\}$ $= {}^\bullet t_o - \{p_o\} = \phi$, it follows from Theorem 6.2(2b) that (N', M_0') is live, bounded and reversible.

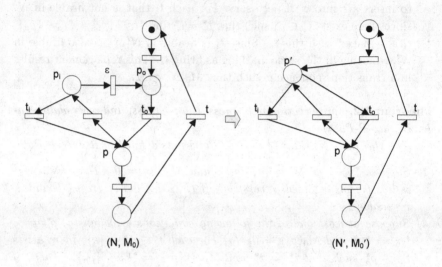

(N, M₀)   (N', M₀')

Fig. 6.6   Both (N, M_0) and (N, M_0) are live, bounded and reversible.

Note that, in Theorem 6.2, each of the three Conditions (1), (2a) or (2b) is sufficient but not necessary. Figures 6.7, 6.8 and 6.9 below show that different results may occur if none of these conditions holds. In N of all these figures, since $^\bullet p_o \neq \{\varepsilon\}$ and $p_i^\bullet \neq \{\varepsilon\}$, neither Condition (1) nor Condition (2a) is satisfied; and, since $t_i^\bullet \neq t_o^\bullet$, Condition (2b) is not satisfied either. Hence, (N', M_0') cannot be concluded to preserve all these three properties of (N, M_0). In fact, in Figure 6.7, (N, M_0) is bounded and reversible but (N', M_0') is unbounded and not reversible. In Figure 6.8, (N, M_0) is live but (N', M_0') is not. After firing $t_1 t_2 t_3 t_i t_i$ in (N', M_0'), transition t is dead. In Figure 6.9, both (N, M_0) and (N', M_0') are live. Note that (N', M_0') is not shown in both Figures 6.8 and 6.9.

Theorem 6.3. *(property-preservation under Reduce-P-Path)*
Let (N, M_0) and (N', M_0') be the two Petri nets involved in Reduce-P-Path (Definition 6.2). Then, the following propositions hold:

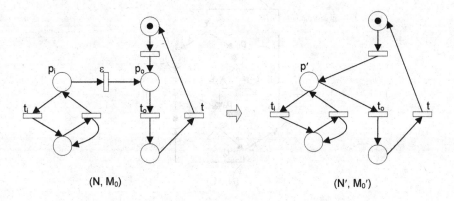

Fig. 6.7 (N, M_0) is bounded and reversible but (N, M_0') is unbounded and not reversible.

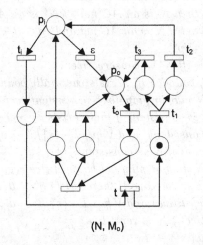

(N, M₀)

Fig. 6.8   A live (N, M_0) that becomes non-live after applying Reduce-P-Path.

1) N is an SM iff N' is an SM.

2) N is an MG iff N' is an MG.

3) (a) In general, N (resp., N') may or may not be an FC net even if N'
     (resp., N) is an FC net. In particular, if ($^\bullet(p_i^\bullet) = \{p_i\}$ or $^\bullet(p_o^\bullet) = \{p_o\}$) in N, then N is an FC net iff N' is an FC net.

   (b) N' is an AC net if N is an FC net.

4) (a) In general, N (resp., N') may or may not be an AC net even if N'

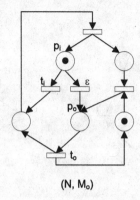

(N, M₀)

Fig. 6.9   A live (N, M_0) that is still live after applying Reduce-P-Path.

(resp., N) is an AC net. In particular, if ($^{\bullet}(p_i^{\bullet}) = \{p_i\}$ or $^{\bullet}(p_o^{\bullet}) = \{p_o\}$) in N, then N is an AC net iff N' is an AC net.

(b) *N' is an AC net if N is an AC net and $\forall p \in P - \{p_o, p_i\}$: $p^{\bullet} \cap p_o^{\bullet} \neq \phi$ implies $p^{\bullet} \subseteq p_o^{\bullet}$.*

5) *N is conservative iff N' is conservative.*

6) *N is structurally bounded iff N' is structurally bounded.*

7) (a) *In general, N may or may not be consistent even if N' is consistent. However, N is consistent if there exists $\beta' > 0$ such that $V'\beta' = 0$ (i.e., N' is consistent) and $V[p_i, T-\{\varepsilon\}]\beta' = a > 0$.*

(b) *N' is consistent if N is consistent.*

8) (a) *In general, N may or may not be repetitive even if N' is repetitive. However, N is repetitive if there exists $\beta' > 0$ such that $V'\beta' \geq 0$ (i.e, N' is repetitive) and $V[p_i, T-\{\varepsilon\}]\beta' = a > 0$.*

(b) *N' is repetitive if N is repetitive.*

9) $Rank(N) = Rank(N') + 1$.

10) *In general, $|C(N)| - 1 \leq |C(N')| \leq |C(N)|$. In particular, $|C(N)| = |C(N')| + 1$ if $p_i^{\bullet} = \{\varepsilon\}$.*

11) *Suppose $p_i^{\bullet} = \{\varepsilon\}$. N satisfies the RC-property iff N' satisfies the RC-property.*

12) (a) *In general, N may or may not have a minimal SM-cover even if N' has a minimal SM cover.*

(b) *N' has a minimal SM-cover if N has a minimal SM cover.*

13) (a) *Suppose D' is a siphon of N'. If $p' \notin D'$, then D' is also a siphon of N; if $p' \in D'$, then $D = D' - \{p'\} \cup \{p_i, p_o\}$ is a siphon of N.*

(b) *Suppose D is a siphon of N. If $p_o \in D$, then $D' = D - \{p_i, p_o\}$*

$\cup \{p'\}$ is a siphon of N'. If p_i, $p_o \notin D$, then $D' = D$ is a siphon of N'. If $p_o \notin D$, $p_i \in D$ and $(\,^\bullet p_o - \{\varepsilon\} \subseteq D^\bullet \cup p_o^\bullet)$ in N, then $D' = D - \{p_i\} \cup \{p'\}$ is a siphon of N'.

14) (a) Suppose S' is a trap of N'. If $p' \notin S'$, then S' is also a trap of N. Otherwise, $S = S' - \{p'\} \cup \{p_i, p_o\}$ is a trap of N.

(b) Suppose S is a trap of N. If $p_i \in S$, then $S' = S - \{p_i, p_o\} \cup \{p'\}$ is a trap of N'. If p_i, $p_o \notin S$, then $S' = S$ is a trap of N'. If $p_i \notin S$, $p_o \in S$ and $(p_i^\bullet - \{\varepsilon\} \subseteq \,^\bullet S \cup \,^\bullet p_i)$ in N, then $S' = S - \{p_o\} \cup \{p'\}$ is a trap of N'.

15) (a) Suppose N is a Petri net process such that the entry place $p_e \neq p_i$ and the exit place $p_x \neq p_o$. Then, N' is also a Petri net process and $Z(N') = Z(N)$.

(b) Suppose N' is a Petri net process such that $p_e \neq p'$, $p_x \neq p'$ and $M_0'(p') = 0$. Then, N is also a Petri net process and $Z(N) = Z(N')$.

16) (a) Suppose N is a Petri net process such that the entry place $p_e \neq p_i$ and the exit place $p_x \neq p_o$. Then, N' is a proper terminated Petri net process if N terminates properly.

(b) Suppose N' is a proper terminated Petri net process such that $p_e \neq p'$, $p_x \neq p'$ and $M_0'(p') = 0$. Then N may or may not terminate properly.

Proof.

1-2) According to the definition of Reduce-P-Path (Definition 6.2), SM and MG, it is obvious that N' is an SM (resp., MG) iff N is SM (resp., MG).

3) (a) It is obvious that N (resp., N') may or may not be an FC net even if N' (resp., N) is an FC net because p' in N' is obtained by merging p_i and p_o in N.

"$\Leftarrow$" For any $p, q \in P$ satisfying $(p^\bullet \cap q^\bullet \neq \phi)$ in N, consider three cases: (1) $p, q \in P - \{p_i, p_o\}$. Then, $p, q \in P' - \{p'\}$. Since N' is an FC net, $(p^\bullet = q^\bullet)$ in N' and thus in N. (2) $p = p_i$ or p_o and $q \in P - \{p_i, p_o\}$. Then $q \in P' - \{p'\}$. Since $(p^\bullet = q^\bullet)$ in N' and $(\,^\bullet(p^\bullet) = \{p\})$ in N, $(p^\bullet = q^\bullet)$ in N. (3) $p = p_i$ and $q = p_o$. By the definition of Reduce-P-Path, $p^\bullet \cap q^\bullet = \phi$, i.e., this case is impossible. Hence, N is an FC net if N' is an FC net and $(\,^\bullet(p_i^\bullet) = \{p_i\}$ or $^\bullet(p_o^\bullet) = \{p_o\})$ in N.

"$\Rightarrow$" Similar to the proof above, we can prove that N' is an FC net if N is an FC net and that $(\,^\bullet(p_i^\bullet) = \{p_i\}$ or $^\bullet(p_o^\bullet) = \{p_o\})$ in N.

(b) For any p, $q \in P'$ satisfying $(p^\bullet \cap q^\bullet \neq \phi)$ in N', consider two cases: (1) p, $q \in P' - \{p'\}$. Then, p, $q \in P - \{p_i, p_o\}$ in N. $(p^\bullet \cap q^\bullet \neq \phi)$ in N' and thus in N. Since N is an FC net, $p^\bullet = q^\bullet$ in N and thus in N'. (2) $p = p'$ and $q \in P' - \{p'\}$. Correspondingly, $(p = p_i$ and $q \in P - \{p_i, p_o\})$ or $(p = p_o$ and $q \in P - \{p_i, p_o\})$ in N. For the first case, since N is an FC net and $(\varepsilon \in p_i^\bullet$ but $\varepsilon \notin q^\bullet)$, the case is impossible. For the second case, $p_o^\bullet = q^\bullet$ in N implies $q^\bullet = p_o^\bullet \subseteq p'^\bullet$ in N'. Hence, N' is an AC net.

(4) (a) It is obvious that N (resp., N') may or may not be an AC net even if N' (resp., N) is an AC net.

"$\Leftarrow$"For any p, $q \in P$ satisfying $(p^\bullet \cap q^\bullet \neq \phi)$ in N, consider three cases: (1) p, $q \in P - \{p_i, p_o\}$. Then, p, $q \in P' - \{p'\}$. Since N' is an AC net, $(p^\bullet \subseteq q^\bullet$ or $q^\bullet \subseteq p^\bullet)$in N' and thus in N. (2) $p = p_i$ and $q \in P - \{p_o\}$. Since N' is an AC net and $(\varepsilon \in p_i^\bullet)$ in N, $(q^\bullet \subseteq p^\bullet)$ in N. (3) $p = p_o$ and $q \in P - \{p_i, p_o\}$. Since N' is an AC net, $(p^\bullet \subseteq q^\bullet$ or $q^\bullet \subseteq p^\bullet)$in N if $(^\bullet(p_i^\bullet) = \{p_i\}$ or $^\bullet(p_o^\bullet) = \{p_o\})$ in N. N is an AC net if N' is an AC net and $(^\bullet(p_i^\bullet) = \{p_i\}$ or $^\bullet(p_o^\bullet) = \{p_o\})$ in N.

"$\Rightarrow$"Similar to the proof above, we can proof N' is an AC net if N is an AC net and $(^\bullet(p_i^\bullet) = \{p_i\}$ or $^\bullet(p_o^\bullet) = \{p_o\})$ in N.

(b) For any p, $q \in P'$ satisfying $(p^\bullet \cap q^\bullet \neq \phi)$ in N'. Consider the following two cases: (1) p, $q \in P' - \{p'\}$. Then p, $q \in P - \{p_i, p_o\}$. Since $(p^\bullet \cap q^\bullet \neq \phi)$ in N', $(p^\bullet \cap q^\bullet \neq \phi)$ in N. If N is an AC net, then $(p^\bullet \subseteq q^\bullet$ or $q^\bullet \subseteq p^\bullet)$in N and thus $(p^\bullet \subseteq q^\bullet$ or $q^\bullet \subseteq p^\bullet)$in N'. (2) $p = p'$ and $q \in P' - \{p'\}$. Then $q \in P' - \{p_i, p_o\}$. Since $(p^\bullet \cap q^\bullet \neq \phi)$ in N', $(p^\bullet \cap q^\bullet \neq \phi)$ in N. If N is an AC net, then $(p^\bullet \subseteq q^\bullet$ or $q^\bullet \subseteq p^\bullet)$ in N. Since $(\varepsilon \in p_i^\bullet)$ in N and $\forall p \in P - \{p_o, p_i\}$: $p^\bullet \cap p_o^\bullet \neq \phi$ implies $(p^\bullet \subseteq p_o^\bullet)$ in N, $(q^\bullet \subseteq p^\bullet)$ in N'. Hence, N' is an AC net if N is an AC net and $\forall p \in P - \{p_o, p_i\}$: $p^\bullet \cap p_o^\bullet \neq \phi$ implies $(p^\bullet \subseteq p_o^\bullet)$ in N.

5) "$\Leftarrow$" (Figure 6.10) If N' is conservative, by Characterization 2.1, $\exists \alpha' = (x_2, x_3, \ldots, x_{|P|}) > 0$ such that $\alpha'V' = 0$. Let $\alpha = (x_2, x_2, x_3, \ldots, x_{|P|})$. Then, $\alpha > 0$ and $\alpha V = 0$. This means that N is conservative. "$\Rightarrow$" If N is conservative, by Characterization 2.1, $\exists \alpha = (x_1, x_2, x_3, \ldots, x_{|P|}) > 0$ such that $\alpha V = 0$. This leads to $x_1 = x_2$ and $x_1 V(p_i, T - \{\varepsilon\}) + x_2 V(p_o, T - \{\varepsilon\}) + \alpha[x_3, \ldots, x_{|P|}]V(P - \{p_i, p_o\}, T - \{\varepsilon\}) = 0$ by ignoring the first column of V. Let $\alpha' = (x_2, x_3, \ldots, x_{|P|})$. Then, $\alpha' > 0$ and the above equality is $\alpha'V' = 0$, i.e., N' is conservative.

6) Similar to the proof of 5).

7) (a) It is obvious that N may or may not be consistent even if N' is consistent. If there exists $\beta' > 0$ such that $V'\beta' = 0$ and $V[p_i, T-\{\varepsilon\}]\beta' = a > 0$, let $\beta = (a, \beta')$. Then $\beta > 0$ and $V\beta = 0$. By Characterization 2.3, N is consistent.

   (b) If N is consistent, then by Characterization 2.3, $\exists \beta = (y_1, y_2, \ldots, y_{|T|}) > 0$ such that $V\beta = 0$. The first, second and remaining rows of this set of equations are: $-y_1 + V[p_i, T - \{\varepsilon\}]\beta[y_2, \ldots, y_{|T|}] = 0$, $y_1 + V[p_o, T - \{\varepsilon\}]\beta[y_2, \ldots, y_{|T|}] = 0$ and $0 + V[P - \{p_i, p_o\}, T - \{\varepsilon\}]\beta[y_2, \ldots, y_{|T|}] = 0$, respectively. Let $\beta' = (y_2, \ldots, y_{|T|})$. Then, $\beta' > 0$ and $V'\beta' = ((V[p_i, T - \{\varepsilon\}] + V[p_o, T - \{\varepsilon\}])\beta', V[P - \{p_i, p_o\}, T - \{\varepsilon\}]\beta') = 0$. Hence, N' is consistent according to Characterization 2.3.

8) Similar to the proof of 7).

9) (Figure 6.10) Adding row 1 to row 2 in V changes the leftmost column to $(-1, 0, \ldots, 0)^T$. Then, all the entries in row 1 except the leftmost one can be reduced to zero by Gaussian column operations without affecting the other rows. This results in a diagonal matrix with diagonal submatrices -1 and V'. Hence, $\text{Rank}(N) = \text{Rank}(N') + 1$.

$$
\begin{array}{c}
\begin{array}{cc}
\quad\quad\quad\quad \varepsilon & \quad\quad\quad\quad T - \{\varepsilon\}
\end{array} \\
\begin{array}{c}
p_i \\
p_o \\
P - \{P_i, P_o\}
\end{array}
\left(
\begin{array}{cc}
-1 & V[p_i, T - T_\varepsilon] \\
1 & V[P_o, T - T_\varepsilon] \\
0 & V[P - \{p_i, p_o\}, T - \{\varepsilon\}]
\end{array}
\right)
\end{array}
$$

$$
\begin{array}{c}
T' = T - \{\varepsilon\} \\
\begin{array}{c}
p' \\
P' - \{p'\}
\end{array}
\left(
\begin{array}{c}
V[p_i, T - \{\varepsilon\}] + V[p_o, T - \{\varepsilon\}] \\
V[P - \{p_i, p_o\}, T - \{\varepsilon\}]
\end{array}
\right)
\end{array}
$$

Fig. 6.10   Matrices V of N and V' of N' (bottom).

10) On fusing p_i with p_o, the clusters $[p_i]$ and $[p_o]$ are merged into the cluster $[p']$ and no other clusters are created or eliminated. Hence, $|C(N)| - 1 \leq |C(N')| \leq |C(N)|$. If $p_i^\bullet = \{\varepsilon\}$, then $[p_i] \neq [p_o]$ and thus $|C(N')| = |C(N)| - 1$.

11) Together 9) and 10), it is known that if $p_i^\bullet = \{\varepsilon\}$, then N satisfies the RC-property iff N' satisfies the RC-property.

12) (a) It is obvious that N may or may not have an SM-cover even if N' has an SM-cover.

(b) Let K be a minimal SM-cover of N. Replace every $S \in K$ containing $\{p_i, \varepsilon, p_o\}$ by $S' = S \cup \{p'\} \cup \{(x, p')| (x, p_i) \in S\} \cup \{(p', x)| (p_o, x) \in S\} - \{p_i, \varepsilon, p_o\} - \{(x, p_i)| (x, p_i) \in S\} - \{(p_o, x)| (p_o, x) \in S\} - \{(p_i, \varepsilon), (\varepsilon, p_o)\}$. (Note that, being strongly connected, every SM-component containing ε must also contain both p_i and p_o) Obviously, all the transitions in S' has the same connectivity as in S. Hence, S' is an SM-component and the resulting K forms an SM-cover of N'.

13) (a) Suppose D' is a siphon of N'. If $p' \notin D'$, then it is obvious that D' is also a siphon of N. If $p' \in D'$, then let $D = D' - \{p'\} \cup \{p_i, p_o\}$. Since $(^\bullet D - \{\varepsilon\})$ in $N = {}^\bullet D'$ in $N' \subseteq (D')^\bullet$ in $N' = (D^\bullet - \{\varepsilon\})$ in N, $(^\bullet D \subseteq D^\bullet)$ in N, i.e., D is a siphon of N.

(b) Suppose D is a siphon of N. If $p_o \in D$, then $p_i \in D$. Let $D' = (D - \{p_i, p_o\}) \cup \{p'\}$. Then, ${}^\bullet(D')$ in $N' = {}^\bullet D$ in $N - \{\varepsilon\}$, $(D')^\bullet$ in $N' = D^\bullet$ in $N - \{\varepsilon\}$. Hence, $(^\bullet D' \subseteq (D')^\bullet)$ in N' provided that $(^\bullet D \subseteq D^\bullet)$ in N. If $p_o \notin D$, there are two subcases: 1) $p_i \notin D$. Let $D' = D$. Then $\varepsilon \notin {}^\bullet D$ in N, $\varepsilon \notin D^\bullet$ in N, $(D')^\bullet$ in $N' = D^\bullet$ in N and ${}^\bullet(D')$ in $N' = {}^\bullet D$ in N. Hence, $(^\bullet D' \subseteq (D')^\bullet)$ in N' provided that $(^\bullet D \subseteq D^\bullet)$ in N. 2) $p_i \in D$. Let $D' = (D - \{p_i\}) \cup \{p'\}$. Then, ${}^\bullet(D')$ in $N' = (^\bullet D$ in $N - \{\varepsilon\}) \cup \{^\bullet p_o\}$ and $(D')^\bullet$ in $N' = (D^\bullet$ in $N - \{\varepsilon\}) \cup \{p_o^\bullet\}$. Since $\varepsilon \notin {}^\bullet D$ and $\varepsilon \in D^\bullet$ in N, $(^\bullet D \subseteq D^\bullet)$ in N implies that $(^\bullet D \subseteq D^\bullet - \{\varepsilon\})$ in N. This means that $(^\bullet D' \subseteq (D')^\bullet)$ in N' provided that $(\{^\bullet p_o\} - \{\varepsilon\} \subseteq \{p_o^\bullet\})$ in N.

14) (a) Suppose S' is a trap of N'. If $p' \notin S'$, then it is obvious that S' is also a trap of N. If $p' \in S'$, then $(S^\bullet - \{\varepsilon\})$ in $N = (S')^\bullet$ in $N' \subseteq {}^\bullet(S')$ in $N' = (^\bullet S - \{\varepsilon\})$ in N. This means that S is a trap of N.

(b) Suppose S is a trap of N. If $p_i \in S$, then $p_o \in S$. Let $S' = (S - \{p_i, p_o\}) \cup \{p'\}$. Then, ${}^\bullet(S')$ in $N' = {}^\bullet S$ in $N - \{\varepsilon\}$, $(S')^\bullet$ in $N' = S^\bullet$ in $N - \{\varepsilon\}$. Hence, $((S')^\bullet \subseteq {}^\bullet(S'))$ in N' provided that $(S^\bullet \subseteq {}^\bullet S)$ in N. If $p_i \notin S$, there are two subcases: 1) $p_o \notin S$. Let $S' = S$. Then, $((S')^\bullet$ in $N' = S^\bullet$ in $N)$ and $(^\bullet(S')$ in $N' = {}^\bullet S$ in $N)$. Hence, $((S')^\bullet \subseteq {}^\bullet(S'))$ in N' provided that $(S^\bullet \subseteq {}^\bullet S)$ in N. 2) $p_o \in S$. Let $S' = (S - \{p_o\}) \cup \{p'\}$. Then, $((S')^\bullet$ in $N' = (S^\bullet$ in $N - \{\varepsilon\}) \cup \{p_i^\bullet\}$ in $N)$ and $(^\bullet(S')$ in $N' = (^\bullet S$ in $N - \{\varepsilon\}) \cup \{^\bullet p_i\}$ in $N)$. Since $((\varepsilon \in {}^\bullet S)$ in $N)$ and $((\varepsilon \notin S^\bullet)$ in $N)$, $((S')^\bullet \subseteq {}^\bullet(S'))$ in N' provided that $(\{p_i^\bullet\} - \{\varepsilon\} \subseteq \{^\bullet p_i\})$ in N.

15) (a) If N is a Petri net process, then $M_0(p_e) = 0$ by the definition of a

Petri net process (Definition 3.1). If the entry place $p_e \neq p_i$ and the exit place $p_x \neq p_o$, then it is obvious that N' is also a Petri net process and $Z(N') = |F'| - |P' \cup T'| + 2 = |F| - 2 - (|P| - 1) - (|T| - 1) + 2 = Z(N)$.

(b) If N' is a Petri net process such that $p_e \neq p'$, $p_x \neq p'$ and $M_0'(p') = 0$, then N is also a Petri net process and $Z(N) = |F| - |P \cup T| + 2 = |F'| + 2 - (|P'| + 1) - (|T'| + 1) + 2 = |F'| - |P' \cup T'| + 2 = Z(N')$.

16) By 15), N' (resp., N) is also a Petri net process if N (resp., N) is a Petri net process under the conditions. This proposition follows from the definitions of proper termination, we know that this result is obvious. $\square$

6.3   Reducing a Place-Bordered Subnet to a Place

This section studies an operator that reduces a subnet N_S within an ordinary Petri net to a single place. This is an extension of the cases studied in Sections 6.2 and 6.3 Conditions for the two-directianal preservation of many properties will be derived.

Definition 6.3. (Reduce-Subnet) (Figure 6.11)
Let $N_S = (P_S, T_S, F_S)$ be a place-bordered (i.e., $({}^\bullet T_S \cup T_S^\bullet) \cap (P - P_S) = \phi$) subnet of an ordinary Petri net $N = (P, T, F)$. Suppose there exists a transition set $T_I \subseteq T - T_S$ such that the subnet generated by P_S and $T_S \cup T_I$ forms a strongly connected SM in N. Reduce-Subnet reduces N_S to a single place p_s by transforming (N, M_0) to (N', M_0'), where $N' = (P', T', F', W')$, as follows:

$P' = P - P_S \cup \{p_s\}$
$T' = T - T_S$
$F' = F - F_S - (\{(t,p), (p,t) | t \in T - T_S, p \in P_S\} \cap F) \cup \{(t, p_s) | t \in T - T_S, t^\bullet \cap P_S \neq \phi\} \cup \{(p_s, t) | t \in T - T_S, {}^\bullet t \cap P_S \neq \phi\}$
$\forall t \in T_A = {}^\bullet P_S \cup P_S^\bullet - T_S - T_I$, $W'(p_s, t) = |{}^\bullet t \cap P_S|$ and $W'(t, p_s) =$
The weight of every other edge in F' remains 1; and $M_0'(p) = M_0(p)$ for $p \in P' - \{p_s\}$ and $M_0'(p_s) = \sum_{p \in P_S} M_0(p)$.

Example 6.2. (Figure 6.11)
In N, the place-bordered subnet N_S lies within the ellipse, $T_I = \{t_i, t_j\}$ and $T_A = \{t_a, t_b\}$. N_S and T_I generate a strongly connected SM. In N', the transitions t_i and t_j form self-loops with p_s and the weight of the arc (t_a, p_s) is 2 because $|t_a^\bullet \cap P_S| = 2$ in N.

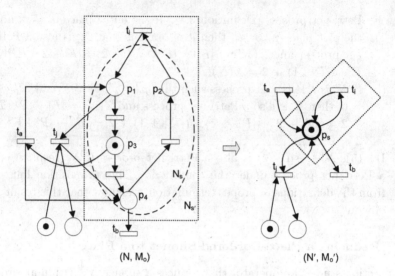

Fig. 6.11   The Petri nets before and after applying Reduce-Subnet.

Some of the property-preservation results described later depend on the following condition.

Definition 6.4. Internal Path Condition (IPC)) (Figure 6.12)
Consider the subnet $N_S = (P_S, T_S, F_S)$ of (N, M_0) in Reduce-Subnet.
$\forall x \in ((T_I \cup T_A)^\bullet \cap P_S) \cup \{p \in P_S | M_0(p) > 0\}$, $\forall y \in {}^\bullet(T_I \cup T_A) \cap P_S$, there exists a path ρ that starts at x and ends at y such that ρ lies entirely within N_S.

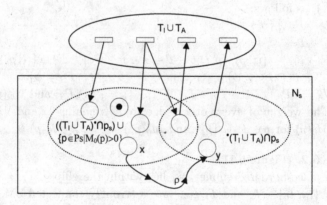

Fig. 6.12   Internal Path Condition (x is any place in $((T_I \cup T_A)^\bullet \cap P_S) \cup p \in P_S | M_0(p) > 0$, y is any place in ${}^\bullet(T_I \cup T_A) \cap P_S$ and γ is a path from x to y within N_S.

Discussion on Reduce-Subnet and Internal Path Condition:

(a) The subnet N_S can model subsystems with multiple entries and exits. For example, the subsystem in N_S of Figure 6.11 has two entries $\{p_1, p_4\}$ and four exits $\{p_1, p_2, p_3, p_4\}$.

(b) Reduce-Subnet takes two practical requirements into consideration in its formulation. It is flexible enough so that it can have a large scope of application. First, the subnet N_S to be reduced is 'almost' a strongly connected SM, meaning that N_S itself is not required to be a strongly connected SM but must become so when combined with a set T_I of 'included' transitions. Second, the choice of T_I and T_A is not unique. For example, for the same N_S in Figure 6.11, one can choose $T_I = \{t_i\}$ and $T_A = \{t_a, t_b, t_j\}$. In particular, if N_S is a strongly connected SM itself, T_I may even be empty. For example, for the subnet $N_{S'}$ within the dotted square in Figure 6.11, one can let $T_I = \phi$ and $T_A = \{t_a, t_b, t_j\}$.

(c) In the definition of Internal Path Condition (Definition 6.4), $(T_I \cup T_A)^\bullet \cap P_S$ denote the set of 'entry' places, $\{p \in P_S | M_0(p) > 0\}$ is the set of initially marked places and $^\bullet(T_I \cup T_A) \cap P_S$ denote the set of 'exit' places. IPC does not require N_S to be strongly connected. It only requires that, within N_S, there exists a directed path from every 'entry' place or initially marked place to every 'exit' place. Obviously, strongly connected N_S automatically satisfies IPC. A weaker condition IPC allows more flexibility in selecting a subnet for reductionduring actual application.

Definition 6.5. (mappings arising from Reduce-Subnet)
Let (N', M_0') and (N, M_0) be the two Petri nets defined in Reduce-Subnet (Definition 6.5). For a firing sequence σ and a marking M of N where $M_0[N, \sigma\rangle M$, the mappings of σ and M from N onto N' are defined as follows:

$f: T^* \to T'^*$:

$f(\lambda) = \lambda$ where λ is the null sequence
$f(\sigma t) = f(\sigma)$
$\quad\ \ = f(\sigma)t$

M' is the restriction of M from P to P':

$M'(P) = M(p)$ if $p \in P' - \{p_s\}$
$M'(p_s) = \sum_{p \in p_s} M(p)$

For the rest of this section, the notations N', M_0', T_A, T_I, σ, $f(\sigma)$ and M' have the same meanings as defined in Definition 6.3 and Definition 6.5. For simplification, the symbols σ and σ' are also used to denote the set of transitions in the sequences σ and σ', respectively.

Lemmas 6.2 and 6.3 below describe the relationships of σ and M with their mappings $f(\sigma)$ and M'.

Lemma 6.2. *Let (N', M_0') and (N, M_0) be the two Petri nets defined in Definition 6.3. For any sequence σ and marking M of N where $M_0[N, \sigma\rangle M$, their mappings $f(\sigma)$ and M' satisfy $M_0'[N', f(\sigma)\rangle M'$.*

Proof. (by induction on the length of σ) For $\sigma = \lambda$, obviously $M = M_0$. By Definition 6.5, $f(\sigma) = \lambda \in L(N', M_0')$ and $M' = M_0'$. Hence, $M_0'[N', f(\sigma)\rangle M'$. Next, assume the proposition holds for every μ, where $|\mu| \leq n$. That is, for such μ and marking M_1, $M_0[N, \mu\rangle M_1$ implies that $M_0'[N', f(\mu)\rangle M_1'$. Let $\sigma = \mu t \in L(N, M_0)$ and marking M satisfy $M_0[N, \mu\rangle M_1[N, t\rangle M$. $\qquad\square$

To show that $f(\sigma)$ is firable, two cases should be considered:

(a) If $t \in T_S$, then $f(\sigma) = f(\mu) \in L(N', M_0')$ by Definition 6.5 and the assumption above.

(b) If $t \in T - T_S$, then $f(\sigma) = f(\mu)t$. By the above assumption $M_0'[N', f(\mu)\rangle M_1'$, it is sufficient to show that t is firable at M_1'. By Definition 6.5, $M_1'(p) = M_1(p)$ for $p \in P - p_s$ and $M_1'(p_s) = \sum_{p \in P_s} M_1(p)$. Since t is firable at M_1 in $N, M_1(p) \geq W(p, t), \forall p \in {}^\bullet t$ in N. Hence, $M_1'(p) \geq W'(p, t), \forall p \in {}^\bullet t$ in N'. This implies that t is firable at M_1' in N'. Hence, $f(\sigma)$ is firable in N'.

Next, consider two cases of t:

(a) If $t \in T_S$, then $f(\sigma) = f(\mu)$. $M'(p_s) = M(P_S) = M_1(P_S) = M_1'(p_s)$ and $M'(p) = M(p) = M_1(p) = M_1'(p)$ for $p \in P - P_S$. Hence, $M' = M_1'$ and $M_0'[N', f(\sigma)\rangle M'$.

(b) If $t \in T - T_S$, then $f(\sigma) = f(\mu)t$. $M'(p) = M(p) = M_1(p)\pm 1 = M_1'(p)\pm 1$ for $p \in {}^\bullet t \cup t^\bullet - p_s$, $M'(p_s) = \sum_{p \in P_s} M(p) = \sum_{p \in (P_s - {}^\bullet t \cup t^\bullet)} M_1(p) + \sum_{p \in (P_s \cap {}^\bullet t \cup t^\bullet)} (M_1(p)\pm 1) = M_1'(p_s) + W(t, p_s) - W(p_s, t)$, and $M'(p) = M(p) = M_1(p) = M_1'(p)$ for $p \in P' - ({}^\bullet t \cup t^\bullet)$. Hence, $M_1'[N', t\rangle M'$ and $M_0'[N', f(\sigma)\rangle M'$.

Lemma 6.3. *Suppose N satisfies the Internal Path Condition in Reduce-Subnet. Then, for any sequence σ' and marking M' of N', where $M_0'[N',*

$\sigma'\rangle M'$, there exist sequence σ and marking M of N such that $\sigma' = f(\sigma)$ and $M_0[N, \sigma\rangle M$, where M' is the mapping of M.

Proof. For any sequence σ' and marking M' of N', where $M_0'[N', \sigma'\rangle M'$, suppose $\sigma' = \sigma_1' t_1 \sigma_2' t_2, \ldots l_i \sigma_i' t_i, \ldots l_j \ldots t_k \sigma_d'$, where every $\sigma_i' \cap (p_s^\bullet \cup {}^\bullet p_s) = \phi$, every $t_i \in {}^\bullet p_s$ and every $l_i \in p_s^\bullet$. Then $t_i, l_i \in T_I \cup T_A$ in N and the Internal Path Condition implies that, $\forall x \in (P_S \cap t_i^\bullet) \cup \{p \in P_S | M_0(p) > 0\}$ and $\forall y \in {}^\bullet l_i \cap P_S, i = 1, 2 \ldots$, there exists a path ρ_i from x to y such that ρ_i lies entirely within N_S in N. Since these paths lie within a connected SM, they are all firable sequences at M_1 if $M_1(p_r) > 0$, where $p_r \in \rho_i$ and $M_1 \in R(N, M_0)$, and every firing will preserve the number of tokens within P_S. In particular, some of them are fired so that every place $y \in {}^\bullet l_i$, $i = 1$, $2, \ldots$, gets a token eventually in N. Let σ_i be such a firing sequence if a sequence in ρ_i is fired and a null sequence otherwise. Hence, the sequence $\sigma = \sigma_1' t_1 \sigma_2' t_2, \ldots \sigma_i l_i \sigma_i' t_i, \ldots \sigma_j l_j \ldots t_k \sigma_d'$ is firable and $f(\sigma) = \sigma'$. Suppose $M_0[N, \sigma\rangle M_2$. Since firing σ_i preserves the number of tokens within P_S, $M_2(P_S) = M'(p_s)$ and $M_2(p) = M'(p)$ for $p \in P - P_S$. Hence $M_2 = M$. $\square$

Example 6.3. (Figure 6.13)
For the place-bordered subnet N_S within the square, $P_S = \{p_a, p_1, p_2\}$, $T_S = \{t_1, t_2, t_5\}$, $T_I = \{t_3, t_4\}$, $T_A = \{t_a\}$, $(T_I \cup T_A)^\bullet \cap P_S = \{p_a, p_2\}$, ${}^\bullet(T_I \cup T_A) \cap P_S = \{p_1, p_2\}$. For N_S, $\{p_a, p_2\}$ is the set of 'entry' places and $\{p_1, p_2\}$ the set of 'exit' places. Since the paths $p_a t_1 p_1$, $p_a t_2 p_2$, $p_2 t_5 p_a t_1 p_1$ and p_2 all lie within N_S, IPC is satisfied. By Lemma 6.3, for the firing sequences $\sigma_1' = t_a t_3$, $\sigma_2' = t_a t_4$, $\sigma_3' = t_a t_3 t_4$ and $\sigma_4' = t_a t_4 t_3$ of (N', M_0'), the firing sequences $\sigma_1 = t_a t_1 t_3$, $\sigma_2 = t_a t_2 t_4$, $\sigma_3 = t_a t_1 t_3 t_4$ and $\sigma_4 = t_a t_2 t_4 t_1 t_3$ of (N, M_0) satisfy $\sigma_i' = f(\sigma_i)$, $i = 1, 2, 3, 4$. The other sequences are just subsequences of these ones.

Without IPC, Lemma 6.3 may be invalid. For example, in Figure 6.13, suppose p_1 is initially marked. Since there is no path from p_1 to p_2 within N_S, IPC is not satisfied. For the firing sequence $\sigma' = t_4$ in N', there is no firing sequence σ in N, such that $\sigma' = f(\sigma)$. Similarly, for the subnet N_S within the ellipse in Figure 6.14, $P_S = \{p_1, p_2, p_3\}$, $T_S = \{t_1, t_2, t_3\}$, $T_I = \{t_4\}$, $T_A = \{t_a, t_b, t_c\}$, $(T_I \cup T_A)^\bullet \cap P_S = \{p_1, p_2, p_3\}$, ${}^\bullet(T_I \cup T_A) \cap P_S = \{p_1, p_2, p_3\}$. Since there is no directed path from p_1 to p_3 within N_S, IPC is not satisfied. For the firing sequence $\sigma' = t_a t_a t_c$ in N', there is no firing sequence σ in N, such that $\sigma' = f(\sigma)$.

Theorem 6.4. *(preservation of boundedness, liveness and reversibility under Reduce-Subnet)*

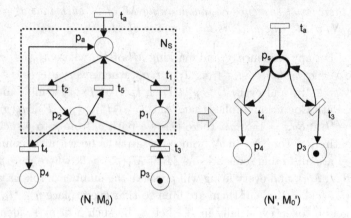

Fig. 6.13   An example for explaining Lemma 6.3.

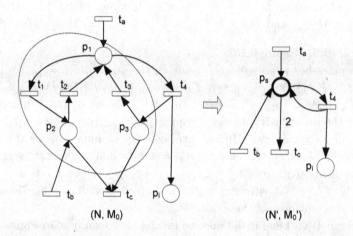

Fig. 6.14   An example for explaining the Internal Path Condition of Lemma 6.3.

Let (N, M_0) and (N', M_0') be the two Petri nets defined in Reduce-Subnet (Definition 6.3). Then, the following propositions hold:

1) If (N', M_0') is bounded, then (N, M_0) is bounded.

2) If (N, M_0) is bounded and either the Internal Path Condition is satisfied or $^\bullet T_I \cap (P - P_S) = \phi$ (i.e., T_I has no input places outside the subnet N_S), then (N', M_0') is bounded.

3) Suppose the Internal Path Condition is satisfied. Then, (N, M_0) is

live (resp., reversible, terminates properly) iff (N', M_0') is live (resp., reversible, terminates properly).

4) *If (N', M_0') is live (resp., reversible, terminate properly) and ${}^\bullet T_I \cap (P - P_S) = \phi$, then (N, M_0) is live (resp., reversible, terminates properly).*

Proof.

1) Suppose (N', M_0') is bounded. For every reachable marking M of (N, M_0), by Lemma 6.2, its mapping M' is a reachable marking of (N', M_0'). By Definition 6.5, for every place p in N, $M(p)$ is bounded by $M'(p)$ or $M'(p_s)$. Hence, (N, M_0) is bounded.

2) Suppose (N, M_0) is bounded. For every $M' \in R(N', M_0')$ obtained by firing $\sigma' = \sigma_1' t_1 \sigma_2' t_2, \ldots l_i \sigma_i' t_i, \ldots l_j \ldots t_k \sigma_d'$ in N', where every $\sigma_i' \cap (p_s^\bullet \cup {}^\bullet p_s) = \phi$, every $t_i \in {}^\bullet p_s$ and every $l_i \in p_s^\bullet$, if the Internal Path Condition is satisfied, by Lemma 6.3, $M'(p)$ is obviously bounded by $M(p)$ or by $M(P_S)$, where M and M' satisfy the mapping relation in Definition 6.5. If the Internal Path Condition is not satisfied, $\forall x \in (P_S \cap t_i^\bullet) \cup \{p \in P_S | M_0(p) > 0\}$ and $\forall y \in {}^\bullet l_i \cap P_S, i = 1, 2 \ldots$, there exists a path ρ_i from x to y such that ρ_i lies entirely within the strongly connected SM generated by N_S and T_I in N. By the assumption ${}^\bullet T_I \cap (P - P_S) = \phi$, the paths are all firable sequences at M_1 if $M_1(p_r) > 0$, where $p_r \in \rho_i$ and $M_1 \in R(N, M_0)$. Some of them are fired so that every place $y \in {}^\bullet l_i$, $i = 1, 2, \ldots$ gets a token eventually in N. Let σ_i be such a firing sequence if a sequence in ρ_i is fired and a null sequence otherwise. Hence, the sequence $\sigma = \sigma_1' t_1 \sigma_2' t_2 \ldots \sigma_i l_i \sigma_i' t_i, \ldots \sigma_j l_j \ldots t_k \sigma_d'$ is firable and $f(\sigma) = \sigma_1' t_1 \sigma_2' t_2, \ldots s_i l_i \sigma_i' t_i, \ldots s_j l_j \ldots t_k \sigma_d'$, where every $s_i \in T_I$. Suppose $M_0[N, \sigma\rangle M_2$. Then, $M_2(P_S) = M'(p_s), M_2(p) = M'(p)$ for $p \notin s_i^\bullet$ and $M_2(p) \geq M'(p)$ for $p \in s_i^\bullet$. Hence, (N', M_0') is bounded.

3) ($\Rightarrow$) Suppose (N, M_0) is live. For every $\sigma' \in L(N', M_0')$ and every $t \in T'$, since the Internal Path Condition is satisfied, by Lemma 6.3, there exists $\sigma \in L(N, M_0)$ such that $\sigma' = f(\sigma)$. Since (N, M_0) is live, there exists $\sigma_1 \in T^*$ such that $\sigma\sigma_1 t \in L(N, M_0)$. By Definition 6.5 and Lemma 6.2, $f(\sigma\sigma_1 t) = \sigma'\sigma_1' t \in L(N', M_0')$. Hence, (N', M_0') is live. ($\Leftarrow$) Suppose (N', M_0') is live. For every $\sigma \in L(N, M_0)$ and every $t \in T$, by Lemma 6.2, there exists $\sigma' \in L(N', M_0')$ such that $\sigma' = f(\sigma)$. Since (N', M_0') is live, there exists $\sigma_1' \in T'^*$ such that $\sigma'\sigma_1' t \in L(N', M_0')$. Since the Internal Path Condition is satisfied, by Lemma 6.3, there exists $\sigma_1 \in T*$ such that $\sigma\sigma_1 t \in L(N, M_0)$ and $f(\sigma\sigma_1 t) = \sigma'\sigma_1' t$. Hence, (N, M_0) is live. The proofs for reversibility and proper termination are similar.

4) Suppose (N', M_0') is live. For every $\sigma \in L(N, M_0)$ and every $t \in T$, by Lemma 6.2, there exists $\sigma' \in L(N', M_0')$ such that $f(\sigma) = \sigma'$. *Case 1.* If $t \in T - T_S$ in N, then $t \in T'$ in N'. Since (N', M_0') is live, there exists $\sigma_1' = \mu_1 a_1 a_2 \ldots \mu_2\, b_1 b_2 \ldots \mu_3 a_3 a_4 \ldots \mu_4 b_3 b_4 \ldots \quad \mu_d t \in T'^*$ such that $\sigma'\sigma_1' \in L(N', M_0')$, where every $\mu_i \cap (^{\bullet}p_s \cup p_s^{\bullet}) = \phi$, $a_i \in {}^{\bullet}p_s = {}^{\bullet}P_S - T_S$ and $b_j \in p_s^{\bullet} = P_S^{\bullet} - T_S$. By the proof of proposition (2), for such a firable sequence in N', there exists a firable sequence $\sigma_1 = \mu_1 a_1 a_2 \ldots \mu_2 \gamma_1 \gamma_2 \ldots b_1 b_2 \ldots \mu_3 a_3 a_4 \ldots \mu_4 \gamma_3 \gamma_4 \ldots b_3 b_4 \ldots \quad \mu_d t$ in N, where each ρ_i is the firable sequence such that firing $\gamma_1 \gamma_2 \gamma_3 \gamma_4 \ldots$ can guarantee that b_j, $j = 1, 2, 3, \ldots$, is still firable in N. Hence, $\sigma\sigma_1 \in L(N, M_0)$. *Case 2.* If $t \in T_S$, then $t \notin T'$ in N'. By Case 1, every $t_j \in T - T_S$ is live. Hence, there exists $\sigma_1 \in T^*$ such that $\sigma\sigma_1 t_j \in L(N, M_0)$, where $t_j \in {}^{\bullet}P_S - T_S$. Let $M_0[N, \sigma\sigma_1 t_j\rangle M_j$. Then, $M_j(P_S) \geq 1$. Since P_S and $T_S \cup T_I$ generate a strongly connected SM and T_I has no input places in $P - P_S$, every $t \in T_S$ is obviously a potentially firable transition in (N, M_j). Hence, (N, M_0) is live. The proofs for reversibility and proper termination are similar.

$\square$

Example 6.4. (Figure 6.15)

In Theorem 6.4, without the condition ${}^{\bullet}T_I \cap (P - P_S) = \phi$, (N, M_0) may be non-live although (N', M_0') is live and (N', M_0') may be unbounded although (N, M_0) is bounded. For example, for the subnet N_S (within the ellipse) of N and $T_I = \{t\}$, ${}^{\bullet}T_I \cap (P - P_S) = \{p\} \neq \phi$, (N', M_0') is live but (N, M_0) is not because N is dead after firing $t_a t_1 t t_2 t_3 t_1$. (N, M_0) is bounded but (N', M_0') is not because p in N' becomes unbounded if the sequence $t_a t_b$ is fired repeatedly.

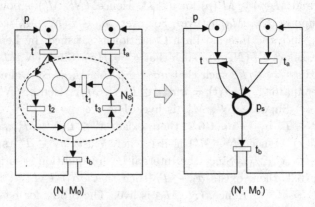

(N, M₀)   (N', M₀')

Fig. 6.15   An example for illustrating the role of ${}^{\bullet}T_I \cap (P - P_S) = \emptyset$ in Theorem 6.4.

Corollary 6.1. *Let (N, M_0) and (N', M_0') be the two Petri nets defined in Reduce-Subnet (Definition 6.3). Suppose the place-bordered subnet N_S is itself a strongly connected state machine. Then, (N, M_0) is live (resp., bounded, reversible) iff (N', M_0') is live (resp., bounded, reversible).*

Proof. In this case, let $T_I = \phi$. It is then always true that ${}^\bullet T_I \cap (P - P_S) = \phi$ and the Internal Path Condition is satisfied. $\qquad\square$

Theorem 6.5. *(property-preservation under Reduce-Subnet)*
Let (N, M_0) and (N', M_0') be the two Petri nets defined in Reduce-Subnet (Definition 6.3). Then, the following propositions hold:

1) *N is an SM iff N' is an SM*

2) (a) *In general, N' (resp., N) may or may not be an MG even if N (resp., N') is an MG.*

   (b) *N is an MG if N' is an MG and $P_S \cup {}^\bullet P_S \cup P_S^\bullet$ generates an MG in N.*

   (c) *N' is an MG if N is an MG and $|{}^\bullet P_S \cap (T - T_S)| = |P_S^\bullet \cap (T - T_S)| = 1$.*

3) (a) *Suppose N' is an FC. Then, N is an FC net provided that $\forall p_1, p_2 \in P_S$, $p_i^\bullet \cap (T - T_S) \neq \phi$ implies ($p_i^\bullet \cap T_S = \phi$ and $p_1^\bullet = p_2^\bullet$ in N), where $i = 1, 2$.*

   (b) *If N is an FC net, then N' is an AC net.*

4) (a) *Suppose N' is an AC net. Then, N is an AC net provided that $\forall p_1, p_2 \in P_S$, $p_i^\bullet \cap (T - T_S) \neq \phi$ implies ($p_i^\bullet \cap T_S = \phi$ and ($p_1^\bullet \subseteq p_2^\bullet$ or $p_2^\bullet \subseteq p_1^\bullet$) in N), where $i = 1, 2$.*

   (b) *If N is an AC net, then N' is an AC net provided that for any $p \in P - P_S$ and $q \in P_S$ satisfying ($p^\bullet \cap q^\bullet \neq \phi$) in N, $p^\bullet \subseteq q^\bullet$.*

5) (a) *If N' is conservative (resp., structurally bounded), then N is conservative (resp., structurally bounded).*

   (b) *If there exists a vector $\alpha = (I_x, \alpha_1) > 0$, where $I_x = (x, x, \ldots, x)$ is a $|P_S|$-vector, such that $\alpha V = 0$ (resp., $\alpha V \leq 0$), where V is the incidence matrix of N, then N' is conservative (resp., structurally bounded).*

6) (a) *If N' is consistent (resp., repetitive), then N is consistent (resp., repetitive) provided that there exist $\beta_1, \beta' > 0$ such that $V'\beta' = 0$ (resp., $V'\beta' \geq 0$) and $\forall p \in P_S$, $V[p, T_S]\beta_1 + V[p, T - \{T_S\}]\beta' = 0$ (resp., $V[p, T_S]\beta_1 + V[p, T - \{T_S\}]\beta' \geq 0$).*

   (b) *If N is consistent (resp., repetitive), then N' is consistent (resp., repetitive).*

7) $Rank(N) = Rank(N') + Rank(N_S)$.

8) $|C(N)| + 1 - |P_S| \leq |C(N')| \leq |C(N)|$.

9) N *(resp., N') may not satisfy the RC-property even if N'(resp., N) satisfies the RC-property.*

10) N' *has a minimal SM-cover iff N has a minimal SM cover.*

11) *For a set of places $D \subseteq P$ of N, suppose either $D \cap P_S = \phi$ or $P_S \subseteq D$. Then, D is a siphon (resp., trap) of N iff D or $(D - P_S) \cup \{p_s\}$ is a siphon (resp., trap) of N'.*

12) N *is a process such that $\{p_e, p_x\} \cap P_S = \phi$ iff N' is also a process and $Z(N') = Z(N) + |P_S| - |T_S| - 1$.*

Proof.

1) Since the connectivity degree of every transition is not changed in Reduce-Subnet, property SM is preserved.

2) (a) Since $|{}^\bullet p_s| = |{}^\bullet P_S \cap (T - T_S)|$ and $|p_s^\bullet| = |P_S^\bullet \cap (T - T_S)|$, N' may not be an MG even if N is an MG. Conversely, the subnet N_S may not be an MG even if N' is an MG. Hence N is not an MG.

   (b) Since $P_S \cup {}^\bullet P_S \cup P_S^\bullet$ generates an MG in N and the connectivity degrees for other places in $P - P_S$ are not changed, N is an MG if N' is an MG.

   (c) By the proof of (a) above, $|{}^\bullet p_s| = |p_s^\bullet| = 1$ and the connectivity degrees of other places in P' are not changed. Hence N' is an MG if N is.

3) (a) For any $p, q \in P$ satisfying $(p^\bullet \cap q^\bullet \neq \phi)$ in N, consider three cases: (1) $p, q \in P_S$. Since N_S is an SM, $p^\bullet \cap q^\bullet \subseteq T - T_S$. By the condition, $(p^\bullet \cup q^\bullet) \cap T_S = \phi$ and $p^\bullet = q^\bullet$ in N. Hence, N is an FC net. (2) $p \in P_S$ and $q \in P - P_S$. Since N_S is not transition-bordered, $p^\bullet \cap q^\bullet \subseteq T - T_S$. By the condition, $p^\bullet \cap T_S = \phi$ in N. Since N' is an FC net and $((p^\bullet \cap q^\bullet \neq \phi)$ in N implies $(p_s^\bullet \cap q^\bullet \neq \phi)$ in N'), $p_s^\bullet = q^\bullet$ in N'. Correspondingly, $p^\bullet = q^\bullet$ in N. Hence, N is an FC net. (3) $p, q \in P - P_S$. Since N' is an FC net, $p^\bullet = q^\bullet$ in N' and thus in N. Hence, N is an FC net.

   (b) Suppose N is an FC net. For any $p, q \in P'$ satisfying $(p^\bullet \cap q^\bullet \neq \phi)$ in N', consider two cases: (1) $p, q \in P' - \{p_s\}$. Then, $p, q \in P - P_S$ in N. $p^\bullet = q^\bullet$ in N and thus in N'. Hence, N' is an FC net. (2) $p = p_s$ and $q \in P' - \{p_s\}$. Then, there exists a place $r \in P_S$ such that $r^\bullet \cap q^\bullet \neq \phi$ in N. Since N is an FC net, $r^\bullet = q^\bullet$ in N. Since $r^\bullet \subseteq P_S^\bullet \cap (T - T_S)$, this implies $q^\bullet \subseteq p_s^\bullet$ in N'. Hence, N' is an AC net.

4) Similar to 3).

5) (a) (Figure 6.16) By the definition of Reduce-Subnet, $\sum_{p \in Ps} V(pT_S)$ $= 0$. If N' is conservative, by Characterization 2.1, $\exists \alpha' = (x, \alpha_1) >$ 0 such that $\alpha'V' = 0$. That is, $x \sum_{p \in Ps} V(p, T - T_S) + \alpha_1 V_1$ $= 0$. Let $\alpha = (I_x, \alpha_1)$, where $I_x = (x, x, \ldots, x)$ is a $|P_S|$-vector. Then, $\alpha > 0$ and $\alpha V = (x \sum_{p \in Ps} V(p, T - T_S) + \alpha_1 V_1,$ $x \sum_{p \in Ps} V(p, T_S)) = 0$. This means that N is conservative. Similarly, structural boundedness can be proved.

$$
\begin{array}{c}
 \quad T_S \qquad\qquad T - T_S \\
\begin{array}{c} P_S \\ P - P_S \end{array}
\left(
\begin{array}{cc}
V[P_S, T_S] & V[P_S, T - T_S] \\
0 & V[P - P_S, T - T_S]
\end{array}
\right)
\end{array}
$$

$$
\begin{array}{c}
 \qquad T' = T - T_S \\
\begin{array}{c} p_s \\ P' - \{p_s\} \end{array}
\left(
\begin{array}{c}
\sum_{p \in PS} V[p, T - -T_S] \\
V[P - P_S, T - T_S]
\end{array}
\right)
\end{array}
$$

Fig. 6.16   Matrices V of N and V' of N' (bottom).

(b) Suppose there exists a vector $\alpha = (I_x, \alpha_1) > 0$, where $I_x = (x, x, \ldots, x)$ is a $|P_S|$-vector, such that $\alpha V = 0$. Then, $x \sum_{p \in Ps} V(p, T - -T_S) + \alpha_1 V_1 = 0$. Let $\alpha' = (x, \alpha_1)$. Then, $\alpha' > 0$ and the above equality is $\alpha'V' = 0$, i.e., N' is conservative. Similarly, structural boundedness can be proved.

6) (a) It is obvious that N may or may not be consistent even if N' is consistent. If there exists β_1, $\beta' > 0$ such that $V'\beta' = 0$ and $\forall p \in P_S$, $V[p, T_S]\beta_1 + V[p, T - \{T_S\}]\beta' = 0$, let $\beta = (\beta_1, \beta')$. Then $\beta > 0$ and $V\beta = 0$. By Characterization 2.3, N is consistent. Proof for repetitiveness is similar.

(b) If N is consistent, then by Characterization 2.3, $\exists \beta = (\beta_1, \beta_2) \geq 1$ such that $V\beta = 0$. That is, $V(P_S, T_S)\beta_1 + V(P_S, T - T_S)\beta_2 = 0$ and $V(P_S, T - T_S)\beta_2 = 0$. For the first equality, adding all the equations together results in the following equality: $\sum_{p \in Ps} V(p, T - T_S)\beta_2 = 0$. Hence, $V'\beta_2 = 0$ and N' is consistent. Similarly, repetitiveness can be proved.

7) (Figure 6.16) For the matrix V, adding the first $|P_S| - 1$ rows to row $|P_S|$ results in a diagonal matrix such that V' and V_S are its diagonal submatrixes, where V_S is the matrix for subnet N_S after applying a Gaussion transformation. Hence, $\text{Rank}(N) = \text{Rank}(N') + \text{Rank}(N_S)$.

8) After applying Reduce-Subnet, the number of clusters decreases by at most $|P_S| - 1$. Hence, $|C(N)| + 1 - |P_S| \leq |C(N')| \leq |C(N)|$.

9) By 7) and 8), it is trivial.

10) By the definition of Reduce-Subnet, N is strongly connected iff so is N'.

  (a) Suppose N' has a minimal SM-cover. For those SM-components N_i such that $p_s \notin N_i$, N_i are also the SM-components of N. For those SM-component N_k such that $p_s \in N_k$, we construct the new SM-components N'_k for N by replacing place p_s in N_k with subnet N_S and keep those edges between p_s and $^\bullet p_s \cup p_s^\bullet$. Hence $N_i \cup N'_k$ is the minimal SM-cover of N.

  (b) Suppose N has a minimal SM-cover. Consider two kinds of SM-components of N: One is those SM-components N_i such that $N_S \cap N_i = \phi$. Such N_i are also SM-components of N'. Another is those components N_k such that $N_S \cap N_k \neq \phi$. We construct a SM-component N'_k for N' based on these N_k as follows: The first step is to union all those N_k, this results in a subnet $N_l = \cup_k N_k$. Since $\cup_{i,k}(N_i \cup N_k)$ is the minimal SM-cover of N, the subnet N_S must be covered by N_l. The second step is to replace the subnet in N_l with the place p_s in N'. Hence, $(\cup_i N_i) \cup N_l)$ is a minimal SM-cover of N'.

11) By the assumption, for any place set $D \subseteq P$ of N, either $D \cap P_S = \phi$ or $P_S \subseteq D$. If $D \cap P_S = \phi$, then $(^\bullet D \text{ in } N) = (^\bullet D \text{ in } N')$ and $(D^\bullet \text{ in } N) = (D^\bullet \text{ in } N')$. Hence, D is a siphon of N iff D is a siphon of N'. If $P_S \subseteq D$, then $(^\bullet D - T_S \text{ in } N) = (^\bullet((D - P_S) \cup \{p_s\}) \text{ in } N')$ and $(D^\bullet - T_S \text{ in } N) = (((D - P_S) \cup \{p_s\})^\bullet \text{ in } N')$. Hence, D is a siphon of N iff $(D - P_S) \cup \{p_s\}$ is a siphon of N'. The proof for a trap is similar.

12) By the definition of Reduce-Subnet, it is obvious that N is a process such that $\{p_e, p_x\} \cap P_S = \phi$ iff N' is a process. $Z(N') = |F'| - |P' \cup T'| + 2 = |F| - 2|T_S| - (|P| - |P_S| + 1 + |T| - |T_S|) + 2 = Z(N) + |P_S| - |T_S| - 1$. $\square$

6.4   Application Examples

Example 6.5. (Figures 6.17, 6.18, 6.19)
This example applies Reduce-T-Path and Reduce-P-Path for simplifying the model of the Automatic Virtual Environment (CAVE) system. CAVE

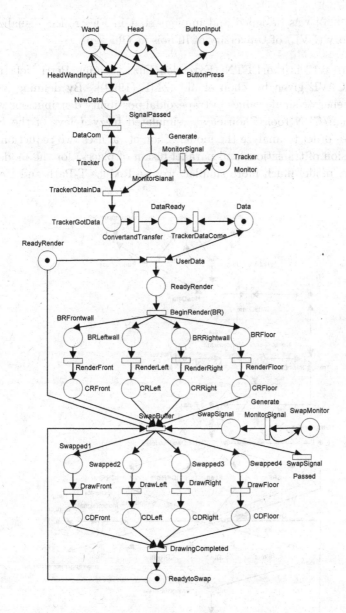

Fig. 6.17   Zhou's Petri net model for the CAVE. (BR means Begin Render, CR means Complete Render and CD means Complete Drawing.)

[PAP (1996)] was designed and implemented in Electronec Visualization Laboratory (EVL) of University of Illinois at Chicago.

Figure 6.17 is the EFTN (Extended Fuzzy-Timing Petri Net) model for the CAVE given by Zhou et al. [ZMF (1999)]. By defining a function to generate single values in trapezoidal possibility distributions within the Design/CPN tool, Zhou et al. simulated fuzzy delays of the EFTN model. In order to analyze the model, Zhou et. al uses two reduction rules (post fusion of transitions and parallel fusion of places) for the model. To make the model much more simpler, we use Reduce-T-Path and Reduce-

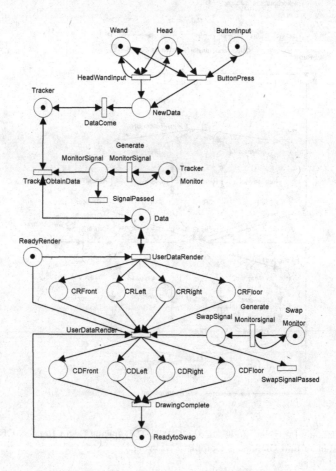

Fig. 6.18   The simplified Petri net model after applying Reduce-P-Path and Reduce-T-Path.

P-Path to the original model first, and then use Zhou's two reduction rules.

Figure 6.18 is the resulting EFTN model after using Reduce-P-Path and Reduce-T-Path several times. Then, the paths 1) *TrackerObtainData-TrackerGotData-CnvertandTransfor-DataReady-TrackerDataCome,* 2) *UseData-ReadyRender-BeginRender,* 3) *BRFrontwall-RenderFrond-CRFrond,* 4) *BRLeftwall-RenderLeft-CRLeft,* 5) *BRRightwall-RenderRight-CRRight,* 6) *BRFlorr-RenderFlorr-CRFlorr,* 7) *Swapped1-DrawFront-CDFrong,* 8) *Swapped2-DrawLeft-CDLeft,* 9) *Swapped3-DrawRight-CDRight* and 10) *Swapped4-DrawFloor-CDFloor* in Figure 6.17 are reduced to single transitions *TrackerObtainData, UseDataRender* and single places *CRFron, CRLeft, CRRight, CRFloor, CDFront, CDLeft, CDRight* and *CDFloor* in FigureFig6.18, respectively. In each reduction step, for the reduced path, either its start transition (or place) has only one output or its end transition (or place) has only one input. Hence, the reduced model can preserve all the nineteen properties described in Theorems 6.1, 6.2 and 6.3. Since our reduction rules fit for general Petri nets, in the EFTN model, the rules can preserve liveness, boundedness, deadlock and timing properties.

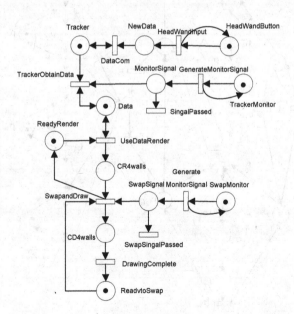

Fig. 6.19   Reduced Petri net model of the CAVE.

Figure 6.19 shows the results of using Zhou's two fusion rules for our simplified model. These two rules can preserve safeness, deadlock and timing properties. Hence, by analyzing the simplest model Figure 6.19, we can get the same simulation results.

Example 6.6. (Figure 6.20 and Table 6.1)

This example illustrates the application of the three operators to verifying a manufacturing system. Although not shown in this chapter, the three operators can also be applied to system specification since they are two-way preserving operators. Chapter 8 and Chapter 9 illustrate the applications to the specification and verification in multi-agent systems and manufacturing systems, respectively.

This manufacturing system consists of three processes: two workstations WS_1 and WS_2 (on the left of Figure 6.20) for assembly work and one machining center (on the right of Figure 6.20) for machining. WS_1 and WS_2 share robot R_2 between themselves and share Robot R_1 with the

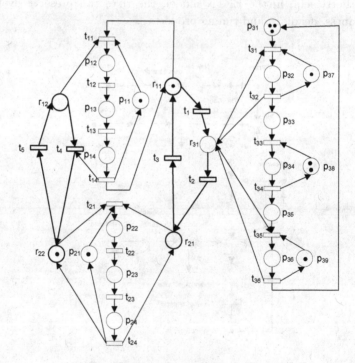

Fig. 6.20   The original system (N, M_0) with two resource subnets (boldfaced).

Table 6.1   The Legend for Figure 6.20.

Places	Transitions
r_{i1} ($i = 1, 2, 3$): Robot R_1 is available	t_{11}: starts acquiring R_1 and R_2
r_{i2}($i = 1, 2$): Robot R_2 is available	t_{12}: starts first step of assembling at WS_1
p_{11}: WS_1 requests R_1 and R_2	t_{13}: starts final step of assembling at WS_1
p_{12}: WS_1 acquires R_1 and R_2	t_{14}: completes assembling at WS_1
p_{13}: first step of assembling at WS_1	t_{21}: starts acquiring R_1 and R_2
p_{14}: final step of assembling at WS_1	t_{22}: starts first step of assembling at WS_2
p_{21}: WS_2 requests R_1 and R_2	t_{23}: starts final step of assembling at WS_2
p_{22}: WS_2 acquires R_1 and R_2	t_{24}: completes assembling at WS_2
p_{23}: first step of assembling at WS_2	t_{31}: starts activity p_{32}
p_{24}: final step of assembling at WS_2	t_{32}: completes activity p_{32} and start activity p_{33}
p_{31}: pallets are available	
p_{32}: machine M_1 loads, fixtures and processes a palleted raw part	t_{33}: completes p_{33} and start the storage activity p_{34}
p_{33}: R_1 unloads an intermediate part to the buffer	t_{34}: completes p_{34} and start activity p_{35}
p_{34}: buffer B stores an intermediate part	t_{35}: completes p_{35} and start p_{36}
p_{35}: machine M_2 loads and processes an intermediate part	t_{36}: completes p_{36}
p_{36}: R_1 unloads a final product from M_2, defixtures and returns the pallet	t_i ($i = 1, 2, 3, 4, 5$): intermediate processing on a robot before passing it from one process to another.
p_{37}: M_1 is available	
p_{38}: B is available	
p_{39}: M_2 is available	

machining center. (Note: The left and right components of Figure 6.20 are extracted from [ZD (1993)]. Zhou et al. used them just for explaining the concepts of mutual exclusions in resource sharing. They are combined here with some modifications to create an example for illustrating our results.) The system runs as follows:

A. In the machining center, parts are machined first by machine M_1 and then by machine M_2. Each part is automatically fixtured to a pallet and loaded into the machine. After processing, robot R_1 unloads the intermediate part from M_1 into buffer B. At machining station M_2, intermediate parts are automatically loaded into M_2 and processed. When M_2 finishes processing a part, the robot R_1 unloads the final product, defixtures it and returns the fixture to M_1.

B. When either WS_1 or WS_2 is ready to execute the assembly task, it requests both robots R_1 and R_2 and acquires them if they are available. When a workstation starts an assembly task, it cannot be interrupted until it is completed. When WS_1 (WS_2) completes, it releases both robots.

C. It is assumed that input parts are always available to be fixtured and that the finished products are removed.

For the specification of the manufacturing system with Petri nets, each operation process is abstracted to a single place and each transition represents the start or/and completion of a process. This is similar to the literature [VAL (1990); ZD (1993)]. For handling resources sharing problems, this chapter has some differences with the literature.

Unlike other systems where the robots are shared among the processes without any modifications, this example considers a more general situation where a robot has to go through some intermediate treatments (e.g., cleaning the oil left from the previous process, adding some parts needed by the next calling process, etc.) when being passed from one process to another. Hence, for the Petri net specification of the system (Figure 6.20), each resource is originally represented by a set of places (called *resource-places* hereafter), one in each of the parts it is involved in. The resource-places may form a connected subnet whose transitions represent the intermediate processes. For example, Robot R_1 is shared by the three parts (WS$_1$, WS$_2$ and the machining center) and needs some intermediate treatments when being passed from one part to another. In Figure 6.20, places r_{11}, r_{21} and r_{31} are resource-places representing robot R_1. Transitions t_1, t_2 and t_3 represent the intermediate processes. The resource-places and these transitions generate a connected subnet (one of the bold-faced subnets in Figure 6.20).

Verification on the final system (Figure 6.20) proceeds in three steps:

Step 1. (N, M_0) (Figure 6.20) is transformed to (N_1, M_1) (Figure 6.21) by using Reduce-Subnet.

In Figure 6.20, the two bold-faced subnets N_{S_1} (generated by $\{r_{11}, r_{21}, r_{31}, t_1, t_2, t_3\}$) and N_{S_2} (generated by $\{r_{12}, r_{22}, t_4, t_5\}$) are strongly connected SMs. By setting $T_{I1} = T_{I2} = \phi$, $T_{A1} = \{t_{11}, t_{14}, t_{21}, t_{24}, t_{32}, t_{33}, t_{35}, t_{36}\}$ and $T_{A2} = \{t_{11}, t_{14}, t_{21}, t_{24}\}$, Reduce-Subnet reduces N_{S_1} and N_{S_2} to places r_1 and r_2, respectively, resulting in (N_1, M_1) (Figure 6.21). By Corollary 6.1, (N, M_0) is live, bounded and reversible iff (N_1, M_1) is.

Step 2. (N_1, M_1) (Figure 6.21) is transformed to (N_2, M_2) (Figure 6.22) by using Reduce-T-Path.

In (N_1, M_1), the transition-bordered paths $s_1 = t_{11}p_{12}t_{12}p_{13}t_{13}p_{14}t_{14}$, $s_2 = t_{21}p_{22}t_{22}p_{23}t_{23}p_{24}t_{24}$, $s_3 = t_{31}p_{32}t_{32}$, $s_4 = t_{33}p_{34}t_{34}$ and $s_5 = t_{35}p_{36}t_{36}$ satisfy the conditions in Theorem 6.1(16b)(17b)(18b). e.g., $|t_{11}^{\bullet}| = |^{\bullet}t_{14}| = 1$ for path s_1 and $|t_{33}^{\bullet}| = |^{\bullet}t_{34}| = 1$ for path s_4. Reduce-T-Path reduces

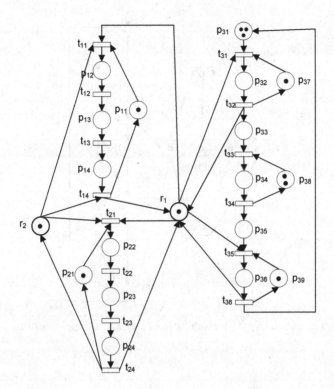

Fig. 6.21   The system (N_1, M_1), resulting from (N, M_0) by applying Reduce-Subnet.

the five paths to transitions s_1, s_2, s_3, s_4, and s_5, respectively, resulting in (N_2, M_2) (Figure 6.22). By Theorem 6.1, (N_1, M_1) is live, bounded and reversible iff (N_2, M_2) is live, bounded and reversible.

Step 3. Deleting all the places p in (N_2, M_2) (Figure 6.22)) satisfying $^\bullet p = p^\bullet$ and $M_2(p) > 0$ results in Petri net (N_3, M_3) (Figure 6.23).

Since those marked places p_{11}, p_{21}, p_{37}, p_{38}, p_{39}, r_1, r_2 consist self-loops with their associated transitons in (N_2, M_2), deleting them and their associated arcs will not affect the firing sequences and token distribution. Hence, (N_2, M_2) is live, bounded and reversible iff (N_3, M_3) is live, bounded and reversible.

Hence, the complex manufacturing model (N, M_0) (Figure 6.20) is live, bounded and reversible if and only if so is (N_3, M_3) (Figure 6.23). Since (N_3, M_3) is an initially marked cycle together with two independently transitions, it is obviously live, bounded and reversible [MUR (1989)].

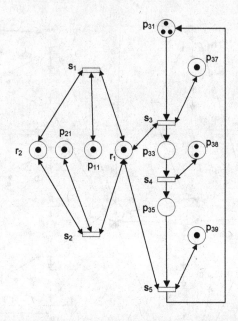

Fig. 6.22   The system (N_2, M_2), resulting from (N_1, M_1) by applying Reduce-T-Path.

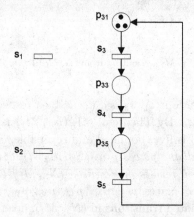

Fig. 6.23   The system (N_3, M_3), resulting from (N_2, M_2) by deleting self-loops.

Hence, the manufacturing system (N, M_0) (Figure 6.20) is live, bounded and reversible.

Based on Petri nets, this chapter has made the following contributions towards solving the resource-sharing and subsystem abstraction problems in system design:

A. *Enhancing the capability for modeling* – In the literature, these problems are described in quite a straightforward manner as exemplified by Chu's AMGs and Zhou's sequential and parallel mutual exclusions. Also, the systems involved are modeled mostly as an SM or MG. This chapter formulates the problems as subnet-reducing operators. Three operators are proposed so that a designer has a lot of flexibility in selecting an appropriate operator for specifying the resources, the system, the subsystems and the problems under investigation. In particular, a resource is now allowed to receive intermediate processing when switching from one user to another.

B. *Formalizing the property-preserving approach for verification* – In the literature, very little has been devoted to the development of formal verification techniques specifically for the resource sharing and system abstraction problems. Usually, just general techniques were suggested. Based on the subnet-reducing operators, this chapter proposes a property-preserving approach for verification. For each of the three operators, conditions are imposed on the structure of the subnets to be reduced so that various properties of the net will be preserved.

C. Besides their applications to system design, the results obtained in this chapter also enrich the theory for property-preserving transformations in Petri nets.

Chapter 7

Place-Merging Operators for Resource Sharing

This chapter presents some more operators, namely, place merging, in PPPA for handling resource-sharing problems in system design. As a result, these operators together with the other operators presented in the previous chapters form a complete algebra PPPA.

Resource sharing is a hard problem arising frequently in system design. We handle this problem by applying place-merging operators. Our approach starts by designing correct components while temporarily ignoring the resource sharing issue. Then, each component will have a *resource-place* representing a duplicate copy of the resource. If these resources are shared, these components are integrated by merging the identical resource places.

In order to formulate the problem, this chapter presents a rather extensive property-preserving methodology for verification and design. Based on the conditions of the shared resources, four classes of place-merging operators are investigated. For each class, conditions and special types of Petri nets are proposed for ensuring the preservation of certain properties. More details follow.

Operator 1. Merging an arbitrary set of places: This is the most general case where several sets of places Q_i are each merged to a single place. No restrictions are assumed on Q_i. Conditions have been obtained for preserving many properties.

Operator 2. Merging a set of non-neighboring places: This is a special case of Operator 1 wherein no two places of every Q_i are connected to a common transition. Various conditions for preserving liveness, boundedness and MST-property etc. are obtained for general Petri nets, free-choice nets and augmented marked graphs.

Operator 3. Composition via merging pairs of places: This is a special case of Operator 2 wherein the net consists of two disconnected components and each Q_i consists exactly of two places, one from each of these components. It is shown that those results for Operators 1 and 2 will hold under less restrictive conditions.

Operator 4. Composition via merging two single places: This is a special case of Operator 3, for which there is only one merge set with two places, one from each of the two disconnected components. It is shown that those results for Operators 1 to 3 will hold under even less restrictive conditions. In addition, the Rank Theorem will be preserved.

From the viewpoint of software transformation, place merging is in fact a kind of the reduction operators. In the case that the resources are shared with independent components, place merging can also be regarded as a kind of the composition operators. With respect to our approach for component-based system design, similar to refinements and reductions, our place-merging techniques also preserve two features of the original Petri net: 1) It is still an element of PPPA so that it can be applied again within the algebra. 2) It preserves the correctness of the original Petri net process. To preserve these two features, we propose conditions on the place merging techniques for the preservation of nineteen properties.

The rest of this chapter presents the technical details of our results as follows. Section 7.1, 7.2, 7.3, 7.4 present the four operators and their preservation of nineteen properties, respectively. In Section 7.5, the above results are applied to two areas: software agent systems and manufacturing systems. Some conclusive remarks are given at last. The results in the chapter are mainly extracted from [HJC (2005); JHC (2005, 2008)]

7.1   Merging Arbitrary Sets of Places

The operator presented in this section merges several arbitrary sets of places of a net each into a single place. This operator is very general. The given net is an ordinary Petri net possibly composed of disconnected components and the resulting net may have self-loops and weights bigger than 1. The merged sets are arbitrary in the sense that no special restrictions are imposed on their places. Conditions will be proposed for preserving the following properties: siphons, traps, ST-property, MST-property, P-invariant, conservativeness, structural boundedness, T-invariant, consistency, repetitiveness and strong connectedness.

Definition 7.1. (Merge-A-Place)

Suppose $N = (P_0 \cup Q_1 \cup \ldots \cup Q_k, T, F)$ is an ordinary net satisfying the condition: $\forall i, j \in \{1, 2, \ldots, k\}$ and $i \neq j$, $P_0 \cap Q_i = \phi$ and $Q_i \cap Q_j = \phi$. Let (N', M_0'), where $N' = (P_0 \cup Q_0, T', F', W')$ and $Q_0 = \{q_1, q_2, \ldots, q_k\}$, be obtained from (N, M_0) by merging the places of each Q_i into q_i as follows:

$T' = T$; F' is obtained from F by replacing each set of arcs of the form $\{(t, p) | p \in Q_i\}$ with (t, q_i) (resp., $\{(p, t) | p \in Q_i\}$ with (q_i, t)).

$\forall q_i \in Q_0$ and $\forall t \in T'$: $W'(q_i, t) = (|Q_i \cap {}^\bullet t|$ in N) and $W'(t, q_i) = (|Q_i \cap t^\bullet|$ in N), where $i \in \{1, 2, \ldots, k\}$. The weights of the other arcs in F' are 1.

$$M_0'(p) = \begin{cases} M_0(p) & p \in P_0 \\ max_{p \in Q_i}\{M_0(p)\} & p = q_i \in Q_0, i = 1, 2, \ldots, k \end{cases}$$

Definition 7.2. (reduction-set-dependent ST-property (resp., reduction-set-dependent MST-property)

N (resp., (N, M_0)) is said to satisfy the reduction-set-dependent ST-property (resp., reduction-set-dependent MST-property) if every siphon D of N satisfies the condition: If $D \subseteq P_0$, then D contains a trap S (resp., a trap S marked by M_0). If $Q_i \subseteq D$ for some i, then there exists a trap S (resp., a trap S marked by M_0) within D such that either $S \subseteq P_0$ or $Q_i \subseteq S$.

Note that, unlike the ordinary ST-property, the above property does not require every siphon to contain a trap but only those siphons that either contain the entire Q_i or are totally contained within P_0 to have a trap that either contains Q_i or is contained within P_0.

Theorem 7.1. *(preservation of siphons, traps, ST-property and MST-property for general Petri nets)*

Let (N', M_0') and (N, M_0) be defined in MERGE-A-PLACE, where $N = (P_0 \cup Q_1 \cup \ldots \cup Q_k, T, F)$ and $N' = (P_0 \cup Q_0, T', F', W')$. Then, the following propositions hold:

1) *For any $D' \subseteq P_0 \cup Q_0$, let $D = \cup\{Q_i | \exists q_i \in D' \cap Q_0\} \cup (D' \cap P_0)$. Then, D' is a siphon (resp., trap) of N' iff D is a siphon (resp., trap) of N.*

2) *Suppose N satisfies the reduction-set-dependent ST-property (resp., (N, M_0) satisfies reduction-set-dependent MST-property). Then, N' satisfies ST-property (resp., (N', M_0') satisfies MST-property).*

Proof.

1) According to MERGE-A-PLACE, we have: $D \cap P_0 = D' \cap P_0$ and $((Q_i \subseteq D)$ in $N)$ iff $((q_i \in D' \cap Q_0)$ in $N')$. Also, since no transitions have been eliminated, $\forall p \in D \cap P_0$, $(p^{\bullet}$ in $N) = (p^{\bullet}$ in $N')$ and $({}^{\bullet}p$ in $N) = ({}^{\bullet}p$ in $N')$. $\forall q_i \in D'$, $(q_i^{\bullet}$ in $N') = (Q_i^{\bullet}$ in $N)$ and $({}^{\bullet}q_i$ in $N') = ({}^{\bullet}Q_i$ in $N)$. Hence, $({}^{\bullet}D$ in $N) = ((\cup^{\bullet}\{Q_i|Q_i \subseteq D\} \cup {}^{\bullet}(D \cap P_0))$ in $N) = ((\cup^{\bullet}\{q_i|\exists q_i \in D' \cap Q_0\} \cup {}^{\bullet}(D' \cap P_0))$ in $N') = ({}^{\bullet}D'$ in $N')$ and $(D^{\bullet}$ in $N) = ((\cup \{Q_i|\exists q_i \in D' \cap Q_0\}^{\bullet}$ in $N) \cup (D' \cap P_0)^{\bullet}$ in $N') = (D'^{\bullet}$ in $N')$. This implies that D' is a siphon (trap) of N' iff D is a siphon (trap) of N.

2) For an arbitrary non-empty siphon D' of N', let $D = \cup\{Q_i|\exists q_i \in D' \cap Q_0\} \cup (D' \cap P_0)$. By Proposition 1), D is a siphon of N. If $D' \cap Q_0 = \phi$, then $D \subseteq P_0$ and thus contains at least one trap S (resp., one trap S marked by M_0) of N and S is also a trap of N'. If $q_i \in D'$, then $Q_i \subseteq D$. By the assumption, D contains a trap S_i (resp., a trap S_i marked by M_0) such that either $S_i \subseteq P_0$ or $Q_i \subseteq S$. Let S be the union of these traps. Then, S is a trap of N. Let $S' = \cup\{q_i|Q_i \subseteq S\} \cup (S \cap P_0)$. Since $S \cap P_0 = S' \cap P_0$, $S = \cup\{Q_i|\exists q_i \in S' \cap Q_0\} \cup (S' \cap P_0) \subseteq D$. Hence, $S' \subseteq D'$. By Proposition (1), S' is a trap of N'. Hence, N' satisfies ST-property (resp., MST-property). $\square$

In general, structural boundedness and conservativeness may not be preserved under MERGE-A-PLACE. For example, the net N in Figure 7.1(a) is structurally bounded but N' in Figure 7.1(b) becomes unbounded by firing t_1 repeatedly. Theorems 7.2 and 7.3 below present the conditions under which these and many other properties are preserved.

Theorem 7.2. *(preservation of P-invariant, conservativeness and structural boundedness for general Petri nets)*
Let N' and N be defined in MERGE-A-PLACE. Let $\alpha = (\alpha_0, \alpha_1, \ldots, \alpha_k)$, where $\alpha_0 = (a_1, a_2, \ldots, a_{|P0|})$ and each α_i is a $|Q_i|$-vector of the form $(d_i, d_i, \ldots, d_i)$, $i = 1, 2, \ldots, k$. Let $\alpha' = (\alpha_0, d_1, d_2, \ldots, d_k)$. *(Note: The orders of the elements of the vectors α and α' conform with the row order of the incidence matrices of N and N', respectively.)* Then, the following propositions hold for such α and α':

1) α' is a P-invariant of N' iff α is a P-invariant of N.
2) If there exists $\alpha > 0$ such that $\alpha V = 0$ (resp., $\alpha V \leq 0$) for N *(Note: This condition requires N to be more than being conservative because*

α has to be in a special pattern.), then N' is conservative (resp., structurally bounded). N is conservative (resp., structurally bounded) if N' is conservative (resp., structurally bounded).

Proof. The incidence matrices V and V' of N and N' have the following forms:

$$
V = \begin{array}{c} \\ P_0 \\ Q_1 \\ Q_2 \\ \cdots \\ Q_k \end{array}
\begin{pmatrix}
\overset{t_1}{V_{01}} & \overset{t_2}{V_{02}} & \cdots & \overset{t_m}{V_{0m}} \\
V_{11} & V_{12} & \cdots & V_{1m} \\
V_{21} & V_{22} & \cdots & V_{2m} \\
\cdots & \cdots & \cdots & \cdots \\
V_{k1} & V_{k2} & \cdots & V_{km}
\end{pmatrix}
$$

and

$$
V' = \begin{array}{c} \\ P_0 \\ q_1 \\ q_2 \\ \cdots \\ q_k \end{array}
\begin{pmatrix}
\overset{t_1}{V_{01}} & \overset{t_2}{V_{02}} & \cdots & \overset{t_m}{V_{0m}} \\
u_{11} & u_{12} & \cdots & u_{1m} \\
u_{21} & u_{22} & \cdots & u_{2m} \\
\cdots & \cdots & \cdots & \cdots \\
u_{k1} & u_{k2} & \cdots & u_{km}
\end{pmatrix}
$$

where

$$
V_{ij} = \begin{array}{c} q_{i1} \\ q_{i2} \\ \cdots \\ q_{i|Q_i|} \end{array}
\begin{pmatrix}
\overset{t_j}{V_{i1,j}} \\
V_{i2,j} \\
\cdots \\
V_{i|Q_i|,j}
\end{pmatrix}
$$

$i = 1, \ldots, k$, and $j = 1, \ldots, m$.

1) According to MERGE-A-PLACE, we have

$$
u_{ij} = \sum_{l=1}^{|Q_i|} v_{il,j} \text{ for } i = 1, \ldots, k \text{ and } j = 1, \ldots, m. \tag{7.1}
$$

Then, it follows from the definitions of α and α' that $\alpha' \geq 0$ and $\alpha'V' = 0$ iff $\alpha \geq 0$ and $\alpha V = 0$. Hence, α' is a P-invariant of N' iff α is a P-invariant of N.

2) If $\exists \alpha \geq 0$ such that $\alpha V = 0$ (resp., $\alpha V \leq 0$). By Equality (7.1), $\alpha'V' = 0$ (resp., $\alpha'V' \leq 0$), i.e., N' is conservative (resp., structurally bounded). If N' is conservative (resp., structurally bounded), then $\exists \alpha' = (a_1, a_2, \ldots, a_{|P0|}, a_{|Q1|}, a_{|Q2|}, \ldots, a_{|Qk|}) > 0$ such that $\alpha'V' = 0$ (resp., $\alpha'V' \leq 0$). Hence, $\alpha V = 0$ (resp., $\alpha V \leq 0$), i.e., N is conservative (resp., structurally bounded). $\qquad\square$

Theorem 7.3. *(preservation of T-invariant, consistency, repetitiveness and strong connectedness for general Petri nets)*
Let N' and N be defined in MERGE-A-PLACE. Then, the following propositions hold:

1) *A T-invariant of N is also a T-invariant of N'.*
2) *N' is consistent (resp., repetitive, strongly connected) if N is consistent (resp., repetitive, strongly connected).*

Proof.

1) If $\beta = (b_1, b_2, \ldots, b_m)$ is a T-invariant of N, then $V\beta = 0$. By Equality 7.1, we have $b_1 u_{i1} + b_2 u_{i2} + \ldots + b_m u_m = \sum_{l=1}^{|Q_i|} \sum_{j=1}^{m} b_j v_{il,j} = 0$ for $i = 1$, $2, \ldots, k$. This implies that $V'\beta = 0$, i.e., β is a T-invariant of N'.
2) Similar to the proof of Proposition (1), we can prove that N' preserves consistency and repetitiveness. It is obvious that N' is strongly connected if N is so.

$\square$

Example 7.1. (Illustration for Theorems 7.1, 7.2 and 7.3)
N (Figure 7.1(a)) is transformed by applying MERGE-A-PLACE to N', N'' and N'''. For N' (Figure 7.1(b)), $Q_1 = \{p_1, p_2\}$. Since $\{q_1, p_4, p_5\}$ is a siphon of N', by Theorem 7.1(1), $\{p_1, p_2, p_4, p_5\}$ is a siphon of N. Similarly, for N'' (Figure 7.1(c)), $Q_1 = \{p_2, p_5\}$. Since $\{q_1, p_1, p_3\}$ is a trap of N'', $\{p_1, p_2, p_3, p_5\}$ is a trap of N.

This example also shows that, in Theorem 7.1(2), the ordinary ST-property is not a sufficient condition for N' to satisfy the ST-property. For example, N' and N'' (Figures 7.1(b)(c)) do not satisfy ST-property even N does. It requires N to satisfy the reduction-set-dependent ST-property. For N''' (Figure 7.1(d)), $Q_1 = \{p_4, p_5\}$ and $P_0 = \{p_1, p_2, p_3\}$. For N, the siphon $\{p_1, p_2, p_3\}$ that does not contain Q_1 is a trap itself, while those siphons $\{p_1, p_4, p_5\}$, $\{p_1, p_2, p_4, p_5\}$ and $\{p_1, p_2, p_3, p_4, p_5\}$ that contain Q_1 all contain the trap $\{p_1, p_4, p_5\}$ that contains Q_1. That is, N satisfies reduction-set-dependent ST-property with respect to Q_1. By Theorem 7.1(2), N''' satisfies ST-property. Since $(2, 1, 1, 1, 1)$ is a positive P-invariant of N, by Theorem 7.2, $(2, 1, 1, 1)$ is also a positive P-invariant of N'' and N'''. This means that N, N' and N''' are conservative (resp., structurally bounded). Moreover, since $(1, 1, 1, 1)$ is a positive T-invariant of N, N is consistent and repetitive. By Theorem 7.3, N', N'' and N''' are also consistent and repetitive.

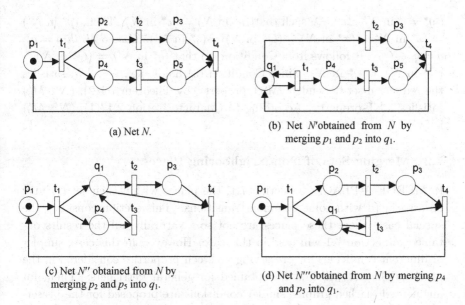

(a) Net N.

(b) Net N' obtained from N by merging p_1 and p_2 into q_1.

(c) Net N'' obtained from N by merging p_2 and p_5 into q_1.

(d) Net N''' obtained from N by merging p_4 and p_5 into q_1.

Fig. 7.1   An example for illustrating Theorems 7.1, 7.2 and 7.3.

The following Theorem 7.4 shows that MERGE-A-PLACE preserves liveness when applied on a free-choice net.

Theorem 7.4. *Let (N, M_0) be an ordinary FC net satisfying the conditions described in MERGE-A-PLACE and the following conditions: (1) $\forall p_1, p_2 \in Q_i$, where $i = 1, 2, \ldots, k$: $p_1^\bullet \cap p_2^\bullet = \phi$. (2) $\forall i, j \in \{1, 2, \ldots, k\}$, if $Q_i^\bullet \cap Q_j^\bullet \neq \phi$, then either $Q_i^\bullet \subseteq Q_j^\bullet$ or $Q_j^\bullet \subseteq Q_i^\bullet$. (3) (N, M_0) satisfies the reduction-set-dependent MST-property. Then, the net (N', M_0') obtained from (N, M_0) by MERGE-A-PLACE is a live AC net satisfying MST-property.*

Proof.   Condition (1) states that all weights of those arcs from places to transitions in N' are 1. Note that, $\forall p \in P_0$, $(p^\bullet$ in $N) = (p^\bullet$ in $N')$ and that, $\forall q_i \in Q_0$, $(Q_i^\bullet$ in $N) = (q_i^\bullet$ in $N')$. We shall first show that N' is an AC net. In N', $\forall p, q \in P_0 \cup Q_0$, where $p^\bullet \cap q^\bullet \neq \phi$ in N', consider the following three cases. *Case 1 (both $p, q \in P_0$):* p and q are in N and their pre-sets and post-sets remain unchanged in N'. Since N is an FC net, $p^\bullet = q^\bullet$ in N and, thus, also in N'. *Case 2 ($p \in P_0$ but $q = q_i \in Q_0$):* Since N is FC and $p^\bullet$

$\cap q_i^\bullet \neq \phi$ in N', $\exists x \in Q_i$ such that $(p^\bullet$ in $N) = (x^\bullet$ in $N)$. Then, $(p^\bullet$ in $N')$ $= (p^\bullet$ in $N) = (x^\bullet$ in $N) \subseteq (Q_i^\bullet$ in $N) = (q_i^\bullet$ in $N')$. *Case 3* (*both $p = q_j$, $q = q_i \in Q_0$*): It follows from Condition (2) that $(q_j^\bullet$ in $N') = (Q_j^\bullet$ in $N) \subseteq (Q_i^\bullet$ in $N) = (q_i^\bullet$ in $N)$. Since Condition (3) holds, i.e., (N, M_0) satisfies the reduction-set-dependent MST-property, by Theorem 7.1(2), (N', M_0') satisfies MST-property. According to Characterization 2.17(1), (N', M_0') is live. $\qquad\square$

7.2   Merging Sets of Non-Neighboring Places

In MERGE-A-PLACE of Section 7.1, the places within each set Q_i may be associated with some common transitions. This section considers the special case where these places are not so. Naturally, all the results obtained in Section 7.1 will hold in this case. However, in this case, simpler or different conditions for preserving nineteen properties considered in the previous chapter will first be obtained for general Petri nets. Then, for augmented marked graphs, similar conditions are proposed for the preservation of liveness, boundedness, reversibility and MST-property. This case includes Zhou's parallel and sequential mutual exclusions [ZD (1993)].

Definition 7.3. (non-neighboring places)
Let $N = (P, T, F)$ be a net. Places p_1 and p_2 are said to be *non-neighboring* iff $(^\bullet p_1 \cup p_1^\bullet) \cap (^\bullet p_2 \cup p_2^\bullet) = \phi$.

MERGE-N-PLACE   (merging a set of non-neighboring places): This place-merging operator is the same as MERGE-A-PLACE except for the additional requirement that, for each of the merged sets Q_i, every two places within Q_i are non-neighboring.

One of the main consequential differences between MERGE-A-PLACE and MERGE-N-PLACE is that, for MERGE-N-PLACE, if N is ordinary (resp., pure), so is N'. (This is why W' does not appear in N'.) This may not be true for MERGE-A-PLACE.

Theorem 7.5. (*property preservation under MERGE-N-PLACE*)
Let (N, M_0) and (N', M_0') be involved in MERGE-N-PLACE, where $N = (P_0 \cup Q_1 \cup \ldots \cup Q_k, T, F)$ and $N' = (P_0 \cup Q_0, T', F')$. Then, the following propositions hold:

1) If N is an SM, then N' is also an SM.
2) Let N be an MG. Three cases may hold for N': (a) In general, N' is not an MG. (b) N' is an FC net if, $\forall p \in P_0$, $\forall i, j \in \{1, 2, \ldots, k\}$,

where $i \neq j$, $p^\bullet \cap Q_i^\bullet = \phi$ and $(Q_i^\bullet \cap Q_j^\bullet \neq \phi$ implies $Q_i^\bullet = Q_j^\bullet)$. *(c)*
N' is an AC net if $\forall i, j \in \{1, 2, \ldots, k\}$, where $i \neq j$, $Q_i^\bullet \cap Q_j^\bullet \neq \phi$
implies either $Q_i^\bullet \subseteq Q_j^\bullet$ or $Q_j^\bullet \subseteq Q_i^\bullet$.

3) *Let N be an FC net. Three cases may hold for N': (a) In general, N'*
   may or may not be an FC net. (b) N' is an FC net if $\forall p \in P_0$, $\forall i, j \in$
   $\{1, 2, \ldots, k\}$, *where $i \neq j$, $p^\bullet \cap Q_i^\bullet = \phi$ and $Q_i^\bullet \cap Q_j^\bullet = \phi$. (c) N' is*
   an AC net if $\forall i, j \in \{1, 2, \ldots, k\}$, where $i \neq j$, $Q_i^\bullet \cap Q_j^\bullet \neq \phi$ implies
   either $Q_i^\bullet \subseteq Q_j^\bullet$ or $Q_j^\bullet \subseteq Q_i^\bullet$.

4) *Let N be an AC net. Two cases may hold for N': (a) In general, N' may*
   or may not be an AC net. (b) N' is an AC net if, $\forall p \in P_0$, $\forall i, j \in \{1,$
   $2, \ldots, k\}$, *where $i \neq j$, $p^\bullet \cap Q_i^\bullet \neq \phi$ implies that $p^\bullet \subseteq Q_i^\bullet$ and $Q_i^\bullet \cap$*
   $Q_j^\bullet \neq \phi$ *implies either $Q_i^\bullet \subseteq Q_j^\bullet$ or $Q_j^\bullet \subseteq Q_i^\bullet$.*

5) *Suppose there exists $\alpha = (\alpha_0, \alpha_1, \ldots, \alpha_k) > 0$, where $\alpha_0 = (a_1, a_2,$*
   $\ldots, a_{|P_0|})$ *and each α_i is a $|Q_i|$-subvector of the form $(d_i, d_i, \ldots, d_i)$,*
   $i = 1, 2, \ldots, k$, *such that $\alpha V = 0$ for N. Then, N' is conservative.*
   (Note: The order of the elements of α conforms with the row order of
   the incidence matrix of N. Also, α is a positive vector with a special
   form.)

6) *If there exists the same α as defined in 5) such that $\alpha V \leq 0$, then N'*
   is structurally bounded.

7) *N' is consistent if N is consistent.*

8) *N' is repetitive if N is repetitive.*

9) $Rank(N) + k - \sum_{i=1}^{k} |Q_i| \leq Rank(N') \leq Rank(N)$.

10) $|C(N)| + k - \sum_{i=1}^{k} |Q_i| \leq |C(N')| \leq |C(N)|$, *where $k = |Q_0|$.*

11) *N' may or may not satisfy the RC-property even if N satisfies the*
    RC-property.

12) *Suppose N has an SM-cover. Two cases may hold for N': (a) In*
    general, N' may not have an SM-cover. (b) N' has an SM-cover if the
    following two conditions hold: (1) For every SM-component N_1 of N,
    $P_1 \cap Q_i \neq \phi$ implies $P_1 \cap Q_j = \phi$, where $j \neq i$. (2) For any two SM-
    components N_1 and N_2 of N satisfying $P_1 \cap Q_i \neq \phi$ and $P_2 \cap Q_i \neq \phi$,
    $^\bullet(P_1 - P_2) \cap^\bullet (P_2 - P_1) = \phi$ and $(P_1 - P_2)^\bullet \cap (P_2 - P_1)^\bullet = \phi$. (Note:
    P_i is the set of places of N_i.)

13) *(a) For any $D' \subseteq P_0 \cup Q_0$, let $D = \cup\{Q_i | \exists q_i \in D' \cap Q_0\} \cup (D' \cap P_0)$.*
    Then, D' is a siphon of N' iff D is a siphon of N. (b) If $D \subseteq P_0$ is a
    minimal siphon of N, then D is also a minimal siphon of N'.

14) *(a) For any $S' \subseteq P_0 \cup Q_0$, let $S = \cup\{Q_i | \exists q_i \in S' \cap Q_0\} \cup (S' \cap P_0)$.*

Then, S' is a trap of N' iff S is a trap of N. (b) If $S \subseteq P_0$ is a minimal trap of N, then S is also a minimal trap of N'.

15) *Suppose N is a process such that the entry place $p_e \notin Q_i$ and the exit place $p_x \notin Q_i$ for $i=1, 2, \ldots, k$. Then, the following properties hold:*

   (a) *N' is a process.* (b) *$Z(N') = Z(N) + \sum_{i=1}^{k} |Q_i| - k$, where $k = |Q_0|$.*
   (c) *(N', M_0') may or may not terminate properly even if (N, M_0) terminates properly.*

16) *In general, (N', M_0') may or may not be bounded even if (N, M_0) is bounded.*

17) *In general, (N', M_0') may or may not be live even if (N, M_0) is live. However, (N', M_0') is live and bounded if the following conditions hold: (a) 12)(b) holds. (b) (N, M_0) satisfies the reduction-set-dependent MST-property.*

18) *In general, (N', M_0') may or may not be reversible even if (N, M_0) is reversible. However, (N', M_0') is reversible if the following conditions hold: (a) 3b) and 6) hold. (b) (N, M_0) satisfies the reduction-set-dependent MST-property.*

Proof.

1) Since every Q_i is non-neighboring, ${}^\bullet t$ in $N = {}^\bullet t$ in N' and $t^\bullet$ in $N = t^\bullet$ in N'. Hence, N' is an SM if N is an SM.

2) Due to the merging, $\forall q_i \in Q_0$: $({}^\bullet q_i$ in $N') = ({}^\bullet Q_i$ in $N)$ and $(q_i^\bullet$ in $N') = (Q_i^\bullet$ in $N)$. We have to consider only those pairs of places at least one of which is in Q_0.

   (a) Since N is an MG and Q_i is non-neighboring, $|{}^\bullet q_i| = |{}^\bullet Q_i| > 1$ or $|q_i^\bullet| = |Q_i^\bullet| > 1$. Hence, N' is not an MG.

   (b) $\forall p \in P_0$, $\forall q_i \in Q_0$, $p^\bullet \cap q_i^\bullet = p^\bullet \cap Q_i^\bullet = \phi$. $\forall q_i, q_j \in Q_0$ in N', where $i \neq j$, $q_i^\bullet \cap q_j^\bullet \neq \phi$ because $Q_i^\bullet \cap Q_j^\bullet \neq \phi$ in N. Hence, N' is an FC net.

   (c) $\forall q_i, q_j \in Q_0$ in N': $q_i^\bullet \cap q_j^\bullet \neq \phi$, we have $Q_i^\bullet \cap Q_j^\bullet \neq \phi$ in N. Suppose $Q_i^\bullet \subseteq Q_j^\bullet$. Then, $q_i^\bullet = Q_i^\bullet \subseteq Q_j^\bullet = q_j^\bullet$. $\forall p \in P_0$, $\forall q_i \in Q_0$: $p^\bullet \cap q_i^\bullet \neq \phi$, there exists $q \in Q_i$ such that $p^\bullet \cap q^\bullet \neq \phi$ in N. Since N is an MG, $p^\bullet = q^\bullet \subseteq Q_i^\bullet = q_i^\bullet$. Hence, N' is an AC net.

3) Similar to 2), we have to consider only those pairs of places at least one of which is in Q_0.

   (a) $\forall p \in P_0$, $\forall q_i \in Q_0$, $p^\bullet \cap q_i^\bullet = p^\bullet \cap Q_i^\bullet$. Hence, $p^\bullet \subset q_i^\bullet$ if

$p^\bullet \cap Q_i^\bullet \neq \phi$. This means that N' may or may not be an FC net even if N is an FC.

(b) $\forall p \in P_0 \forall q_i \in Q_0$, $p^\bullet \cap q_i^\bullet = p^\bullet \cap Q_i^\bullet = \phi$. $\forall q_i, q_j \in Q_0$ in N', where $i \neq j$, $q_i^\bullet \cap q_j^\bullet = \phi$ because $Q_i^\bullet \cap Q_j^\bullet = \phi$ in N. Hence, N' is an FC net.

(c) Suppose $\forall i, j \in \{1, 2, \ldots, k\}$ where $i \neq j$: $Q_i^\bullet \cap Q_j^\bullet \neq \phi$ implies either $Q_i^\bullet \subseteq Q_j^\bullet$ or $Q_j^\bullet \subseteq Q_i^\bullet$. Note that, $\forall p \in P_0$, $(p^\bullet$ in $N) = (p^\bullet$ in $N')$ and that, $\forall q_i \in Q_0$, $Q_i^\bullet = q_i^\bullet$. In N', $\forall p, q \in P_0 \cup Q_0$, where $p^\bullet \cap q^\bullet \neq \phi$ in N', consider the following three cases. *Case 1 (both p, $q \in P_0$):* p and q are in P_0 and thus their pre-sets and post-sets remain unchanged in N'. Since N is an FC net, $(p^\bullet = q^\bullet)$ in N and, thus, also in N'. *Case 2 ($p \in P_0$ but $q = q_i \in Q_0$):* Since N is FC and $p^\bullet \cap q_i^\bullet \neq \phi$ in N', $\exists x \in Q_i$ such that $(p^\bullet$ in $N) = (x^\bullet$ in $N)$. Then, $(p^\bullet$ in $N') = (p^\bullet$ in $N) = (x^\bullet$ in $N) \subseteq (Q_i^\bullet$ in $N) = (q_i^\bullet$ in $N')$. *Case 3 (both $p = q_j, q = q_i \in Q_0$):* It follows that $(q_j^\bullet$ in $N') = (Q_j^\bullet$ in $N) \subseteq (Q_i^\bullet$ in $N) = (q_i^\bullet$ in $N')$. Hence, N is an AC net.

4) (a) Similar to the proof of 3a).

(b) Consider that $p^\bullet \cap Q_i^\bullet \neq \phi$ implies that $p^\bullet \subseteq Q_i^\bullet$, this proof is similar to that of 3c).

5-6) It follows from Theorem 7.2.

7-8) It follows from Theorem 7.3.

9) For the matrix V, adding the rows in Q_i to a single row and doing some row exchanges will result in a new matrix such that V' is its submatrix. Hence, $\text{Rank}(N) + k - \sum_{i=1}^{k} |Q_i| \leq \text{Rank}(N')$, $\text{Rank}(N') \leq \text{Rank}(N)$ is obvious.

10) After applying MERGE-N-PLACE, the number of clusters decreases by at most $\sum_{i=1}^{k} |Q_i| - k$, where $k = |Q_0|$. Hence, $|C(N)| + k - \sum_{i=1}^{k} |Q_i| \leq |C(N')| \leq |C(N)|$.

11) By (9) and (10), it follows that N' may or may not satisfy the RC-property even if N satisfies the RC-property.

12) Note that $\forall q_i \in Q_0$: $(^\bullet q_i$ in $N') = (^\bullet Q_i$ in $N)$ and $(q_i^\bullet$ in $N') = (Q_i^\bullet$ in $N)$. (a) By Definition 2.24 about SM-cover, it is obvious that N' may not have an SM-cover even if N has an SM-cover. (b) We distinguish between two kinds of SM-components of N. For those SM-component $N_1 = (P_1, T_1, F_1)$, where $P_1 \subseteq P_0$, N_1 is

still an SM-component of N'. For those SM-components, say $N_k = (P_k, T_k, F_k)$, of N, such that $P_k \cap Q_i \neq \phi$ for some Q_i, we can combine all of them to form a subnet called $N_i' = (P_i', T_i')$ because of Condition (1). It will be shown below that N_i' is an SM-component of N'. Without loss of generality, we may assume that N_i' is formed from two SM-components $N_1 = (P_1, T_1, F_1)$ and $N_2 = (P_2, T_2, F_2)$ of N. Obviously, N_i' is strongly connected and $T_i' = T_1 \cup T_2 = ({}^\bullet P_1 \cup P_1{}^\bullet) \cup ({}^\bullet P_2 \cup P_2{}^\bullet) = {}^\bullet P_i' \cup P_i'{}^\bullet$. Lastly, $|{}^\bullet t| = |t^\bullet| = 1 \; \forall t \in {}^\bullet(P_1 - P_2) \cup {}^\bullet(P_2 - P_1) \cup (P_1 - P_2)^\bullet \cup (P_2 - P_1)^\bullet$ because of Condition (2) and $|{}^\bullet t| = |t^\bullet| = 1 \; \forall t \in {}^\bullet(P_1 \cap P_2) \cup (P_2 \cap P_1)^\bullet \subseteq T_1 \cap T_2$ because both N_1 and N_2 are SM. Hence, N_i' is an SM.

13-14) (a) According to MERGE-N-PLACE, we have: $D \cap P_0 = D' \cap P_0$ and ($Q_i \subseteq D$ in N) iff ($q_i \in D' \cap Q_0$ in N'). Also, since no transitions have been eliminated, $\forall p \in D \cap P_0$, ($p^\bullet$ in N) = ($p^\bullet$ in N') and (${}^\bullet p$ in N) = (${}^\bullet p$ in N'); $\forall q_i \in D'$, ($q_i^\bullet$ in N') = ($Q_i^\bullet$ in N) and (${}^\bullet q_i$ in N') = (${}^\bullet Q_i$ in N). Hence, (${}^\bullet D$ in N) = (($\cup {}^\bullet \{Q_i | Q_i \subseteq D\} \cup {}^\bullet(D \cap P_0)$)) in N) = (($\cup {}^\bullet \{q_i | \exists q_i \in D' \cap Q_0\} \cup {}^\bullet(D' \cap P_0)$)) in N') = (${}^\bullet D'$ in N') and ($D^\bullet$ in N) = (($\cup \{Q_i | \exists q_i \in D' \cap Q_0\}^\bullet \cup (D' \cap P_0)^\bullet$) in N) = ($D'^\bullet$ in N'). This implies that D' is a siphon (resp., trap) of N' iff D is a siphon (resp., trap) of N. (b) If $D \subseteq P_0$ is a minimal siphon (resp., trap) of N, then by the result above, it is obvious that D is also a minimal siphon (resp., trap) of N'.

15) Suppose N is a process such that $p_e \notin Q_i$ and $p_x \notin Q_i$ for $i = 1$, $2, \ldots, k$. Then, after MERGE-N-PLACE, $p_e \notin Q_0$ and $p_x \notin Q_0$. (a) By the definitions of MERGE-N-PLACE and a process, it is obvious that N' is also a process. (b) $Z(N') = |F'| - |P' \cup T'| + 2 = |F| - |P \cup T| + (\sum_{i=1}^{k} |Q_i| - k) + 2 = Z(N) + \sum_{i=1}^{k} |Q_i| - k$. (c) If N is a process and (N, M_0) terminates properly, then it is obvious that (N', M_0') may or may not terminate properly. For example, refer to the figures obtained from Figure 7.2(a) and 7.2(c) by deleting the transition t_5 and the two corresponding arcs and moving the tokens in p_2, p_3, p_5 in Figure 7.2(a) and in q_1, p_3 in Figure 7.2(c).

16) Figure 7.2 shows that MERGE-N-PLACE may transform a bounded net N (Figure 7.2(a)) to an unbounded N' or to a bounded N''. N' (Figure 7.2(b)) is obtained by merging $Q_1 = \{p_1, p_3\}$ to q_1. By firing t_1 and t_2 repetitively, p_4 becomes unbounded. N'' (Figure 7.2(c)) is obtained by merging $Q_1 = \{p_2, p_5\}$ to q_1. It is obvious that N'' is bounded.

17) The Petri net N in Figure 7.2(a) and N' in Figure 7.2(b) are live but N'' in Figure 7.2(c) is not live. Hence, (N', M_0') may or may not be live even if (N, M_0) is live. Obviously, N' is also ordinary and pure. By Proposition 12) Condition (a) means that N' is SM-coverable. By Theorem 7.1, Condition (b) implies that (N', M_0') satisfies MST-property. By Characterization 2.18, (N', M_0') is live and bounded.

18) It is obvious that N in Figure 7.2(a) is reversible but N' in Figure 7.2(b) and N'' in Figure 7.2(c) are not. Hence, (N', M_0') may or may not be reversible even if (N, M_0) is reversible. If 3b) holds, then N' is an FC net. If 6) holds, then N' is structurally bounded. Condition (b) implies that (N', M_0') satisfies MST-property. By Characterization 2.19, (N', M_0') is a live and bounded FC net. Since (N, M_0) is reversible, every trap of N is marked. By 14), every trap of N' is marked by M_0'. Hence, (N', M_0') is reversible. □

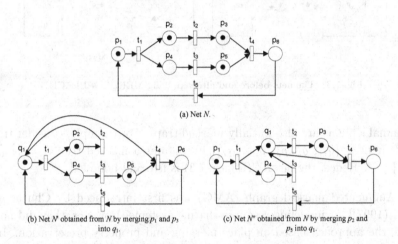

(a) Net N.

(b) Net N' obtained from N by merging p_1 and p_3 into q_1.

(c) Net N'' obtained from N by merging p_2 and p_3 into q_1.

Fig. 7.2   An example for illustrating Propositions 15) - 18).

Example 7.2. This example is used to illustrate Theorem 7.5(17) Figure 7.3 shows a Petri net N with $P_0 = \{p_1, p_2, \ldots, p_{23}, p_{24}\}$, $Q_1 = \{q_{11}, q_{12}\}$, $Q_2 = \{q_{21}, q_{22}, q_{23}, q_{24}\}$ and $Q_3 = \{q_{31}, q_{32}\}$. They satisfy the conditions required in MERGE-N-PLACE. N satisfies Condition (a) of Theorem 7.5(17) because each of the minimal SM-component containing a resource place is unique. For instance, the SM-component containing q_{11} is $q_{11}t_1p_2 (t_2, t_3) q_{11}$ Condition (b) of Theorem 7.5(17) is satisfied since all the

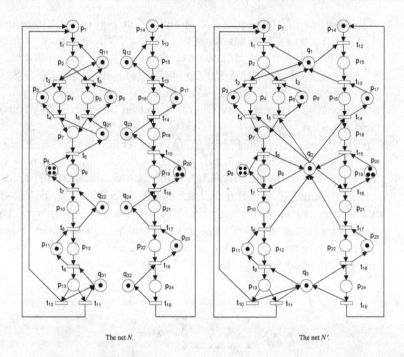

The net N.   The net N'.

Fig. 7.3   The nets before and after applying MERGE-N-PLACE.

minimal siphons are also initially marked traps. For example, consider the siphon $D = \{q_{11}, q_{12}, p_2, p_{16}, p_{17}\}$, it contains a marked trap $S = \{p_{16}, p_{17}\} \subset P$. Hence, the net N' (Figure 7.3) is live and bounded.

Augmented marked graph (AMG) was first introduced by Chu et al [CX (1997)] for modeling resource-sharing systems. However, they did not take the approach based on place merging and property preservation. In this chapter, we assume that the set of resources R is created by merging the resource-places. If there are k resources, then R can be partitioned into k non-overlapping subsets $Q_1, Q_2, \ldots, Q_k$, each Q_i consisting of those non-neighboring places to be merged into a single place q_i.

Before providing some property preserving results, we present a rank property for AMG.

Theorem 7.6. *For any augmented marked graph N, $Rank(N) = Rank(G)$, where G is the corresponding marked graph obtained from N by deleting all the resources R.*

Proof. Suppose resource place $r \in R$ associates with k pairs of transitions $\{(a_i, b_i), i = 1, 2, \ldots, k\}$. Then, by Definition 2.8, r is located at k elementary cycles $C_1, C_2, \ldots, C_k$ which cover the k transition pairs, respectively. For each place $p \in P$ in N, if p appears at n of these cycles, then in the incidence matrix V of N, add n times row p to row r. After doing the addition operations for all places in P, row r becomes zero in V. Similarly, every other resource place row becomes zero after doing this Gaussian Elimination. Hence, $Rank(N) = Rank(G)$. $\qquad\square$

Example 7.3. (for explaining Theorem 7.6)
In order to make the above proof more clearly, we use an example to explain how to eliminate the rows of resource places to zero in the incidence matrix of an augmented marked graph. Figure 7.4 is an augmented marked graph and its incidence matrix, where r_1, r_2 are resource places. r_1 is associated with a transition pair (t_1, t_3) and the elementary path from t_1 to t_3 is $t_1 p_2 t_2 p_3 t_3$, by Gaussion-elimination, adding $\text{row}(p_2) + \text{row}(p_3)$ to $\text{row}(r_1)$ can eliminate $\text{row}(r_1)$ to zero. Similarly, r_2 is associated with two transition pairs (t_1, t_3) and (t_2, t_4), the elementary path from t_1 to t_3 is $t_1 p_2 t_2 p_3 t_3$ and the elementary path from t_2 to t_4 is $t_2 p_3 t_3 p_4 t_4$. By Gaussion-elimination, adding $\text{row}(p_2) + 2\text{row}(p_3) + \text{row}(p_4)$ to $\text{row}(r_2)$ can eliminate $\text{row}(r_2)$ to zero.

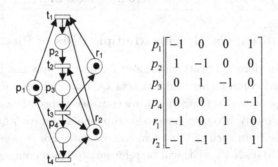

Fig. 7.4 An augmented marked graph and its incidence matrix.

The following theorem extends the results published by the authors in [HJC (2003)].

Theorem 7.7. *Let (N', M_0') and (N, M_0) be defined in MERGE-N-PLACE. Let (N, M_0) be an AMG and $Q_i \subseteq R$(the set of resources), $i = 1, 2, \ldots, k$. Then, the following propositions hold:*

1) N' is an AMG.

2) Suppose, for every siphon D of N, if $Q_i \subseteq D$, there exists a trap S marked by M_0 within D such that $S \subseteq P_0 \cap D$ or $Q_i \subseteq S$. Then, (N', M_0') is live and reversible if (N, M_0) is live.

3) $Rank(N') = Rank(N)$.

Proof.

1) Since only the places of R are merged, Conditions in Definition 2.8 are preserved and hence (N', M_0') is also an AMG.

2) Let D' be a siphon of N'. By Theorem 7.1(1), $D = \cup\{Q_i | \exists q_i \in D' \cap Q_0\} \cup (D' \cap P_0)$ is a siphon of N. If $D' \cap Q_0 = \phi$, then $D' = D$. Since $(N, M_0|_P)$ is a live marked graph, D and thus D' contains a trap S of N marked by M_0 [MUR (1989)]. Moreover, for every siphon D of N containing Q_i, there exists a trap S marked by M_0 such that $S \subseteq P_0 \cap D$ or $Q_i \subseteq S$. That is, (N, M_0) satisfies the reduction-set-dependent MST-property. By Theorem 7.1(2), (N', M_0') satisfies MST-property. It follows from Characterization 2.20 that (N', M_0') is live. By Characterization 2.21, (N', M_0') is reversible.

3) Since $G = G'$, by Theorem 7.6, $Rank(N') = Rank(G') = Rank(G) = Rank(N)$, where G (resp., G') is the marked graph obtained from N (resp., N') by removing all the resource places in R. $\square$

7.3   Composition via Merging Multiple Pairs of Places

The net N considered in Sections 7.1 and 7.2 may or may not be composed of disconnected components and the sets Q_i may belong to the same or different components. In this section, we consider a special case where N is composed of exactly two disconnected components and each Q_i consists of two places, one from each of these two components. For convenience in application, MERGE-N-PLACE will be reformulated as a composition of two disjoint Petri nets via merging several pairs of their places to single places. Some of the results of Sections 7.1 and 7.2 will be restated as corollaries for this composition and new results specifically for this composition will be derived.

COMPOSITION-MP (composition via merging multiple pairs of places)

Consider two disconnected ordinary Petri nets $N_1 = (P_1 \cup C, T_1, F_1)$ and $N_2 = (P_2 \cup C, T_2, F_2)$, where $P_1 \cap P_2 = \phi$, $P_1 \cap C = \phi$, $P_2 \cap C = \phi$, T_1

$\cap T_2 = \phi$, $F_1 \cap F_2 = \phi$ and C is a set of places appearing in both N_1 and N_2. Let (N, M_0) be composed from (N_1, M_1) and (N_2, M_2) by merging the pairs of places in C. That is, $N = (P, T, F)$, where $P = P_1 \cup P_2 \cup C$, $T = T_1 \cup T_2$, $F = F_1 \cup F_2$ and

$$M_0(p) = \begin{cases} M_1(p) & p \in P_1 \\ M_2(p) & p \in P_2 \\ max\{M_1(p), M_2(p)\} & p \in C \end{cases}$$

Corollary 7.1. *(based on Theorem 7.1(1))* (preservation of siphons and traps for general Petri nets). *Let $N = (P, T, F)$, $N_1 = (P_1 \cup C, T_1, F_1)$ and $N_2 = (P_2 \cup C, T_2, F_2)$ be defined in COMPOSITION-MP. Then, the following propositions hold:*

1) *Suppose D is a siphon (resp., trap) of N. Let $D_i = D \cap (P_i \cup C)$ for $i = 1, 2$. Then, (a) D_i is a siphon (resp., trap) of N_i for $i = 1, 2$; (b) $D = D_1 \cup D_2$ and $D_1 \cap C = D_2 \cap C$.*
2) *Suppose D_i is a siphon (resp., trap) of N_i for $i = 1, 2$. Then, (a) For $i = 1$ or 2, D_i is a siphon (resp., trap) of N if $({}^\bullet(D_i \cap C) \cap T_j \subseteq (D_i \cap C)^\bullet \cap T_j)$ in N (resp., $((D_i \cap C)^\bullet \cap T_j \subseteq^\bullet (D_i \cap C) \cap T_j)$ in N) where $i \neq j$; (b) $D_1 \cup D_2$ is a siphon (resp., trap) of N if $D_1 \cap C = D_2 \cap C$.*

Proof.

1) (a) By the definition of COMPOSITION-MP, ${}^\bullet D_i$ in $N_i = {}^\bullet D \cap T_i$ in $N \subseteq D^\bullet \cap T_i$ in $N = D_i^\bullet$ in N_i. Hence, D_i is a siphon of N_i. (b) $D_1 \cup D_2 = (D \cap (P_1 \cup C)) \cup (D \cap (P_2 \cup C)) = D \cap (P_1 \cup P_2 \cup C) = D$. $D_1 \cap C = D \cap (P_1 \cup C) \cap C = D \cap C = D \cap (P_2 \cup C) \cap C = D_2 \cap C$. The proof for a trap is similar.
2) Since D_i is a siphon of N_i, $({}^\bullet D_i$ in $N_i) \subseteq (D_i^\bullet$ in $N_i)$. (a) Suppose $i = 1$. If $({}^\bullet(D_1 \cap C) \cap T_2$ in $N_2) \subseteq ((D_1 \cap C)^\bullet \cap T_2$ in $N_2)$, then $({}^\bullet D_1 \cap T$ in $N)$ $= (({}^\bullet D_1 \cap T_1)$ in $N_1) \cup (({}^\bullet D_1 \cap T_2)$ in $N_2) = (({}^\bullet D_1 \cap T_1) \cup ({}^\bullet(D_1 \cap C) \cap T_2)$ in $N) \subseteq ((D_1^\bullet \cap T_1) \cup ((D_1 \cap C)^\bullet \cap T_2)$ in $N) = (D_1^\bullet \cap T$ in $N)$. This means that D_1 is a siphon of N. (b) If $D_1 \cap C = D_2 \cap C$, then $({}^\bullet(D_1 \cup D_2)$ in $N) = ({}^\bullet D_1$ in $N_1) \cup ({}^\bullet D_2$ in $N_2) \subseteq (D_1^\bullet$ in $N_1) \cup (D_2^\bullet$ in $N_2) = ((D_1 \cup D_2)^\bullet$ in $N)$. Hence, $D_1 \cup D_2$ is a siphon of N. The proof for a trap is similar. $\square$

Example 7.4. (for explaining Corollary 7.1)
In Figure 7.5, for the set of places $D = \{r_1, p_{12}, p_{14}, p_{25}\}$ and $C = \{r_1, r_2\}$. Let $D_1 = D \cap (\{p_{11}, p_{12}, p_{14}\} \cup C) = \{r_1, p_{12}, p_{14}\}$ and $D_2 = D \cap$

$(\{p_{21}, p_{23}, p_{25}\} \cup C) = \{r_1, p_{25}\}$. By Corollary 7.1, since D_1 is a siphon of N_1 and D_2 is a siphon of N_2, D is a siphon of N. The reverse is also true. Similarly, the set of places $S = \{r_2, p_{12}, p_{23}, p_{25}\}$ is a trap of N if and only if $S_1 = \{r_2, p_{12}\}$ is a trap of N_1 and $S_2 = \{r_2, p_{23}, p_{25}\}$ is a trap of N_2.

Note that, for the particular case $D \cap C = \phi$, i.e., no resource places are involved, D_i is also a siphon (or trap) in N after the composition. For example, siphon $D_1 = \{p_{11}, p_{12}, p_{14}\}$ of N_1 is also a siphon of N and trap $S_2 = \{p_{21}, p_{23}, p_{25}\}$ of N_2 is also a trap of N.

When applied to COMPOSITION-MP, the reduction-set-dependent ST-property (resp., reduction-set-dependent MST-property) can be rewritten as Condition A (resp., B) as follows.

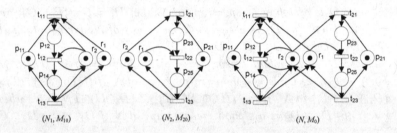

(N_1, M_{10}) (N_2, M_{20}) (N, M_0)

Fig. 7.5   Two augmented marked graphs (N_1, M_{10}) and (N_2, M_{20}) and their composite Petri net (N, M_0).

Condition A: Petri nets $N_1 = (P_1 \cup C, T_1, F_1)$ and $N_2 = (P_2 \cup C, T_2, F_2)$ are said to satisfy Condition A if, for any siphons D_1 of N_1 and D_2 of N_2, the following conditions are satisfied: (1) If $D_i \subseteq P_i$, then D_i contains a trap S_i of N_i. (2) If $D_1 \cap D_2 \neq \phi$, then either (a) for $i = 1$ or 2, D_i contains a trap S_i of N_i such that $S_i \cap C = \phi$, or (b) for both $i = 1$ and 2, D_i contains a trap S_i of N_i such that $S_1 \cap C = S_2 \cap C$.

Condition B: Petri nets (N_1, M_1) and (N_2, M_2) are said to satisfy Condition B if N_1 and N_2 satisfy Condition A where the trap S_i is marked by M_i.

Corollary 7.2. *(based on Theorem 7.1(2))* (preservation of ST-property and MST-property for general Petri nets) *Let (N, M_0), (N_1, M_1) and (N_2, M_2) be defined in COMPOSITION-MP. Then, the following propositions hold:*

1) If N_1 and N_2 satisfy Condition A, then N satisfies ST-property.

2) If (N_1, M_1) and (N_2, M_2) satisfy Condition B, then (N, M_0) satisfies MST-property.

Proof. Consider any siphon D of N. By Corollary 7.1(1), there exist a siphon D_1 of N_1 and a siphon D_2 of N_2 such that $D = D_1 \cup D_2$, where $D_i = D \cap (P_i \cup C)$ for $i = 1$ and 2.

1) If N_1 and N_2 satisfy Condition A, then consider two cases: *Case 1* $(D \cap C = \phi)$: Then, $D_1 \subseteq P_1$ and $D_2 \subseteq P_2$. Hence, there exist a trap S_1 of N_1 and a trap S_2 of N_2 such that $S_1 \subseteq D_1$ and $S_2 \subseteq D_2$. According to Corollary 7.1(2), S_1 and S_2 are traps of N. *Case 2* $(D \cap C \neq \phi)$: Then, $D_1 \cap D_2 \neq \phi$. Hence, either (a) for $i = 1$ or 2, D_i contains a trap S_i of N_i such that $S_i \cap C = \phi$, or (b) for both $i = 1$ and 2, D_i contains a trap S_i of N_i such that $S_1 \cap C = S_2 \cap C$. By Corollary 7.1(2), in case (a), S_1 or S_2 is a trap of N, and in case (b), $S_1 \cup S_2$ is a trap of N.

2) Similar to the proof of Proposition 1). $\qquad\square$

Corollary 7.3. *(based on Theorems 7.2 and 7.3)* (preservation of P-invariant, conservativeness, structural boundedness, T-invariant, consistency, repetitiveness and strong connectedness for general Petri nets) *Let N, N_1 and N_2 be defined in COMPOSITION-MP. Let $\alpha_1 = (x_1, \ldots, x_{|P_1|}, z_1, \ldots, z_{|C|})$, $\alpha_2 = (y_1, \ldots, y_{|P_2|}, z_1, \ldots, z_{|C|})$ and $\alpha = (x_1, \ldots, x_{|P_1|}, y_1, \ldots, y_{|P_2|}, z_1, \ldots, z_{|C|})$. Then, the following propositions hold (Note that α_i has a special pattern.):*

1) α_i *is a P-invariant of N_i for $i = 1$ and 2 iff α is a P-invariant of N.*
2) N *is conservative (resp., structurally bounded) if $\alpha_i > 0$ such that $\alpha_i V_i = 0$ (resp., $\alpha_i V_i \leq 0$) for $i = 1$ and 2. N_1 and N_2 are conservative if N is conservative.*
3) $\beta = (\beta_1, \beta_2)$ *is a T-invariant of N if the $|T_i|$-vector β_i is a T-invariant of N_i for $i = 1$ and 2.*
4) N *is consistent (resp., repetitive, strongly connected) if N_1 and N_2 are consistent (resp., repetitive, strongly connected).*

Proof. The incidence matrices V, V_1 and V_2 of N, N_1 and N_2 have the forms:

$$V = \begin{array}{c} \\ P_1 \\ Q_2 \\ C \end{array} \begin{array}{c} T_1 \quad\ T_2 \\ \begin{pmatrix} V_{11} & 0 \\ 0 & V_{22} \\ V_{31} & V_{32} \end{pmatrix} \end{array}$$

$$V_1 = \begin{matrix} & T_1 \\ P_1 \\ C \end{matrix} \begin{pmatrix} V_{11} \\ V_{31} \end{pmatrix}$$

and

$$V_2 = \begin{matrix} & T_2 \\ P_2 \\ C \end{matrix} \begin{pmatrix} V_{22} \\ V_{32} \end{pmatrix}$$

1) This follows from the definitions of α_1, α_2 and α that $\alpha_1 \geq 0$, $\alpha_2 \geq 0$, $\alpha_1 V_1 = 0$ and $\alpha_1 V_1 = 0$ iff $\alpha \geq 0$ and $\alpha V = 0$.
2) Similar to the proof of Proposition 1).
3) Since β_i is a T-invariant of N_i for i = 1, 2, $V_1\beta_1 = 0$ and $V_2\beta_2 = 0$, implying that $V_{11}\beta_1 = 0$, $V_{31}\beta_1 = 0$, $V_{22}\beta_2 = 0$ and $V_{32}\beta_2 = 0$. Let $\beta = (\beta_1, \beta_2)$, then $V\beta = (V_{11}\beta_1, V_{22}\beta_2, V_{31}\beta_1 + V_{32}\beta_2) = 0$. That is, β is a T-invariant of N.
4) Similar to the proof of Proposition 3).

$\square$

Theorem 7.2(2) states that MERGE-A-PLACE may not preserve structural boundedness. However, as shown in Theorem 7.8 below, it is always preserved under COMPOSITION-MP.

Theorem 7.8. (preservation of structural boundedness under COMPOSITION-MP for general Petri nets)
Let N, N_1 and N_2 be defined in COMPOSITION-MP. Then, N is structurally bounded iff N_1 and N_2 are both structurally bounded.

Proof.   Let the incidence matrices V, V_1 and V_2 of N, N_1 and N_2 have the same forms as in the proof of Corollary 7.3. Since N is structurally bounded, $\exists \alpha = (x_1, \ldots, x_{|P_1|}, y_1, \ldots, y_{|P_2|}, z_1, \ldots, z_{|C|}) > 0$ such that $\alpha V \leq 0$. Let $\alpha_1 = (x_1, \ldots, x_{|P_1|}, z_1, \ldots, z_{|C|})$ and $\alpha_2 = (y_1, \ldots, y_{|P_2|}, z_1, \ldots, z_{|C|})$. Then, $\alpha_1 > 0$, $\alpha_2 > 0$, $\alpha_1 V_1 \leq 0$ and $\alpha_2 V_2 \leq 0$. Hence, N_1 and N_2 are structurally bounded. Next, suppose N_1 and N_2 are structurally bounded. If $\exists M_0$ such that (N, M_0) is unbounded, then (N_1, M_1) or (N_2, M_2) is unbounded, where $M_i = M_0|_{P_i \cup C}$ for $i = 1$ and 2, contradicting with the fact that both N_1 and N_2 are structurally bounded. $\square$

Theorem 7.9. (preservation of boundedness under COMPOSITION-MP for general Petri nets)
Let (N, M_0), (N_1, M_1) and (N_2, M_2) be defined in COMPOSITION-MP. Then, (N_1, M_1) and (N_2, M_2) are bounded if (N, M_0) is bounded.

Proof. If (N_i, M_i) is unbounded for $i = 1$ or 2, then, for any positive integer k, $\exists p \in P_i \cup C$ and $\exists \sigma \in L(N_i, M_i)$ such that $M_i[N_i, \sigma\rangle M_i'$ and $M_i'(p) > k$. By the definition of COMPOSITION-MP, $\sigma \in L(N, M_0)$. This means that $\exists M' \in R(N, M_0)$ such that $M[N, \sigma\rangle M'$ and $M'(p) > k$. That is, p is unbounded in (N, M_0) – a contradiction. $\square$

Corollary 7.4. *(based on Theorem 7.5)* *(preservation of liveness and boundedness under COMPOSITION-MP for general Petri nets). Let (N, M_0), (N_1, M_1) and (N_2, M_2) be defined in COMPOSITION-MP. Suppose (N_1, M_1) and (N_2, M_2) satisfy Condition B and both are SM-coverable. Then, (N, M_0) is live and bounded.*

Proof. Suppose (N_1, M_1) and (N_2, M_2) satisfy Condition B. By Corollary 7.2 (2), (N, M_0) satisfies MST-property. If both (N_1, M_1) and (N_2, M_2) are SM-coverable, then the union of their SM-components is obviously a set of SM-components covering N. Hence N is SM-coverable. By Characterization 2.18, (N, M_0) is live and bounded. $\square$

For completeness, some important results about the preservation of properties of AMGs under COMPOSITION-MP is stated below.

Lemma 7.1. *Suppose AMG (N, M_0) is obtained by applying Composition MP from AMGs (N_i, M_{i0}), $i = 1, 2$. Then, every firing sequence σ_i of (N_i, M_{i0}) is a firing sequence of (N, M_0). Furthermore, if $M_{10}[\sigma_1\rangle M_1$ in N_1, then $M_0[\sigma_1\rangle(M_1 + M_{20}(P_2))$ in N. If $M_{20}[\sigma_2\rangle M_2$ in N_2, then $M_0[\sigma_2\rangle(M_2 + M_{10}(P_1))$ in N.*

Proof. In N, if we refrain from firing any transitions in the part N_2, then σ_1 is obviously firable in N. Similarly, σ_2 is a firing sequence of N. Furthermore, if $M_{i0}[\sigma_i\rangle M_i$, then $M_0[\sigma_1\rangle(M_1 + M_{20}(P_2))$ and $M_0[\sigma_2\rangle(M_2 + M_{10}(P_1))$ follows directly. $\square$

Lemma 7.2. *Suppose AMG $(N, M_0) = (P \cup R, T, F, M_0)$ is obtained by applying Composition MP from AMGs $(N_i, M_{i0}) = (P_i \cup R, T_i, F_i M_{i0})$, $i = 1, 2$, and $\sigma = \sigma_{11}\sigma_{21}\sigma_{12}\sigma_{22} \ldots \sigma_{1j}\sigma_{2k}$ is a firing sequence of (N, M_0), where $\sigma_{11}, \sigma_{12} \ldots \sigma_{1j}$ are transition sequence in T_1 and $\sigma_{21}, \sigma_{22} \ldots \sigma_{2k}$ are transition sequence in T_2. Then, $\sigma_1 = \sigma_{11}\sigma_{12}\ldots\sigma_{1j}$ is a firing sequence of (N_1, M_{10}) and $\sigma_2 = \sigma_{21}\sigma_{22}\ldots\sigma_{2k}$ is a firing sequence of (N_2, M_{20}). Furthermore, if $M_0[\sigma\rangle M$ and $M_{i0}[\sigma_i\rangle M_i$, then $M(P_i) = M_i(P_i)$.*

Proof. (by induction on the length of the firing sequence σ.) For $\sigma = t$, without lose of generality, suppose $t \in T_1$. Let $\sigma_1 = t$ and $\sigma_2 = \phi$. Since

$M_0 = M_{10} + M_{20}(P_2)$, $\forall p \in {}^\bullet t$, $M_{10}(p) = M_0(p) > 0$, t is enabled at M_{10}. Hence, σ_i is a firing sequence of (N_i, M_{i0}) such that $M_{i0}[\sigma_i\rangle M_i$. Suppose $M_0[N, \sigma\rangle M$. Then, by Lemma 7.1, $M = M_1 + M_2(P_2) = M_1(P_1) + M_2(P_2) + M(R)$, where $M_2 = M_{20}$ because $\sigma_2 = \phi$. Next, suppose σ, σ_1 and σ_2 are firing sequences of (N, M_0), (N_1, M_{10}) and (N_2, M_{20}), respectively such that $M_{i0}[\sigma_i\rangle M_i$ and $M = M_1(P_1) + M_2(P_2) + M(R)$. Consider the firing sequence $\sigma' = \sigma t$, $\forall p \in {}^\bullet t$, $M(p) > 0$. Since the resource places are safe and "consumed" by both components, $\forall r \in R$, $M(r) \le M_i(r)$. By the assumption, $M_i(P_i) = M(P_i)$. Hence, $\forall p \in {}^\bullet t$, $M_i(p) > 0$, t is enabled at the reachable marking M_i and $\sigma'_i = \sigma_i t$ is a firing sequence of (N_i, M_{i0}), $i = 1$ or 2. Suppose $M[t\rangle M'$ and $M_i[t\rangle M'_i$, similar to the base part, $M' = M'_1(P_1) + M'_2(P_2) + M'(R)$ follows.   □

Theorem 7.10. *(preservation of liveness and reversibility)*
Suppose AMG (N, M_0) is obtained by applying Composition MP from AMGs (N_i, M_{i0}), $i = 1$, 2, where $N = (P \cup R, T, F)$ and $N_i = (P_i \cup R, T_i, F_i)$. Then, the following propositions hold:

1) *(N, M_0) is live and reversible if, for every siphon D_i in N_i containing the resource place set $R_r \subseteq R$, D_i contains an initially marked trap S_i such that either $R_r \subseteq S_i$ or $R \cap S_i = \phi$, $i = 1$, 2.*
2) *Both (N_1, M_{10}) and (N_2, M_{20}) are live if (N, M_0) is live.*
3) *Both (N_1, M_{10}) and (N_2, M_{20}) are reversible if (N, M_0) is reversible*
4) *$Rank(N) = Rank(N_1) + Rank(N_2)$.*

Proof.

1) For every siphon D in N that does not contain any resource place set $R_r \subseteq R$, by Corollary 7.1, D is the union of siphon D_1 in N_1 and D_2 in N_2 and both D_1 and D_2 do not contain any shared resource places. Since both (N_1, M_{10}) and (N_2, M_{20}) are live, by Characterization 2.23, $F(D_i) > 0$. By Lemmas 7.1 and 7.2, for every reachable marking M in N, there exists a reachable marking M_i in N_i such that $M(D) = M_1(D_1) + M_2(D_2)$. Hence, $F(D) > 0$. If D contains a resource place set $R_r \subseteq R$, by Corollary 7.1, there exists a siphon D_i in N_i such that $R_r \subseteq D_i$ and $D = D_1 \cup D_2$. By the assumption, D_i contains an initially marked trap S_i such that either $R_r \subseteq S_i$ or $R \cap S_i = \phi$, $i = 1$, 2. By Corollary 7.1, $S = S_1 \cup S_2$ is an initially marked trap in N which is contained in D. Hence, $F(D) > 0$ and (N, M_0) is live. By Characterization 2.21, (N, M_0) is reversible.

2) For any marking M_1 and firing sequence σ_1 such that $M_{10}[N_1, \sigma_1\rangle M_1$ and any transition $t \in T_1$, we want to prove there exists a transition sequence σ_1' such that $\sigma_1\sigma_1't$ is a firing sequence of (N_1, M_{10}). By Lemma 7.1, σ_1 is a firing sequence of (N, M_0) and $M_0[N, \sigma_1\rangle M$, where $M = M_1 + M_{20}(P_2)$. Since (N, M_0) is live, $\exists \sigma' = \sigma_{11}\sigma_{21}\sigma_{12}\sigma_{22}\dots$ $\sigma_{1i}\sigma_{2j}$, where $\sigma_{11}\sigma_{12}\dots \ \sigma_{1i}$ are transition sequences of T_1, such that $\sigma_1\sigma't$ is a firing sequence of (N, M_0). Let $\sigma_1' = \sigma_{11}\sigma_{12}\dots \ \sigma_{1i}$. By Lemma 7.2, $\sigma_1\sigma_1't$ is a firing sequence of (N_1, M_{10}). Hence, (N_1, M_{10}) is live. Similarly, (N_2, M_{20}) is live.
3) Similar as the proof of Proposition 2).
4) By the proof of Theorem 7.6, in the incidence matrix

$$C = \begin{bmatrix} C[P_1 \quad T_1] & 0 \\ 0 & C[P_1 \quad T_1] \\ C[R \quad T_1] & C[R \quad T_2] \end{bmatrix}$$

of N, the left bottom will be eliminated to zero. Hence, $Rank(N) = Rank(G_1) + Rank(N_2) = Rank(N_1) + Rank(N_2)$. □

Example 7.5. (for illustrating Theorem 7.10)
Petri nets (N_1, M_{10}) and (N_2, M_{20}) in Figure 7.6 are live and reversible. Since every siphon in these two component nets satisfies Condition 1) of Theorem 7.10, the net N in Figure 7.6 is also live and reversible. Condition 1) of Theorem 7.10 is needed for guaranteeing the liveness of N. For example, in Figure 7.7, the siphon $D_1 = \{r_1, r_2, p_{14}\}$ in N_1 contains an initially marked trap $S_1 = \{r_2, p_{14}\}$ which does not contain the resource place r_1, (N, M_0) is non-live.

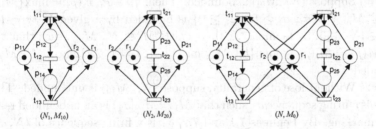

Fig. 7.6   Two augmented marked graphs (N_1, M_{10}) and (N_2, M_{20}) and the composite Petri net (N, M_0).

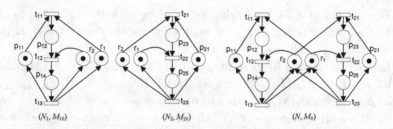

Fig. 7.7   Two augmented marked graphs (N_1, M_{10}) and (N_2, M_{20}) and their composite Petri net (N, M_0).

Note that, by Characterization 2.23, if every minimal siphon D in an AMG (N, M_0) satisfies $F(D) > 0$, then (N, M_0) is live. But, by Composition MP, even if every siphon D_i in N_i satisfies $F(D_i) > 0$, the siphon D computed from D_i by Corollary 7.1 may not satisfy $F(D) > 0$. For example, in Figure 7.7, $D_1 = \{r_1, r_2, p_{14}\}$ and $D_2 = \{r_1, r_2, p_{25}\}$ are siphons of N_1 and N_2 and satisfy $F(D_1) > 0$ and $F(D_2) > 0$, respectively. But in N, siphon $D = \{r_1, r_2, p_{14}, p_{25}\}$ does not satisfy $F(D) > 0$ since after firing transition sequence $t_{11}t_{21}$, D becomes empty. This is why the property "liveness" is not preserved under Composition MP.

Theorem 7.11. *(preservation of boundedness)*
Suppose AMG (N, M_0) is obtained from AMGs (N_i, M_{i0}) by applying Composition MP, where $N = (P \cup R, T, F)$, $N_i = (P_i \cup R, T_i, F_i)$, $i = 1, 2$. Then, (N, M_0) is bounded iff both (N_1, M_{10}) and (N_2, M_{20}) are bounded.

Proof.   By Characterization 2.24 [HJC (2003)], the resource places are safe. In the following, we only need to consider the boundedness of those non-resource places.

($\Leftarrow$) Suppose (N_i, M_{i0}) is bounded. Then, $\forall p \in P_i$, $b_i(p) = \max\{M_i(p) | M_i = M_{i0} + V_i \sigma_i, \sigma_i \geq 0, M_i \geq 0\}$ is bounded by a given K_i, $i = 1, 2$. $\forall M \in [N, M_0\rangle$, by Lemma 7.2, there exists $M_i \in [N_i, M_{i0}\rangle$, such that $\forall p \in P_i$, $M(p) = M_i(p) \leq \max\{b_i(p)\} \leq \max\{K_i\}$, $i = 1, 2$. Hence, (N, M_0) is bounded.

($\Rightarrow$) Without lose of generality, suppose (N_1, M_{10}) is unbounded. Then, there is a firing sequence σ_1 such that $M_1 = M_{10}[\sigma_1\rangle$ is an unbounded reachable marking. By Lemmas 7.1 and 7.2, σ_1 is a firing sequence of (N, M_0) and $M = M_1 + M_{20}$ is a reachable marking by firing σ_1. Hence, M is an unbounded reachable marking. This contradicts with the assumption.

$\square$

7.4   Composition via Merging Two Single Places

This section considers a special case (denoted as COMPOSITION-SP) of COMPOSITION-MP, where only one single pair of places is merged, i.e., $|C| = 1$. It will be shown that some of the conditions in Corollaries 7.1-7.4 and Theorem 7.8 are satisfied automatically and that the Rank Theorem is preserved for this special case only.

Theorem 7.12. *(preservation of conservativeness, strong connectedness, ST-property, MST- property, liveness and boundedness for general Petri nets)*
Let (N, M_0), (N_1, M_1) and (N_2, M_2) be defined in COMPOSITION-SP. Then, the following propositions hold:

1) *N is conservative (resp., strongly connected) iff N_1 and N_2 are conservative (resp., strongly connected).*
2) *N satisfies ST-property (resp., (N, M_0) satisfies MST-property) if N_1 and N_2 satisfy ST-property (resp., (N_1, M_1) and (N_2, M_2) satisfy MST-property).*
3) *(N, M_0) is live and bounded if both N_1 and N_2 are ordinary, pure and SM-coverable, and both (N_1, M_1) and (N_2, M_2) satisfy MST-property.*

Proof.

1) If N_1 and N_2 are conservative and c is their unique shared place, then there exist $\alpha_1 = (x_1 \ldots x_{|P1|}, z') > 0$ and $\alpha_2 = (y_1 \ldots y_{|P2|}, z'') > 0$ such that $\alpha_1 V_1 = 0$ and $\alpha_2 V_2 = 0$. Let $\alpha = (z''x_1 \ldots z''x_{|P1|}z'y_1 \ldots z'y_{|P2|}, z'z'')$. Then, $\alpha > 0$ and $\alpha V = (z''\alpha_1 V_1\ z'\alpha_2 V_2) = 0$, i.e., N is conservative. By Corollary 7.3, N_1 and N_2 are conservative if N is conservative. Since N_1 and N_2 share only a place c, it is obvious that N is strongly connected iff N_1 and N_2 are strongly connected.
2) Let D_i be an arbitrary siphon of N_i for $i = 1$ and 2. If $D_1 \cap D_2 \neq \phi$, then $D_1 \cap D_2 = \{c\}$. This means that N_1 and N_2 satisfy Condition A (resp., (N_1, M_1) and (N_2, M_2) satisfy Condition B). By Corollary 7.2, N satisfies ST-property (resp., (N, M_0) satisfies MST-property).
3) Since N_1 and N_2 are ordinary and pure, N is ordinary and pure. By Corollary 7.4, N is SM-coverable. By Proposition (2), (N, M_0) satisfies MST-property. It follows from Characterization 2.18 that (N, M_0) is live and bounded. $\square$

In general, even if both (N_1, M_1) and (N_2, M_2) are live and only one pair of places is merged, the net (N, M_0) obtained by COMPOSITION-SP may not be live. However, as shown in Theorem 7.13 below, (N, M_0) is live if both N_1 and N_2 satisfy a special property.

Theorem 7.13. *Let (N, M_0), (N_1, M_1) and (N_2, M_2) be defined in COMPOSITION-SP. Suppose, for $i = 1, 2$, (N_i, M_i) is live if it satisfies MST-property. Then, (N, M_0) is live if (N_1, M_1) and (N_2, M_2) satisfy MST-property.*

Proof. For any transition $t \in T = T_1 \cup T_2$ and any marking $M' \in R(N, M_0)$, we will show that $\exists M'' \in R(N, M')$ such that t can be enabled at M''. Without loss of generality, we can assume that $t \in T_1$. For $i = 1$ and 2 and any siphon D_i of N_i, let S_i be the largest trap of N_i within D_i. By assumption, S_i is marked by M_i and also by M_0. It will be shown that either $M'(S_1) > 0$ or $M'(S_2) > 0$. Consider three cases: *Case 1* (S_1 *does not contain* c): By Corollary 7.1 (1a), S_1 is a trap of N. Hence, S_1 is marked by M', i.e., $M'(S_1) > 0$. *Case 2* (S_2 *does not contain* c): By similar argument as in Case 1, $M'(S_2) > 0$. *Case 3* (*Both S_1 and S_2 contain* c): By Corollary 7.1(2), $S_1 \cup S_2$ is a trap of N. Since $M_0(S_1 \cup S_2) > 0$, $M'(S_1 \cup S_2) > 0$, implying either $M'(S_1) > 0$ or $M'(S_2) > 0$. Now, consider the two cases: *Case 1* (If $M'(S_1) > 0$). Then, S_1 is marked by M' in N and by $M_1' = M'|_{P_1 \cup \{c\}}$ in N_1. Since $N_1 \in$ MST, (N_1, M_1') is live. That is, $\exists \sigma_1 \in T_1^*$ and a marking $M_1'' \in R(N_1, M_1')$ such that $M_1'[N_1, \sigma_1\rangle M_1''[N_1, t\rangle$. Hence, $\exists M''$ such that $M'[N, \sigma_1\rangle M''[N, t\rangle$, where $M_1'' = M''|_{P_1 \cup \{c\}}$. *Case 2* (If $M'(S_2) > 0$). Then, (N_2, M_2') is live, where $M_2' = M'|_{P_2 \cup \{c\}}$. Hence, $\exists \sigma_2 \in T_2^*$ and a marking $M_2''' \in R(N_2, M_2')$ such that $M_2'[N_2, \sigma_2\rangle M_2'''$ and $M_2'''(c) > 0$ because $^\bullet c \cup c^\bullet \neq \phi$. That is, $\exists M'''$ and σ_2 such that $M'[N, \sigma_2\rangle M'''$ and $M'''(c) > 0$. This implies that $M''' (S_1) > 0$ and leads to Case 1. Hence, $\exists M''$ and σ_1 such that $M'[N, \sigma_2\rangle M'''[N, \sigma_1\rangle M''[N, t\rangle$. $\square$

Note that there are many Petri nets that satisfy the condition required in Theorem 7.13, such as free choice nets, extended non-self controlled nets [BCD (1995)] and augmented marked graphs. There is some difference in applying Theorem 7.12(3) and Theorem 7.13 to determine the liveness of the net obtained from COMPOSITION-SP. Theorem 7.12(3) requires N_i to be conservative and (N_i, M_i) to satisfy MST-property; whereas Theorem 7.13 does not require N_i to be conservative but requires (N_i, M_i) to be live as a consequence to its satisfaction of MST-property.

As shown in the following theorem, COMPOSITION-SP preserves liveness, boundedness and reversibility for some special classes of Petri nets without additional constraints.

Theorem 7.14. *(preservation of liveness, boundedness and reversibility under COMPOSITION-SP for AMGs, FC nets and AC nets)*
Let (N, M_0), (N_1, M_1) and (N_2, M_2) be the three Petri nets and c be the common place defined in COMPOSITION-SP. Then, the following propositions hold:

1) Suppose (N_1, M_1) and (N_2, M_2) are AMGs. Then, (a) (N, M_0) is live and reversible iff both (N_1, M_{10}) and (N_2, M_{20}) are live. (b) (N, M_0) is reversible iff both (N_1, M_{10}) and (N_2, M_{20}) are reversible.
2) Suppose N_1 and N_2 are FC nets and $M_i(c) > 0$ for $i = 1$ and 2. Then, (N, M_0) is a live and bounded ordinary AC net satisfying ST-property iff (N_i, M_i) is live and bounded.

Proof.

1) (a) ($\Rightarrow$) This follows from Theorem 7.10(2).
   ($\Leftarrow$) Similar as the proof of Theorem 7.10(1), for every minimal siphon D in N, if $c \notin D$, then $F(D) > 0$. If $c \in D$, by Corollary 7.1, there exist siphons D_1 in N_1 and D_2 in N_2 such that $D = D_1 \cup D_2$. Since (N_1, M_{10}) and (N_2, M_{20}) are live, $F(D_i) > 0$. Since there is only one resource place, at any time, the resource place is "consumed" at one component net. By the proof of Lemma 7.2, for every reachable marking M in N, there exists a reachable marking M_i in N_i such that either $M(c) = M_1(c)$ or $M(c) = M_2(c)$. Hence, $M(D) = M_1(D_1) + M_2(D \cap P_2)$ or $M(D) = M_1(D_1 \cap P_1) + M_2(D_2)$. For both cases, $F(D) \geq F(D_i) > 0$. By Characterization 2.23, (N, M_0) is live.

   (b) ($\Rightarrow$) This follows from Theorem 7.10(3).
   ($\Leftarrow$) Since there is only one resource place, by Lemma 7.2, for every reachable marking M and firing sequence σ such that $[M_0, \sigma\rangle M$, there is a reachable marking M_i of (N_i, M_{i0}) such that $M = M_1 + M_2(P_2)$ or $M = M_2 + M_1(P_1)$. Without loss of generality, suppose $M = M_1 + M_2(P_2)$. This implies that the resource is "consumed" by N_1. Since (N_1, M_{10}) is reversible, there exists a firable transition sequence σ_1 such that $M_1[\sigma_1\rangle M_{10}$. By Lemma 7.1, σ_1 is a firable sequence of N such that $M[\sigma_1\rangle(M_{10} + M_2(P_2))$,

where $(M_{10} + M_2(P_2)) = M_{10}(P_1) + M_2$. Since (N_2, M_{20}) is reversible, there exists a firable sequence σ_2 such that $M_2[\sigma_2\rangle M_{20}$. By Lemmas 7.1 and 7.2, $\sigma\sigma_1\sigma_2$ is a firable sequence of N such that $M[\sigma_1\sigma_2\rangle(M_{10}(P_1) + M_{20}) = M_0$. Hence, (N, M_0) is reversible.

2) ($\Leftarrow$): By the definition of COMPOSITION-SP, N is an AC net if N_1 and N_2 are FC nets. For $i = 1$ and 2, since (N_i, M_i) is a live and bounded FC net, (N_i, M_i) satisfies MST-property and N_i is structurally bounded [DE (1995)]. By Theorem 7.12(2), (N, M_0) satisfies MST-property. According to Characterization 2.19, (N, M_0) is live. By Theorem 7.8, N is structurally bounded. ($\Rightarrow$): Since (N, M_0) is a live and bounded ordinary AC net satisfying ST-property, N is covered by minimal siphons and every minimal siphon of N is a trap marked by M_0 [JCL (2004)]. For $i = 1$ and 2, consider any minimal siphon D_i of N_i. If $c \notin D_i$, by Corollary 7.1, D_i is a siphon of N. It is obvious that D_i is also a minimal siphon of N and thus is a trap marked by M_0. By Corollary 7.1 and the definition of COMOPOSITION-SP, D_i is a trap of N_i marked by M_i. Suppose $c \in D_i$. Let D be a minimal siphon of N such that $c \in D$. According to Corollary 7.1(1), $(D \cap P_i) \cup \{c\}$ is a siphon of N_i. By Corollary 7.1(2), $D_i \cup ((D \cap P_j) \cup \{c\})$ is a siphon of N, where $i \neq j$. If $D_i \cup ((D \cap P_j) \cup \{c\})$ is not minimal, then $\exists D' \subset D_i \cup ((D \cap P_j) \cup \{c\})$ such that D' is a minimal siphon of N, this contradicts with the fact that D is minimal in N or that D_i is minimal in N_i. Hence, $D_i \cup ((D \cap P_j) \cup \{c\})$ is a minimal siphon of N and also is a trap of N marked by M_0. By Corollary 7.1(1), D_i is a trap of N_i. Since $M_i(c) > 0$, D_i is marked by M_i. That is, (N_i, M_i) satisfies MST-property and thus (N_i, M_i) is live. Since (N, M_0) is a live and bounded ordinary AC net satisfying ST-property, N is structurally bounded [JCL (2004)]. By Theorem 7.8, N_i is structurally bounded, implying that (N_i, M_i) is bounded. $\square$

Next, it will be shown that Rank Theorem [DE (1995)] is preserved under COMPOSITION-SP.

Theorem 7.15. *(preserving the Rank Theorem under COMPOSITION-SP for general Petri nets)*
Let N, N_1 and N_2 be defined in COMPOSITION-SP. Then, the following propositions hold:

1) N satisfies the Rank Theorem if both N_1 and N_2 satisfy the Rank Theorem.

2) If N satisfies the Rank Theorem, then either both N_1 and N_2 satisfy the Rank Theorem or none of them does.

Proof. By Corollary 7.3(1), existence of positive P-invariants is preserved. It remains to consider T-invariants and the RC-property. Since the only change in clustering after COMOPOSITION-SP is that the two clusters ($[c]$ in N_1) and ($[c]$ in N_2) are merged to one, we have $|C(N)| = |C(N_1)| + |C(N_2)|-1$. The incidence matrices of N, N_1 and N_2 have the forms:

$$V = \begin{array}{c} \\ P_1 \\ P_2 \\ c \end{array} \begin{array}{cc} T_1 & T_2 \\ \begin{pmatrix} V_{11} & 0 \\ 0 & V_{22} \\ V_{31} & V_{32} \end{pmatrix} \end{array}$$

$$V_1 = \begin{array}{c} \\ P_1 \\ c \end{array} \begin{array}{c} T_1 \\ \begin{pmatrix} V_{11} \\ V_{31} \end{pmatrix} \end{array}$$

and

$$V_2 = \begin{array}{c} \\ P_2 \\ c \end{array} \begin{array}{c} T_2 \\ \begin{pmatrix} V_{22} \\ V_{32} \end{pmatrix} \end{array}$$

1) By Corollary 7.3(3), N has a positive T-invariant if N_i for $i = 1$ and 2 has a positive T-invariant. For $i = 1$, 2, since N_i satisfy RC-property, $Rank(V_i) = |C(N_i)|- 1$ By Theorems 7.6 and 7.10, $Rank(V) = Rank(V_1) + Rank(V_2) = Rank(V_{11}) + Rank(V_{22}) = (|C(N_1)|- 1) + (|C(N_2)|- 1) = |C(N)|- 1$.

2) By similar argument as 1), $Rank(V_{11}) + Rank(V_{22}) = Rank(V) = |C(N)|- 1 = |C(N_1)| + |C(N_2)|- 2$. That is, $Rank(V_{11}) - |C(N_1)| + 1 = -(Rank(V_{22}) - |C(N_2)| + 1)$. This implies either both or none of N_1 and N_2 satisfies the RC-property. If $\beta = (\beta_1, \beta_2)$ is a positive T-invariant of N, then $V\beta = 0$, implying $V_{11}\beta_1 = 0$, $V_{22}\beta_2 = 0$ and $V_{31}\beta_1 + V_{32}\beta_2 = 0$. If $Rank(V_1) = |C(N_1)|- 1 < |P_1|$, then the row of V_{31} is a linear combination of the other rows of V_{11}. Then, $V_{11}\beta_1 = 0$ implies that $V_{31}\beta_1 = 0$ and $V_1\beta_1 = 0$. In turns, this implies $V_{32}\beta_2 = 0$ and $V_2\beta_2 = 0$. Hence, both N_1 and N_2 have a positive T-invariant. $\square$

7.5   Application

It is beyond the scope of this chapter to describe in detail how to compose
an agent or manufacturing system in general. Our aim is just to illustrate
how to apply our methodology presented in Sections 7.2 and 7.3 to design
a correct system through examples in the Multi-agent systems and Manufacturing systems.

Application to agent system design:

An agent is a software module that provides specialized services by coor-
dinating with various resources (e.g., plans, knowledge-bases, environments,
utilities, other agents, etc.) via message communications. Very often, these
resources are shared among agents. When a service request enters an agent,
a token is deposited into the entry place of its Petri net representation.
Processing a message may pass through several agents. Structurally, an
agent is composed of various kinds of modules, such as *internal message
process* (IMP), *external message process* (EMP), *internal resource handler*
(IRH) and *external resource handler* (ERH). An IRH manipulates resources
used within an agent and an ERH manipulates resources shared by several
agents.

Example 7.6. Consider an agent composed of two basic modules - an IMP
(N_1 in Figure 7.8(a)) and an ERH (N_2 in Figure 7.8(b)). The two modules
have very similar structure. In particular, both have four resource places:
OMP, GPK, UD and *ME*. In general, these resources do not have to be
shared. In this example, merging the *OMP*s, *GPK*s and *UD*s indicates
that these resources are now shared in usage. The merging of *ME*s ren-
ders the three functions *make-decision*, *gpk-sensing* and *e-sensing* mutually
exclusive.

Verification: It may be assumed that place merging has occurred in two
stages because two different place-merging operators have been applied.

In the first stage, COMPOSITION-MP merges the common set of places
$C = \{ME, UD, GPK\}$. Before merging, (N_1, M_1) and (N_2, M_2) are both
live, bounded and reversible. N_1 and N_2 have the sets of minimal siphons
$K_1 = \{\{entry1, p_1, OMP1, exit1\}, \{ME, p_1, UD\}, \{GPK\}\}$ and $K_2 =
\{\{entry2, p_2, OMP2, exit2\}, \{ME, p_2, UD\}, \{GPK\}, \{EN\}\}$, respectively.
(Note: The last digit indicates the net it belongs to.) Both N_1 and N_2
satisfy Condition (1) of Corollary 7.4. For example, for the siphons $D_1 =
\{entry1, ME, p_1, UD, OMP1, exit1\}$ of N_1 and $D_2 = \{entry2, ME, p_2, UD,$

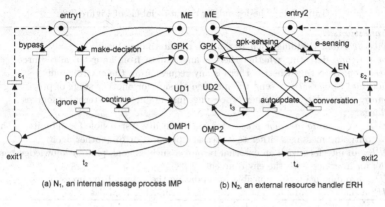

(a) N_1, an internal message process IMP   (b) N_2, an external resource handler ERH

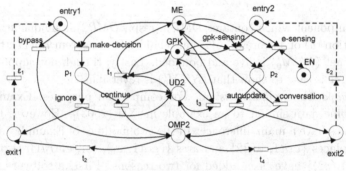

(c) N', the integrated net of the agent

Fig. 7.8   An example of applying resource-place merging techniques to agent system design.

OMP2, *exit2*} of N_2, where $D_1 \cap D_2 \neq \phi$, there exist traps $S_1 = \{ME, p_1, UD\} \subseteq D_1$ and $S_2 = \{ME, p_2, UD\} \subseteq D_2$ such that $S_1 \cap C = \{ME, UD\} = S_2 \cap C$. N_1 and N_2 also satisfy Condition 2) of Corollary 7.4 because both N_1 and N_2 are SM-coverable. Hence, by Corollary 7.4, the integrated net (N, M_0) (not shown) is live and bounded.

In the second stage, MERGE-N-PLACE is applied to N with $Q_1 = \{OMP1, OMP2\}$. Since all the minimal siphons of N, i.e., $D_1 = \{GPK\}$, $D_2 = \{EN\}$, $D_3 = \{ME, p_1, UD, p_2\}$ and $D_4 = \{entry1, p_1, OMP1, exit1, entry2, p_2, OMP2, exit2\}\}$ are also traps and that, every siphon D satisfying $Q_1 \subseteq D$ contains at least one of these traps. Hence, N satisfies Condition (b) of Theorem 7.5(17). For example, for the siphon $D = \{entry1, p_1, OMP1, exit1, OMP2, p_2, ME, p_1, UD\}$ of N satisfying $Q_1 \subseteq D$, there exists the trap $S = \{ME, p_1, UD, p_2\}$ such that $Q_1 \subseteq S$. The

Table 7.1   The legend for essential labels of Figure 7.8.

Transitions:
make-decision: deciding what to do with an internal message
gpk-sensing: detecting whether there is any request from an external resource
e-sensing: detecting whether there is any request from the environment
continue; conversation: send the message outside for another stage of processing
ignore; autoupdate: no further processing, just update the external resources
Places:
OMPx (*Outgoing message process*): a public utility responsible for submitting the outgoing messages either to a message protocol or to another agent
GPK: an interface with an external resource, such as goal, plan or knowledge-base
EN: an interface with the environment
UDx: a public utility responsible for updating the external resources
ME: a place for realizing mutual exclusion with other modules

SM-component which contains $OMP1$ (resp., $OMP2$) is unique and hence Condition (a) of Theorem 7.5(17) also holds. By Theorem 7.5(17), the produced net (N', M_0') is live and bounded. Lastly, though not supported by theory, it is easy to show that (N', M_0') is reversible.

Comparison and discussion: For comparison purposes, Example 7.6 uses some data similar to an example in [XU (2003)]. However, the two examples have a main difference: In the two modules of Example 7.6, some extra places $\{exit1, \ exit2\}$ and arcs $\{(exit1, \ \varepsilon_1), \ (\varepsilon_1, \ entry1), \ (exit2, \ \varepsilon_2), \ (\varepsilon_2, \ entry2)\}$ have been added for two reasons. First, a software module needs an exit to indicate the end of an execution cycle. Second, a Petri net can never be live without these additions.

Let us compare the design approaches adopted in the two papers. In [XU (2003)], without any specific rules, a design similar to Figure 7.8(c) but with a deliberate error is first submitted to a tool for verification and the error is found after traversing many firing sequences. The design is then modified and rechecked. Note that this step may have to be repeated and is very time-consuming. In Example 7.6, we start by creating two correct modules, an IMP and an ERH. These modules are quite primitive and their liveness and boundedness can be easily verified. An agent is then created by merging their resource places. Since the place-merging processes preserve these properties, the agent satisfies these properties without the need of verification.

Application to manufacturing system design:

Example 7.7. Figure 7.9 shows a manufacturing system consisting of two disconnected subsystems N_1 and N_2 for handling raw materials F-parts and

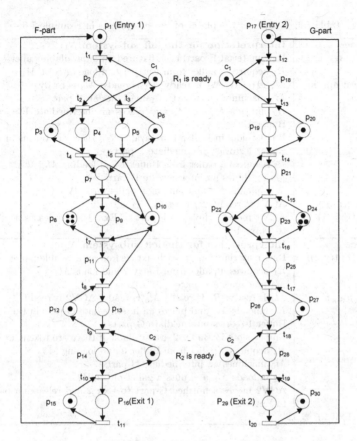

Fig. 7.9   Sharing Robot R_1 (at c_1) and Robot R_2 (at c_2); merging at places c_1 and c_2.

G-parts, respectively. Robots R_1 and R_2 are shared between N_1 and N_2. Robot R_3 is shared within N_1 and Robot R_4 within N_2. COMPOSITION-MP is applied to merge their resource places $C = \{c_1, c_2\}$. The data in this example is extracted from [ZD (1993)] after some modifications. However, Zhou used it for a different purpose and verification is done by direct proof.

Subsystem N_1 runs as follows (N_2 runs in a similar way): Depositing a token in p_1 indicates that an F-part is available for processing. Robot R_1, if available, loads this part onto machine M_1 or M_2 which then produces an intermediate F-part. When such a part is available and there is an empty slot in Buffer 1, Robot R_3, if available, will unload M_1 or M_2 and place it into the slot. Next, if Machine M_3 is ready, Robot R_3, if available, will acquire the finished intermediate part and load it into M_3. M_3 will machine

Table 7.2 Labels for the places of the subsystems in Example 7.6.

Place	Interpretation for the left subsystem N_1:
p_1 (Entry 1)	Raw material F-part, each fixtured on an available pallet
p_2	R_1 acquires a pallet from Entry 1 and loads M_1 or M_1
p_3; p_6; p_{10}; p_{12}	M_1 is ready; M_1 is ready; R_1 is ready; M_3 is ready
p_4; p_5	M_1 machines an F-part; M_2 machines an F-part
p_7	R_3 unloads M_1 or M_2 and places an intermediate F-part in Buffer 1
p_8	Empty slots in Buffer 1 are available if there are tokens in it
p_9	Buffer 1 stories intermediate F-parts
p_{11}	R_3 acquires a pallet from Buffer 1 and loading M_3
p_{13}	M_3 machines an intermediate F-parts
p_{14}	M_2 unloads M_3 and putting a pallet on AGV1
p_{15}; c_1; c_2	AGV1 is ready; R_1 is ready; R_2 is ready
p_{16} (Exit 1)	AGV1 moves a finished F-part to Exit 1 and releases a pallet to Entry 1
Place	**Interpretation for the left subsystem N_2:**
p_{17} (Entry 2)	Raw material G-part, each fixtured on an available pallet
p_{18}	R_1 acquires a pallet from Entry 2 and loads M_4
p_{19}	M_4 machines a G-part
p_{20}; p_{22}; p_{27}	M_4 is ready; R_4 is ready; M_5 is ready; AGV2 is ready
p_{30}	R_4 unloads M_4 and places an intermediate G-part in Buffer 2
p_{21}	Buffer 1 stories intermediate G-parts
p_{23}	Empty slots in Buffer 2 are available if there are tokens in it
p_{24}	R_4 acquires a pallet from Buffer 2 and loading M5
p_{25}	M_5 machines an intermediate G-parts
p_{26}	R_2 unloads M_5 and puts a pallet on AGV2
p_{28}	AGV2 moves a finished G-part to Exit 2 and releases a pallet to Entry 2
p_{29} (Exit 2)	

this intermediate F-part. Then, Robot R_2, if available, will unload M_3 and put a pallet on AGV1. When ready, AGV1 will move a finished F-part to Exit 1 and release a pallet for the next F-part.

In Figure 7.9, N_1 and N_2 have the sets of minimal siphons $K_1 = \{\{p_2, c_1\}, \{p_3, p_4\}, \{p_5, p_6\}, \{p_7, p_{10}, p_{11}\}, \{p_8, p_9\}, \{p_{12}, p_{13}\}, \{p_{14}, c_2\}, \{p_{15}, p_{16}\}, \{p_1, p_2, p_4, p_5, p_7, p_9, p_{11}, p_{13}, p_{14}, p_{16}\}\}$ and $K_2 = \{\{p_{17}, p_{18}, p_{19}, p_{21}, p_{23}, p_{25}, p_{26}, p_{28}, p_{29}\}, \{p_{18}, c_1\}, \{p_{19}, p_{20}\}, \{p_{21}, p_{22}, p_{25}\}, \{p_{23}, p_{24}\}, \{p_{26}, p_{27}\}, \{p_{28}, c_2\}$ and $\{p_{29}, p_{30}\}\}$, respectively. N_1 and N_2 satisfy Condition (1) of Corollary 7.4 because, $\forall D_i \in K_i$, $i = 1$ and 2, D_i is also a trap itself and, if $D_1 \cap D_2 \neq \phi$, there exist a trap S_1 in D_1 and a trap S_2 in D_2 such that $S_1 \cap C = S_2 \cap C$. N_1 and N_2 also satisfy Condition (2) of Corollary 7.4 because both of them are SM-coverable. Hence, by Corollary 7.4, the integrated net is live and bounded.

Example 7.8. The manufacturing system (Figure 7.10, Table 7.3) consists of three processes: two workstations WS_1 and WS_2(Process 1 and Process 2 on the left of Figure 7.10) and one machining centre (Process 3 on the right of Figure 7.10). Processes 1 and 2 share robot R_2 between themselves and share Robot R_1 with Process 3. Machining is done in Process 3; whereas assembly work is done in WS_1 and WS_2. (Note: The left and right components of Figures 7.10 are used in [ZHO (1996)] for explaining the concepts of vertical and horizontal mutual exclusions in resource sharing. We combine them to create an example for illustrating our results.) The description of running the system was presented in Example 6.6.

In manufacturing system modeling, a place usually represents the status of a resource or an operation. A transition represents the start or completion of an event or operation. Our method has two steps:

(1) Represent each process by an augmented marked graph (Figure 7.10).
(2) Merge the processes by Composition MP (Figure 7.11)

Figure 7.10 shows the three augmented marked graphs of the independent processes after Step 1. All three component nets are live, bounded and

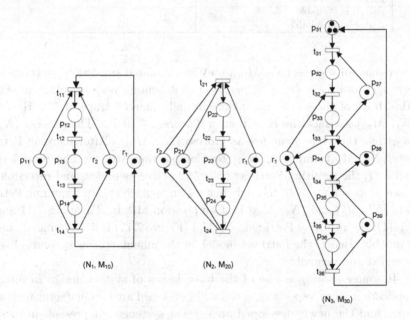

Fig. 7.10   Three independent processes created in Step 1.

Table 7.3   The legend for Figure 7.10.

Place	Transitions
r_1: Robot 1 is available	t_{11}: start acquiring R_1 and R_2
r_2: Robot 2 is available	t_{12}: start first step of assembling
p_{11}: WS_1 requests R_1 and R_2	at WS_1
p_{12}: WS_1 acquires R_1 and R_2	t_{13}: start final step of assembling
p_{13}: first step of assembling at WS_1	at WS_1
p_{14}: final step of assembling at WS_1	t_{14}: complete assembling at WS_1
p_{21}: WS_2 requests R_1 and R_2	t_{21}: start acquiring R_1 and R_2
p_{22}: WS_2 acquires R_1 and R_2	t_{22}: start first step of assembling
p_{23}: first step of assembling at WS_2	at WS_2
p_{24}: final step of assembling at WS_2	t_{23}: start final step of assembling
p_{31}: pallets are available	at WS_2
p_{32}: machine M_1 loads, fixtures and	t_{24}: complete assembling at WS_2
processes a palleted raw part	t_{31}: start activity p_{32}
p_{33}: R_1 unloads an intermediate part	t_{32}: complete activity p_{32} and
to the buffer	start activity p_{33}
p_{34}: buffer B stores an intermediate	t_{33}: complete p_{33} and start the
part	storage activity p_{34}
p_{35}: machine M_2 loads and processes	t_{34}: complete p_{34} and start
an intermediate part	activity p_{35}
p_{36}: R_1 unloads a final product from	t_{35}: complete p_{35} and start p_{36}
M2, defixtures and returns the	t_{36}: complete p_{36}
pallet	
p_{37}: M_1 is available	
p_{38}: B is available	
p_{39}: M_2 is available	

reversible. In process (N_1, M_{10}), only two minimal siphons $S_1 = \{r_1, p_{12}, p_{13}, p_{14}\}$ and $S_2 = \{r_2, p_{12}, p_{13}, p_{14}\}$ that contain resource place subset. Also, both of these two siphons are initially marked traps of N_1. Hence, (N_1, M_{10}) satisfies the condition in Theorem 7.10(1). The process (N_2, M_{20}) has the same structure as process (N_1, M_{10}). After merging Petri nets (N_1, M_{10}) and (N_2, M_{20}) by Composition MP, by Theorems 7.10 and 7.11, the resulting Petri net (N', M_0') is live, bounded, and reversible. Since process (N_3, M_{30}) has only one resource place r_1, after merging Petri nets (N', M_0') and (N_3, M_{30}) by Composition MP, by Theorems 7.11 and 7.14(1), the resulting Petri net (N, M_0) (Figure 7.11) is live, bounded, and reversible. Hence, the Petri net model for the manufacturing system is live, bounded and reversible.

Resource sharing is one of the basic issues of system design in many application areas, especially in the well-developed area, manufacturing systems, and the newly-developed area, agent systems. At present, in most cases, this problem is dealt with by verifying the design directly by an ad

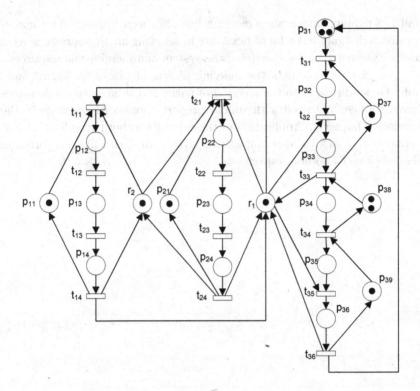

Fig. 7.11   The Petri net model created in Step 2.

hoc approach. If errors are found, the design has to be modified and re-verified. This chapter proposes a new approach which includes two steps: Step 1 creates a set of the subsystems while assuming the resources are not shared. Step 2 creates a design by integrating these subsystems via merging their resource-places. This approach is based on two principles: (1) Most subsystems are primitive and thus easy to design and verify. (2) If the subsystems are correct (i.e., possessing the desirable properties) and the place merging process preserves these properties, then the design is correct. This approach avoids the lengthy process of verifying the design.

An extensive methodology has been developed for supporting this approach. Briefly, depending on the structure of the resources with respect to the subsystems, four classes of place-merging operators are investigated. For each class, conditions on the merging of the Petri nets adopted for modeling are proposed whereby the produced design will automatically be correct. Preservation of many properties other than liveness, boundedness and reversibility under place-merging has also been proved. As a consequence, a designer has a lot of flexibility in selecting an appropriate process and a right net type for specifying the system and merging the resources.

This chapter considers the merging of sets of places to single places only. Generalizations to the merging of paths and subnets to single places have also been obtained with similar property-preservation results in the previous Chapter 4. Application of these results to both specification and verification of MAS and manufacturing system designs will be reported in the next two Chapters, respectively.

Chapter 8

Application of PPPA to Component-Based Design of Manufacturing Systems

This Chapter presents the application of PPPA to the specification and verification of manufacturing system design. It illustrates the application by designing a Manufacturing System (MS) with component-based approach. PPPA handles three problems arising in the component-based approach: 1) Integrating the primitive modules by using composite operators specifed in PPPA to create the final model; 2) Handling resource sharing by using place merging operators; and 3) Specifying the machining and/or assembly operations by using place refinement operators. Since all the operators in PPPA preserve many properties, among other features, PPPA does not need to verify composite components. In other words, if the primitive modules satisfy the desirable properties, each of the composite components, including the system itself, also satisfies these properties.

8.1 Background and Motivation

Recently, researchers pay more attentions on the following features for manufacturing system (MS) design:

Resource sharing: In a manufacturing system, resources such as robots, workers, machines, assembly lines, buffers etc. sometimes are shared by several processes. It becomes a very important aspect to handle resource-sharing during MS design. Note that 'sharing' does not imply simultaneous usage in MS. Simultaneous usage can be handled by re-enterable codes in software systems but is not allowed in manufacturing systems. In MS, 'sharing' requires a resource to be occupied by some part(s) exclusively during utilization and is released afterwards. For multiple-resource systems, deadlocks or overflow may be caused by a wrong order in occupying and releasing these resources. In the literature, handling resource-sharing problems focuses on mutual exclusion [ZD (1993)] and resource allocation [COL (2003); REV (1998)]. They assume that the resources will not be modified when they are switched from one process to another. In this book, besides the unchangeable case in the literature [HJC (2003); JHC (2005, 2008); ZD (1993)], we consider the case that the resources are modified when they leave one process for another. A property-preserving place-merging operator is applied to handle the resource sharing problem. Property-preserving means, if the original system satisfies some properties, after the operators in the algebra were applied, the resultant system still has the same properties.

Component-based architecture: Component-based architecture allows implemented components to be constructed from some existing ones. This architecture not only enlarges the reusability of the system modules but also increases the system running speed. However, it may face difficulties during integration when applied to operation system, interfaces, communication etc. In the literature, the net-based approaches concerning integration include place merging [AC (1978); HJC (2003); NV (1985); SOU (1990)], transition merging [VMS (1988)], path merging [AV (1988); KB (1986)] and modular synthesis [DG (1984); GH (1986); ZDD (1989)]. In this book, the integration problem is handled by applying property-preserving composition operators (Chapter 4) such as enable, choice, interleave and iteration, etc.

Property-preserving algebra: In the manufacturing system design, the designed properties are often destroyed when processes are modified or constituted for creating new processes. Hence, researchers pay many efforts on searching for the property-preserving operators for MS design in order to preserve the desirable properties. For example, [LF (1987); SM (1983); HJC (2005)] found many property-preserving reduction rules, [GV (2003); HAM (2003); HCM (2004); SM (1983)] considered refinement methodologies and [HJC (2003); MAK (2001); NV (1985); SOU (1990); JHC (2005, 2008)] paid attention on composition rules. [AAL (1997); BES (2001); CZ (1998); MAK (2001); HUA (2004)] included reduction, refinement and composition. Many earlier references can be found in the survey papers [BGV (1990); JD (1993)]. Although there are many synthesis and reduction results for Petri nets, the following perspectives are still need to be improved: (a) The approaches should be applied not only to special Petri nets, but also general Petri nets; (b) Many more system properties such as siphon, trap, reversibility, proper termination and so on should be preserved when the modules are merged and/or linked or stepwise refinements are applied; and (c) Property-preserving should be verified in the approaches.

The Property-Preserving Petri Net Process Algebra (PPPA) applied in this book overcomes the difficulties mentioned above.

For designing MS, the component-based approach starts by specifying the first level primitive modules of the MS as PNPs. Then, according to the control flow of the system, the second level MS are created by integrating the primitive modules based on the integration operators of PPPA. Finally, the implementation level MS is obtained by applying refinements in PPPA. Among other features, this approach has two desirable features for the design of MS: (1) It eliminates the need of verifying the composite components since all the operators are required to preserve the desirable properties. Hence, if the primitive PNPs satisfy these properties (correct), each of the composite components, including the MS itself, also satisfies these properties (correct). (2) It avoids such errors as deadlock and overflow by using a property-preserving place-merging technique. These errors often arise when handling resource-sharing. At the same time, considering changeable resources makes the resource sharing problem more flexible.

It should be mentioned that, in the literature, there are two basic approaches for the design of Petri net based manufacturing systems: bottomup and top-down. Bottom-up approach begins with the primitive modules. The final system was constructed by merging and /or linking of all these primitive modules. Top-down approach begins with the first level Petri net model. Then the implementation level was reached by the stepwise refinement for the first level model. The component-based approach applied in this book is in fact a combination of bottom-up and top-down approaches. Hence, the results concerning both bottom-up [AV (1988); DG (1984); GH (1986); HJC (2003); KB (1986); SOU (1990); VMS (1988); HJC (2005)] and top-down approaches [GV (2003); HAM (2003); HCM (2004); LF (1987); SM (1983)] are also available in the approach of this book.

For the rest of this chapter, Section 8.2 outlines how to apply PPPA to a component-based approach for MS design. Based on PPPA, the approach for specifying and verifying the design of MSs is illustrated in detail with an example. Section 8.3 is devoted to the specification of the primitive modules and the creation of the composite components of MS and the verification of these components. Some concluding remarks are given at last.

The results in this chapter are mainly extracted from the literature [HCW (2007)].

8.2 Summary of Applying PPPA to Component-Based Approach

Briefly speaking, the component-based approach combines bottom-up and top-down approaches. It starts with the creation of the top-leveled primitive modules of the system. Then, similar to bottom-up approach, the top-leveled system model is constructed by combining the modules. Finally, the implementation-leveled model is obtained by stepwise refinement. PPPA is applied mainly to the last two steps for specification and verification. In the following, we describe how PPPA is applied to each step of the component-based approach.

Step 1. (Creating the primitive modules):

The subsystems and resources are described as autonomous primitive modules. Each module is specified as a Petri net process (PNP) (see Chapter 3).

Usually, for different systems, the modeling methods for the primitive modules may be different. For example, in a multi-agent system, the primitive modules such as message checker, resource-request handler, protocol selector etc. and the resources such as knowledge bases, environment, plan etc. are modeled according to their functions (Chapter 9). In the workflow system, the primitive modules are modeled according to the workflow of the system [MAK (2001)]. In this chapter, the primitive modules for a manufacturing system are modeled according to the relations between the basic operations and the resources used in the operations in the system. In order to simplify the verification process and provide a correct system model, each operation of the system is modeled as a single place. Hence, the structure for each primitive module is very simple and the verification for each module is trivial.

Step 2. (Creating the top-leveled system model):

According to the relations between the primitive modules, the PNPs created in Step 1 are integrated by some composition operators such as Choice, Enable, etc. or modified by applying some refinement and reduction operators. Resource sharing, if being part of the requirements, will be resolved by place merging. The resultant system is also a PNP (Chapters 3-7).

In this step, PPPA are mainly applied to solve the following problems:

1) *Integrating the modules by using composition operators and reduction operators.*
Previous research integrates the components by merging places [AC (1978); NV (1985); SOU (1990); HJC (2003)], merging transitions [VMS (1988)], merging paths [AV (1988); KB (1986)] or modular synthesis [DG (1984); GH (1986); ZDD (1989)]. By the integration methods, except P-invariant, liveness and boundedness, no other properties can be verified. In this chapter, composition operators such as Choice, Enable, etc in the PPPA are applied to the integration of components. Reduction operators are used in order to simplify the structure of the system models and preserve the functions of the system. Composition operators and reduction operators can preserve liveness, boundedness and reversibility. They can also preserve or conditionally preserve many other properties such as proper termination, P-invariant, siphon, trap and so on.

2) *Handling resource sharing by place merging.*
Handling resource sharing is a difficult problem in the system design. The literature [ZD (1993)] solves this problem by using parallel and sequential exclusion techniques. Although the technique is an efficient tool for verifying liveness, boundedness and reversibility, it itself is too complex to be used for checking many other properties of the system. Place merging is found to be very efficient for handling resource sharing problems in systems design. In the literature, conclusions for place merging are limited to preserving P-invariant [AC (1978)], T-invariant [NV (1985)] and liveness [SOU (1990)]. Recently, preserving many other properties is proved to be possible for place merging operators [JHC (2005, 2008)](Chapter 7).
In this chapter, the resources used in different modules are considered to be different. As will be seen in Section 8.3, these resource places representing the same resources are merged into single ones. As described in Chapter 7, the place-merging technique for handling resource sharing will preserve many desirable properties.

Step 3. (Creating the implementation-leveled system model):
In step 1, each module is modeled as the first-leveled model. Hence, the model for the system after integrating the components is still at the high level. In order to obtain the implementation-leveled model, some places or

transitions are needed to be refined with their detailed functional models. In this chapter, since each operation is modeled as a single place, place refinement is applied to specify the operations. In the previous research, resources are considered to be unchangeable during proceeding in the system. In this chapter, resources are allowed to be modified when they leave one process for another. This assumption makes the models more flexible. Hence, after place merging for handling resources sharing, place refinement should be applied to the resource places in order to specify the modification for the resources.

In fact, this step is similar to the top-down approach. Hence, besides place refinement operators in this chapter, other synthesis and reduction techniques [LF (1987); SM (1983); HCM (2004); HJC (2005); GV (2003); HAM (2003); SM (1983)] in the top-down approaches are applicable in this step when necessary.

Step 4. (Verifying the system model):

No much effort is needed for verification in the approach because of the property-preserving characteristics of all the operators. In fact, if the constituent PNPs satisfy such properties as liveness, boundedness, reversibility and proper termination, the intermediate PNPs obtained by applying these operators also satisfy these properties. However, the burden is shifted to making sure that the initial PNPs satisfy the specified conditions.

8.3   Application of PPPA to Manufacturing System Design

This section illustrates the application of PPPA with an example of designing a manufacturing system. The organization of this section is according to the steps described in Section 8.2.

8.3.1   *Creating the primitive modules*

This subsection specifies the various primitive modules of the manufacturing system as a set of PNPs. It is Step 1 of the approach as outlined in Section 8.2.

The manufacturing system constructs the final product from three primitive parts by using four machining centers MC_i (each MC_i contains a machine M_i), $i = 1, 2, 3, 4$, two assembly stations A_1 and A_2, two robots R_1 and R_2 and a buffer B.

The production is produced as follows:

1) Part 1 is machined by M_1. Part 2 is machined by M_2. Each part is automatically fixtured to a pallet and loaded into the machine.

2) After processing, Parts 1 and 2 enter assembly station A_1 for producing Part S. When either A_1 or A_2 is ready to execute the assembly task, it requests both robots R_1 and R_2 and acquires them if they are available. When A_1 (A_2) completes, it releases both robots.

3) Part 3 is machined first by M_3 and then by M_4. In M_3, the part is automatically fixtured to a pallet and loaded into the machine. After processing, robot R_1 unloads the intermediate part from M_1 into buffer B and M_1 is released.

4) At machining center MC_4, intermediate part is automatically loaded into M_4 and processed. When M_4 finishes processing a part, robot R_1 unloads the final product Part T and releases M_4.

5) Assembly station A_2 assembles Parts S and T to produce the final products.

6) It is assumed that input parts are always available to be fixtured and that the finished products are removed.

7) Robot R_1 needs to be modified when it is switched from one user to another. When R_1 is switched from A_1 to A_2, it needs to be oiled; Before entering M_3 or M_4 from A_2, robot R_1 needs the addition of some parts; Before R_1 is switched from M_3 or M_4 to A_1, the oil left from M_3 or M_4 should be cleaned.

We assume that once the system is executed, it cannot be interrupted. To avoid mixing two independent execution cycles of a PNP, one either has to use colored Petri nets or control the procedure of entering into the process. We adopt the second approach in order to simplify the illustration. In other words, the system cannot begin a new iteration before termination.

Modeling the primitive modules in the manufacturing system (Figure 8.1):

In this approach, each module is specified as a marked process. When the entry place is marked, the process is executable. When the exit place has a token, the process terminates immediately (Chapter 3). A place represents a resource status or an operation. A transition represents either start or completion of an operation process.

Machining Center MC_1 (Figure 8.1(a)): This module has one operation for machining Part 1. When the machine M_1 is available (i.e., resource place m_1 has a token), Part 1 begins to be machined by firing transition t_1. p_1 is the operation place, when it has a token, the machining operation is performed. Transition t_2 represents the completion of the machining operation. After firing t_2, the machine M_1 is available again.

Machining Center MC_2 (Figure 8.1(b)): The Petri net process for this module is similar to process MC_1 except that it is for machining Part 2.

Assembly Station A_1 (Figure 8.1(c)): This module assembles Part 1 and Part 2 to produce Part S. When both the robots are available (i.e., resource places r_{11} and r_{21} have tokens), the assembly can begin by firing transition t_5. When the assembly is completed, i.e., firing transition t_6, both robots are released.

Machining Center MC_3 (Figure 8.1(f)): This module has two operations, i.e., machining Part 3 and unloading the intermediate part. When the machine M_3 is available, i.e., place m_3 has tokens, Part 3 can be machined by firing transition t_{11}. Place p_6 is the operation place, representing performing of the machining operation when it has tokens. After machining, the intermediate part is unloaded if robot R_1 is available. Transition t_{12} represents the completion of machining operation and the start of the unloading operation. Place p_7 is the operation place, representing the process of unloading operation. After unloading, both robot R_1 and machine M_3 are released.

Buffer B (Figure 8.1(e)): This module stores the intermediate part produced from module MC_3. When the buffer is available (i.e., resource place b has a token), the intermediate part can be stored by firing transition t_9. When the intermediate part is removed by firing transition t_{10}, the buffer is released.

Machining Center MC_4 (Figure 8.1(g)): The intermediate part produced by MC_3 and stored in buffer B is processed in this module to produce Part T. The process is similar to that in module MC_3 and thus the Petri net process for this module is similar to MC_3.

Assembly Station A_2 (Figure 8.1(d)): Part S and Part T are assembled in this module. The process is similar to that in module A_1 and thus the Petri net process for this module is similar to A_1.

Machining Operation *MO* (Figure 8.1(h)): This is a Petri net process specifying the operations in Machining Center MC_1. When the conveyor C is available, raw materials for Part 1 are moved from the storage to the Machining Center MC_1. Transitions t_{23} and t_{24} represent the start and completion of the movement operation, respectively. Place p_{13} is the operation place. After the materials enter the machine, Part 1 begins to be machined. Place p_{14} represents the machining process and its associated transitions represent the start and completion of the machining process, respectively.

8.3.2   Creating the system model

This subsection describes how to apply the operators of PPPA for creating MS from the PNPs obtained in Section 8.3.1. This is Step 2 and Step 3 of the component-based approach for system design outlined in Section 8.2.

Step 2.1 (Integrating the Modules by using Composition and Reduction):

This sub-step applies the composition operators and reduction operators on the PNPs of Figure 8.1. The intermediate PNPs are shown in Figure 8.2.

(a) Modules MC_1 and MC_2 are integrated by applying the operator Interleave, resulting in the PNP MC_{12} (Figure 8.2(a)). This integration is motivated by the fact that MC_1 and MC_2 handle Parts 1 and 2 asynchronously.

(b) Components MC_{12} and A_1 are integrated by applying the operator Enable, resulting in the PNP P_1 (Figure 8.2(b)). This integration is motivated by the fact that, after machining Parts 1 and 2 in MC_1 and MC_2, respectively, the next step is to assemble these parts in A_1.

(c) Modules MC_3, B and MC_4 are integrated with the operator Enable, resulting in the process P_2 (Figure 8.2(c)). This integration is motivated by the fact that, Part 3 is first machined in M_3, then stored in the buffer B and at last machined in M_4.

(d) Components P_1 and P_2 obtained in (b) and (c), respectively, are first integrated with the operator Interleave. Then, the paths (MC_{12X}, ε_3, A_1), (MC_{3X}, ε_4, B) and (B_X, ε_5, MC_4) are reduced to single places A_1, B and MC_4, respectively, by using the operator Reduce-P-Path. The result is the PNP P_{12} shown in Figure 8.2(d). This integration is motivated by the fact that P_1 and P_2 handle Parts S and T asynchronously.

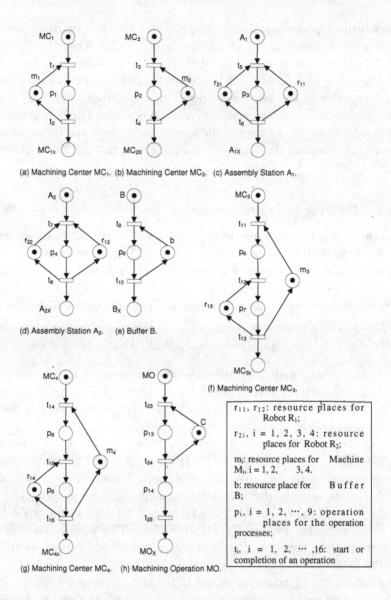

(a) Machining Center MC₁. (b) Machining Center MC₂. (c) Assembly Station A₁.

(d) Assembly Station A₂.   (e) Buffer B.

(f) Machining Center MC₃.

(g) Machining Center MC₄.   (h) Machining Operation MO.

r_{11}, r_{12}: resource places for Robot R_1;

r_{2i}, i = 1, 2, 3, 4: resource places for Robot R_2;

m_i: resource places for Machine M_i, i = 1, 2, 3, 4.

b: resource place for Buffer B;

p_i, i = 1, 2, ⋯, 9: operation places for the operation processes;

t_i, i = 1, 2, ⋯, 16: start or completion of an operation

Fig. 8.1   Petri net Processes for the primitive modules of MS.

(e) Components P_{12} and A_2 are first integrated with the operator Enable. Then, the path $(P_{12X}, \varepsilon_8, A_2)$ (not shown) is reduced to place A_2 by using the operator Reduce-P-Path. The result is the PNP MS shown in Figure 8.2(e). The Petri net representation of process MS is shown in Figure 8.3(a).

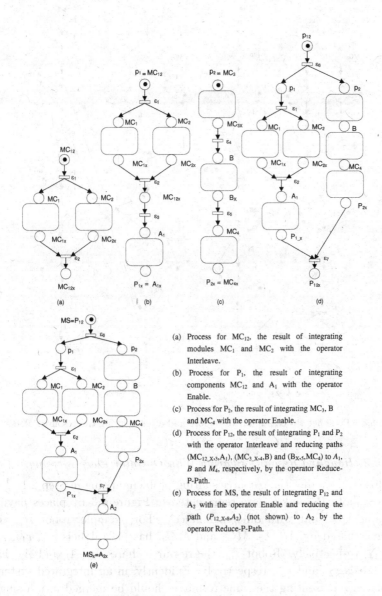

(a) Process for MC_{12}, the result of integrating modules MC_1 and MC_2 with the operator Interleave.

(b) Process for P_1, the result of integrating components MC_{12} and A_1 with the operator Enable.

(c) Process for P_2, the result of integrating MC_3, B and MC_4 with the operator Enable.

(d) Process for P_{12}, the result of integrating P_1 and P_2 with the operator Interleave and reducing paths $(MC_{12\_X,3},A_1)$, $(MC_{3\_X,4},B)$ and $(B_{X,5},MC_4)$ to A_1, B and M_4, respectively, by the operator Reduce-P-Path.

(e) Process for MS, the result of integrating P_{12} and A_2 with the operator Enable and reducing the path $(P_{12\_X,8},A_2)$ (not shown) to A_2 by the operator Reduce-P-Path.

Fig. 8.2   Step 2.1 for constructing the manufacturing system MS.

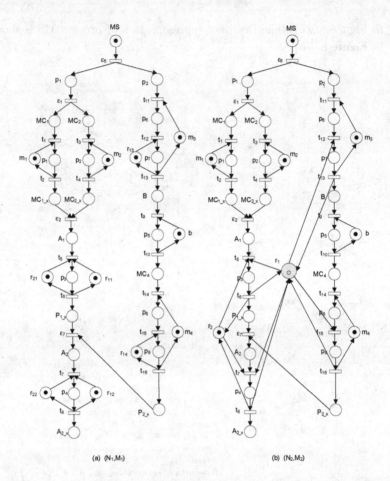

(a) (N_1, M_1) (b) (N_2, M_2)

Fig. 8.3   Petri net Processes for MS before and after merging the resource places.

Step 2.2 (Handling Resource Sharing With Operator Place Merging):

This step handles the resource sharing problem arising in Step 2.1. The shared robots R_1 and R_2 are represented in Figure 8.3 by places having the same label but with different suffices. For example, robot R_1, the resource shared by A_1, A_2, MC_3 and MC_4, has place labels r_{11}, r_{12}, r_{13} and r_{14}, respectively. Robot R_2, the resource shared by A_1 and A_2, has place labels r_{21} and r_{22}, respectively. Evidently, in an integrated system, the places representing the same resource should be merged into a single place. Such merging is done according to MERGE-N-PLACE (Chapter 7). The resulting PNP is shown in Figure 8.3(b).

Step 3 (Creating the Implementation-Leveled System Model by Stepwise Refinement)

In the modeling approach specified in this book, each operation is represented by an operation place. In order to provide a more detailed specification of the operation processes, the operation places can be refined with their detailed Petri net processes. The PNP Machining Operation MO in Figure 8.1(h) is an example for the operation place p_1 in the module MC_1 (Figure 8.1(a)) and the process MS (Figure 8.3(b)). The refinement process is done by using the operator Place-Refinement (Chapter 5). To simplify the appearance of the Figure, refining place p_1 in Figure 8.3(b) is not shown.

The 'place merging' process in Step 2.2 shows only that robot R_1 is simply shared by several processes. However, R_1 is assumed to be able to accommodate the shift differences between those modules. Hence, place r_1 in Figure 8.3(b) represents a function call to a subnet called Robot Modification RM shown in the ellipse in Figure 8.4. This subnet handles three operations, i.e., oiling, adding parts and cleaning oil. Places p_{10}, p_{11} and p_{12} represent these three operations, respectively. Their associated transitions represent the start or completion of the corresponding operations, respectively. After refinement for Figure 8.3(b), the resulting PNP is shown in Figure 8.4.

8.3.3   Verifying the system model

In this subsection, it will be shown that all the primitive modules created in Step 1 and all the composite components created in Steps 2 and 3 are correct. This is Step 4 of the component-based approach for manufacturing system design as outlined in Section 8.2. For this example, a component is considered to be *correct* if it is almost live, bounded and almost reversible and terminates properly. Generally, it is up to the designer to select the set of properties for defining the meaning of correctness.

A *Correctness of the primitive modules created in Step 1 of Section 8.3.2:*
A component-based approach begins with a set of correct primitive modules. For MS, they are PNPs MC_1, MC_2, MC_3, MC_4, A_1, A_2, B and MO as shown in Figure 8.1. Since their structures are quite simple, it is not difficult to see that every one of them is a PNP and is correct. For example, all of them cannot self-start. Since every associated process in Figure 8.1 is a strongly connected marked graph

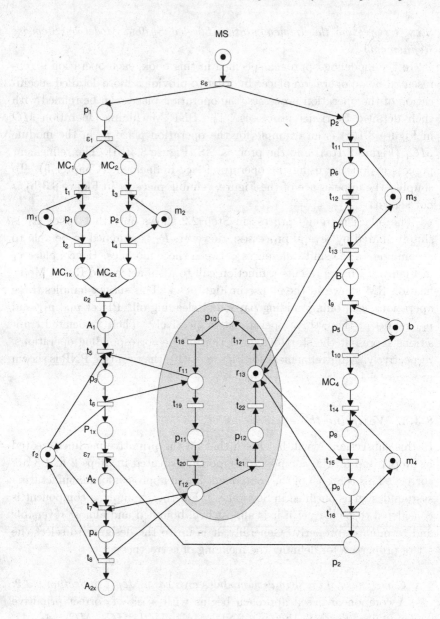

Fig. 8.4   Process MS after applying the place refinements.

and each cycle has exactly one token, the PNPs in Figure 8.1 are almost live and bounded [MUR (1989)]. Another example is the proper termination property of the module MC_3 in Figure 8.1(f). This PNP has only one possible firing sequence $\sigma = t_{11}t_{12}t_{13}$ such that, a token is deposited into the exit place MC_{3X} after firing the sequence. By the firing rule of marked Petri net process (Definition 3.1), execution should terminate. At this moment, initial token distribution is resumed except that the token in the entry place is now in the exit place, meaning that termination is proper.

B *Correctness of the PNPs after composition and reduction in Step 2.1 of Section 8.3.2:*

It will be shown that every PNP produced in Step 2.1 of Section 8.3.2 is correct. In Step 2.1, MC_1 and MC_2 are first integrated by using the operator Interleave, resulting in MC_{12} (Figure 8.2(a)). Next, MC_{12} and A_1 are integrated by using the operator Enable, resulting in P_1 (Figure 8.2(b)). MC_3, B and MC_4 are integrated by using the operator Enable again, resulting in P_2 (Figure 8.2(c)). Then, components P_1 and P_2 are integrated by using Interleave again, resulting in P_{12} (Figure 8.2(d)). Since MC_1, MC_2, A_1, MC_3, B and MC_4 are all correct, it follows from Theorems 4.1 and 4.3 that P_{12} is correct. Step 2.1(d) also reduces the paths $(MC_{12X}, \varepsilon_3, A_1)$, $(MC_{3X}, \varepsilon_4, B)$ and $(B_X, \varepsilon_5, MC_4)$ to places A_1, B and MC_4, respectively. Since ${}^\bullet MC_{12X} \neq \phi$, $MC_{12X}{}^\bullet = \{\varepsilon_3\}$ and $M_0(A_1) = 0$, ${}^\bullet MC_{3X} \neq \phi$, $MC_{3X}{}^\bullet = \{\varepsilon_4\}$ and $M_0(B) = 0$, ${}^\bullet B_X \neq \phi$, $B_X{}^\bullet = \{\varepsilon_5\}$ and $M_0(MC_4) = 0$, these reductions satisfy Theorem 6.2(1) and Theorem 6.3(16a) and hence preserve the properties. Hence, P_{12} is correct. In Step 2.1(e), P_{12} and A_2 are integrated by using the operator Enable and the path $(P_{12\_X}, \varepsilon_8, A_2)$ is reduced to place A_2. Since both P_{12} and A_2 are correct, by Theorem 4.1, the produced process MS_0 (not shown) obtained by integrating P_{12} and A_2 is correct. In MS_0, since $A_2{}^\bullet \neq \phi$, ${}^\bullet A_2 = \{\varepsilon_8\}$ and $M_0(A_2) = 0$, Theorem 6.2(2a) and Theorem 6.3(16a) are satisfied. Hence, reducing the path preserves the correctness and the process MS is correct. That is, the Petri net process (N_1, M_1) for MS shown in Figure 8.3(a) is almost live, bounded, almost reversible and terminates properly.

C *Correctness of the PNPs after place merging in Step 2.2 of Section 8.3.2:*

In Step 2.2, two sets of resource places $Q_1 = \{r_{11}, r_{12}, r_{13}, r_{14}\}$ and $Q_2 = \{r_{21}, r_{22}\}$ in (N_1, M_1) (Figure 8.3(a)) are merged by MERGE-N-PLACE to r_1 and r_2, respectively, resulting in Petri net process (N_2, M_2) (Figure 8.3(b)). In the following, it will be shown that $(N_2,$

M_2) is almost live, bounded and almost reversible by applying Theorem 7.5(17) and Theorem 7.7(2) on the associated process of (N_1, M_1). Furthermore, it can be shown without theoretical support that (N_2, M_2) terminates properly.

Before applying Theorem 7.5(17), let us make two observations: (a) Theorem 7.5(17) has to be applied on the associated process of (N_1, M_1) instead of (N_1, M_1). This is because Condition (b) can never be satisfied by the latter. In fact, the entry place of a process forms a siphon by itself in P_0 that contains no traps, where $P_0 = \{MS, P_1, P_2, MC_1, MC_2, MC_4, m_1, m_2, m_3, m_4, MC_{1X}, MC_{2X}, A_1, P_{1X}, P_{2X}, A_2, A_{2X}, B, b, p_1, p_2, p_3, p_4, p_5, p_6, p_7, p_8, p_9\}$. (b) The set of all the minimal siphons of N_2 are: $D_1 = \{r_{11}, p_3\}$, $D_2 = \{r_{12}, p_4\}$, $D_3 = \{r_{13}, p_7\}$, $D_4 = \{r_{14}, p_9\}$, $D_5 = \{r_{21}, p_3\}$, $D_6 = \{r_{22}, p_4\}$, $D_7 = \{m_1, p_1\}$, $D_8 = \{m_2, p_2\}$, $D_9 = \{m_3, p_6, p_7\}$, $D_{10} = \{m_4, p_8, p_9\}$, $D_{11} = \{b, p_5\}$, $D_{12} = \{MS, A_{2X}, p_4, A_2, p_{1X}, p_3, A_1, MC_{1X}, p_1, MC_1, P_1\}$, $D_{13} = \{MS, A_{2X}, p_4, A_2, p_{1X}, p_3, A_1, MC_{2X}, p_2, MC_2, P_1\}$ and $D_{14} = \{MS, A_{2X}, p_4, A_2, p_{2X}, p_5, p_6, p_7, p_8, p_9, MC_4, B, P_2\}$. These siphons are also traps initially marked by M_2. Furthermore, it can be shown that any siphon containing r_{11} (resp., r_{12}, r_{13}, r_{14}) also contains D_1 (resp., D_2, D_3, D_4) and that any siphon containing r_{21} (resp., r_{22}) also contains D_5 (resp., D_6). For example, let siphon D contain r_{11}. Then, $t_6 \in {}^\bullet D$ implies $p_3 \in D$. Hence, $D_1 = \{r_{11}, p_3\} \subseteq D$.

We are now going to show that the two conditions of Theorem 5.5(17) are satisfied for the associated process of (N_1, M_1). For Condition (a), it is easy to verify that the associated process of N_1 is SM-coverable. All the SM-covers in N_1 that contain resource places in Q_1 are $r_{11}t_5p_3t_6r_{11}$, $r_{12}t_7p_4t_8r_{12}$, $r_{13}t_{12}p_7t_{13}r_{13}$, $r_{14}t_{15}p_9t_{16}r_{14}$ and they have no transitions in common. All the SM-covers in N_1 that contain resource places in Q_2 are $r_{21}t_5p_3t_6r_{21}$ and $r_{22}t_7p_4t_8r_{22}$, they have no common transitions either.

For Condition (b), for example, the siphon D_{12}, where $D_{12} \subseteq P_0$, contains the trap $S = D_{12}$ marked by M_1. For the siphon $D = \{r_{11}, r_{12}, r_{13}, r_{14}, A_2, p_{1X}, p_{2X}, B, p_3, p_4, p_7, p_9\}$, Observation (b) in the last paragraph implies that $D \supseteq D_1 \cup D_2 \cup D_3 \cup D_4 \supseteq Q_1$. Since for $i = 1, 2, 3, 4$, each D_i is an initially marked trap, $S = D_1 \cup D_2 \cup D_3 \cup D_4$ is also an initially marked trap satisfying $Q_1 \subseteq S$.

The above analysis implies that (N_2, M_2) is almost live and bounded. In the following, we will show that the Petri net (N_2, M_2) is almost reversible. Firstly, it will be shown that the associate process of (N_1,

M_1) is a live augmented marked graph by checking the conditions in Definition 2.8(5) one by one. a) $(N_1, M_1) \mid (Q_1 \cup Q_2)$ is almost live and bounded. After deleting some resource places, the correctness of each process in Figure 8.1 is not changed. By Step 2.1, (N_1, M_1) is still correct even if the resource places are deleted. b) Each resource place is associated with a transition pair (a_i, b_i) and there exists an elementary path O_i from a_i to b_i. For example, resource place r_{11} is associated with (t_5, t_6) and there exists an elementary path (t_5, p_3, t_6) from t_5 to t_6. c) Every resource place is marked by M_1 and no elementary path O_i is marked by M_1.

Secondly, since the two conditions in Theorem 7.5(17) are satisfied and that, satisfaction of Condition (b) in Theorem 7.5(17) implies satisfaction of Theorem 7.7(2), (N_2, M_2) is almost reversible (By Characterization 2.20 and 2.21, liveness and reversibility can also be verified in another way).

Finally, it can be observed that (N_2, M_2) terminates properly.

D *Correctness of the PNPs after the refinements in Step 3 of Section 8.3.2:*

Since process MO is correct and place p_1 in MC_1 (therefore in the process (N, M_0)) has input transition t_1, it follows from Theorem 5.6(12)(13)(14) that refining place p_1 with the process MO shown in Figure 8.1(h) preserves the correctness. For simplicity, refining place p_1 in (N, M_0) is not shown.

Since refinement subnet RM is a strongly connected state machine, by Corollary 6.1, refining place r_1 with subnet RM preserves the correctness of (N_2, M_2). Hence, the Petri net process (N, M_0) in Figure 8.4 is correct.

This chapter applies PPPA for MS design. The approach has three main features: (1) It is component-based. The algebra provides the means for representing the MS modules as Petri net processes, integrating simple modules into composite ones. Though only three composition operators, namely, Enable, Choice, and Interleave, one refinement operator and some reduction operators are described in this chapter, this algebra in fact includes many other operators [AAL (1997); BDK (2001); CMR (2003); CZ (1998); GV (2003); HAM (2003); KMR (2001, 2003); MAK (2001); XU (2003)]. Hence, very complex modules and MS architectures can be created. (2) PPPA guarantees the correctness in every step of the integration process. Manufacturing systems with a great variety of property requirements can be designed. (3) By using a property-preserving place-merging technique, it avoids such errors as deadlock and overflow that often arise when handling resource sharing. The assumption about changeable resources makes the resource-sharing problem more flexible.

Chapter 9

Application of PPPA to Multi-Agent System Design

This chapter presents the application of PPPA to a property-preserving and component-based methodology for the specification and verification of multi-agent system (MAS) designs. The idea is similar to Chapter 8. It is assumed that an MAS is composed of software modules. The method starts by specifying these modules as Petri net processes (PNPs). Then, according to the control flow requirements of the system, the components of the MAS are created by integrating these PNPs with various operations, such as composition (e.g., enable, choice, interleaving), refinement and reduction. Among other features, the methodology overcomes two difficulties arising in the design of MASs: (1) It eliminates the requirement of verifying composite components because all the operations preserve many properties. Hence, if the primitive PNPs satisfy these properties, each of the composite components, including the agent itself, also satisfies these properties. (2) It avoids the errors due to resource-sharing by using a property-preserving place-merging technique.

9.1 Background and Motivation

In the last few decades, many expert systems and knowledge bases have been developed independently. The subsequent problem is how to coordinate these systems and databases into a cooperative system for providing information-oriented services to the users. This motivated the birth of multi-agent systems (MAS).

A multi-agent system (MAS) coordinates many resources into a cooperative system for providing information-oriented services to the users. These resources may be expert systems, knowledge bases, ongoing project plans, goals, sub-agents systems, etc. Research and development in MAS have two different goals:

A. *Enriching the features of agents.* The goal of researchers is to explore what kinds of features (e.g., operations, adaptability, etc.) should be available in a useful agent system. The point is not the realization of these features but their qualities. A large amount of knowledge engineering methodologies has been used for modeling the basic features, such as plans, goals, protocols, sensors, effectors, knowledge-bases, etc. [CDS (2002)]. DESIRE and METATEM, two logic-based methodologies to be discussed later, have also been involved in developing the functional aspects of information-related properties. Since this is not the research area of this book, further review will be omitted.

B. *Advancing design methodologies for realizing agent systems.* How to convert some required features into a real system is a complex software development issue. A design process bridges the gap between requirement analysis and implementation. In the case of agent systems, for example, the features mentioned in Part A above may be considered as requirements. When described in a research paper, these features may not have a 'computational' solution yet. A design specification represents a computational solution for realizing all the required features of a system before being coded into a computer program. It not only describes the architecture and operational flows of the system but also serves as a guideline for programming. Hence, it must be correct.

It is not a straightforward process to develop a correct design specification. First, a designer has to determine a brief but overall plan about the architecture, operational flow and functional flow of the system. This plan will be gradually developed into a full computational solution by trial and error. A specification may be modified and verified (i.e., proved for correctness) many times before becoming final. For large systems, it is very difficult for a designer to manipulate such a complex process without the assistance of a design methodology.

The goal of research in design methodologies is to advance the techniques for modification, specification and verification. Many models, some object-oriented, have been reported [KGR (1996); KMJ (1996)]. A partial review can be found in [HL (2005)].

An important and ever-existing objective in the research of agent systems is to acquire a useful design methodology. This methodology must be versatile but flexible for specification and verification under the architecturally modular and functionally resource-prone nature of agent systems. Currently, two major formal approaches are being developed for this purpose: *logic-based* and *net-based*. In the logic-based approach, two representative methods have been proposed. One method applied Concurrent METATEM [FW (1997); WOO (1998)], a language based on temporal belief logic for modeling reactive systems. Another method applied DESIRE [BDJ (1997); CJT (2001); JT (2002)], a means based on order-sorted predicate logic for specifying complex knowledge-based agent systems. In the net-based approach, several kinds of nets have been used, including Cooperative Nets [MK (1999)], Agent-Oriented Colored Petri Nets [MW (1997)], Reference Nets [CMR (2003); KMR (2001, 2003)], Component Protocol Net [HAM (2003)] and Multi-agent Interaction Protocol Net [LL (2003)]. A few recent developments are the applications of Predicate-Transition nets [XVI (2002); XYD (2003)] or G-nets [XU (2003)].

Problems addressed in this chapter:

As far as we know, the methodology adopted in this book is the first formal model for the modular design of MAS in the net-based approach. As explained more clearly below, unlike other net-based papers which just illustrate the ideas of using nets for specifying MAS, our model has a complete set of property-preserving operators for developing error-free design specifications level by level. It should also be pointed out that, though depending on two quite different techniques for computational developments, the logic-based approach of DESIRE/METATEM and the net-based approach to be developed in this book have several concepts and objectives in common, such as component integration, property validation, etc. At present, however, since there are many open issues in this fast-developing field, these approaches have each focused only on some special issues. For example, while the logic-based approach has done a lot of study on the functional specification of information-related properties., such as *information providing correct, information acquisition effective* [JT (2002)], etc., our book focuses more on the issues of resource-sharing and automatic error-freeness during the design process.

This chapter focuses on the following typical problems arising in the design of MAS:

- *What should be the approach for specifying the architecture of a MAS.* A MAS is composed of many components.These components may run in parallel, in sequential order, in either-or mode, etc. They may also be built by different unrelated parties and at different times. Hence, in order to assure uninterrupted developments of the individual components, it is strongly suggested that a complex MAS should adopt a modular and component-based architecture and that a modular approach should be used to specify and develop such an architecture.

- *How to get rid of errors arising in the process of modular design:* In modular design, the modules are integrated via various operators, such as choice, sequential, parallel, disable, etc., or via asynchronous message exchanges and resource sharing. Naturally, errors may be brought on by such modifications. In designing a system as complex as MAS, re-verifying the correctness of the system after every modification is a very heavy burden.

- *How to get rid of errors arising due to resource sharing.* In general, since users' requests for services may be of very diversified nature, it is hardly feasible for a single agent to install sufficient resources for supporting all kinds of requests. In other words, a MAS heavily relies on sharing various resources with others. As we know, resource sharing may induce many errors, such as deadlock or overflow. Therefore, handling resource sharing is a particularly important issue in the design of MAS.

For the rest of this chapter, a methodology based on PPPA for specifying and verifying the design of MASs is illustrated in detail with an example called Health-Care-System (HC-MAS). Section 9.2 is devoted to the specification of the primitive modules and the creation of the composite components of HC-MAS and Section 9.3 to the verification of these components. Some concluding remarks are given in Section 9.4.

9.2 PPPA-Based Methodology for Designing Multi-Agent Systems

In this section, PPPA is applied for the development of design specifications of a health-care system called HC-MAS. The figures to be generated in this example should not be simply considered as a pure architectural description of a system. Rather, they demonstrate how PPPA facilitates the development process.

HC-MAS has many typical features of a MAS. However, since this book is concerned for the technical aspects of design methodologies rather than the feature development aspects of MAS, the actual meanings and specifications of some low-level features of HC-MAS will not be shown. For comparison purposes, this example has been deliberately made similar to several others being used in net-based research for MAS. HC-MAS bears some resemblance with an earlier example about applying agent methodology to medical informatics [HUA (1995)]. However, that example uses an agent-oriented programming language rather than Petri nets for specification.

Under PPPA, composing a MAS from its agents is similar to composing an agent from its components. Hence, it is enough to describe the process of designing a single agent. A top-to-down modular approach will be used, meaning that a specification will be developed level by level. Generally, the designer starts by specifying the top level as a relatively simple PNP containing some 'undefined' transitions/places. An undefined transition/place represents an interface to a not-yet-designed module. In general, at each level, some of the transitions/places may still be undefined. When these elements are refined with some newly-designed modules, the specification will go forward to a lower level. In this way, the level of a specification may become lower and lower until the designer wants it to stop.

In our example, HC-MAS consists of five agents: PATIENT, HOSPITAL, CLINIC, DOCTOR and NURSE. According to our explanation above, we will develop specifications of just the first two levels for CLINIC. This is sufficient for demonstration purposes.

9.2.1 *Summary of our methodology for designing MAS*

Before starting, a designer should know how CLINIC operates. Basically, CLINIC accepts three types of messages, namely, *unchecked messages* (u-message), *protocol-invocation messages* (p-messages) and *resource-triggered messages* (r-messages). Hence, on entering CLINIC, a message will be dispatched to one of the three modules MC (Message Checking), EPS (External Protocol Selector) and RH (Resource Handler) for processing. Each message submits a request for solving some problem. It may find a specific method or a protocol for solution or asks CLINIC to propose one. There is a little difference between a method and a protocol: A method is mostly private. Its interface usually involves just a simple call. On the other hand, a protocol's interface involves several rounds of message exchanges designed according to a standard or an agreed contract. To solve the problems requested in the messages, CLINIC may consult its own resources or other agents. This means that messages may come externally from the users or other agents, or internally from the agent itself. Hence, CLINIC needs two more modules, namely, IPS (Internal Protocol Selector) and PP (Protocol Processor).

Overall planning is probably the most challenging stage in the entire design process. At the beginning, the designer is given a requirement statement about CLINIC's features based on which she/he should first make a rough design plan. The plan may not contain all the details at this stage. However, if a modular approach is taken, it should provide at least an architecture and a distribution of modules/features for the top level of the specification. For CLINIC, for instance, the plan requires five modules MC, RH, EPS, IPS and PP to be available at the top level. It includes also all the problem-oriented and computation-oriented features and their representative places and transitions (See Table 9.1). The designer has to decide how to distribute these features to the five modules in order to develop the next level of the specification.

An agent consists of many modules and resources which may be designed and installed in different locations by unrelated parties. Since it is too diversified to adopt a rigid and universal architecture, it is highly desirable to use a component-based approach for its design.

The methodology proposed in this chapter consists of the following three steps:

Step 1 (Creation of an individual agent):

> *Substep 1.1 (Specification of the primitive modules):* After the requirement analysis phase, an agent's behaviors and/or resources should be available as a set of primitive modules. These modules are then specified as Petri net processes (PNP) (Definition 3.1).

> *Substep 1.2 (Specification of each agent):* According to the message processing requirement of an agent, the PNPs created in Substep 1.1 for each agent are either integrated via some composition operators such as Choice, Enable, etc. or modified by applying some refinement and reduction operators. Resource sharing, if being part of the requirements, will be resolved by place merging. The resulting agent system is also a PNP.

Step 2 (Creation of an MAS): Since each agent is also a PNP and thus can be regarded as the components of the MAS, the creation of the MAS is similar as Step 1.

Step 3 (Verification of an MAS): It is not necessary to spend much more effort for verification because all the operators are property-preserving. That is, if the constituent PNPs satisfy such properties as almost *liveness*, *boundedness*, almost reversibility and proper termination, the PNP obtained by applying any of the operators satisfies these properties as well. (Refer to the theoretical part of this book for details.) However, the burden has been shifted to making sure that the conditions required for the operators are satisfied.

9.2.2 Specifying the modules as Petri net processes (Substep 1.1 in Section 9.2.1)

This subsection specifies the various primitive modules of CLINIC as a set of PNPs.

Five primitive modules of CLINIC:

Message Checker (**MC**, Figure 9.1(c)) - MC handles u-messages. A u-message is sent by a user or another agent that expects the called agent to provide a protocol or method for resolving a problem. MC will first analyze the problem and then make one of the following two decisions: (1) *Invoke-a-method*. If the problem can be resolved by invoking some methods available within this agent, MC will proceed to apply these methods by consulting the various local resources. For example, if a diagnosis plan is available from a DK (place *DK1*), the plan triggers CK (place *MC.ck*) to apply some decision rules in IK (place *MC.ik*), some medical domain knowledge in DK (place *DK1*), and the patient's data from WM (place *WM1*). At the transition *make-decision*, MC has to make a decision whether to *end* this treatment by updating the relevant resources or to *continue* with another round of treatment by invoking an internal protocol from IPS (introduced later). (Note that the difference between *end* and *continue* is not shown in the figures but lies in the message passed to IPS for processing. *end* passes a 'complete' message and *continue* requests an internal protocol.) (2) *Invoke-a-protocol*. If the problem can be handled by invoking a protocol available within this agent or in other agents, MC will create and switch a p-message to IPS for further processing.

Resource-Request Handler (**RH**, Figure 9.1(d)) - An RH handles r-messages. In RH, each resource is represented by a place, such as *DK2*, *EN*, *ME*, *RH.ck* and *RH.ik*, etc. When a resource has a problem to solve, it will deposit a token into its representative place and send an r-message to the agent. Processing an r-message may involve consultation with various expert systems, knowledge bases, etc. At the end, RH will decide whether it has to update these resources (shown by transition *(data-update)* or not (shown by transition *no-change*). In order to detect which of the resources trigger a service request, several kinds of functional tasks called detectors have been added into the RH: a *k-detector* for detecting such resources as plans, goals and knowledge base; an *e-detector* for detecting the environment. They are similar to the sensors used in [XU (2003)].

Protocol Selector (PS, Figure 9.1(a)) - A p-message designates a specific protocol for resolving a problem. On entering an agent, it is dispatched to PS. PS will select the protocol and pass the message together with the selected protocol to the Protocol Processor PP (introduced later) for processing. Three types of protocols are available for selection: *communication protocols* (c-protocols) for the exchange of messages between two agents, *application protocols* (a-protocols) for the exchange of messages between an agent and an application package, and *non-available protocols* (n-protocols). If the called agent does not have the protocol (usually an a-protocol) specified in the incoming p-message, it will attach a non-available sign to it, indicating that the specified application protocol is not available in this agent.

As an illustration of the various ways of creating composite components in a design, CLINIC has installed two types of PS: *Implicit Protocol Selector* (IPS) and *Explicit Protocol Selector* (EPS). An IPS contains some 'implicit' protocols that are not published to the users or other agents. An EPS handles p-messages with an explicitly-specified protocol. If the specified protocol is available, the protocol will be selected and the message is passed to PP for processing. If the protocol is not available, the message will be switched to another agent through PP.

Protocol Processor (PP, Figure 9.1(e)) - When a p-message is dispatched to PP from PS, PP will process the various tasks needed by the attached protocol such as setting up an interface with the relevant application package for an a-protocol, setting up a communication with the relevant agent for a c-protocol, and searching for another agent that has a specified n-protocol.

Note that while different protocols may be selected from different PSs, they are all processed in a common PP. This conforms with the current technology of having one common communication server for interfacing with other servers and various application packages in a distributed environment.

Subtask Manager (SM, Figure 9.1(b)) - The module SM within PP is responsible for processing the various subtasks according to the protocol submitted to PP.

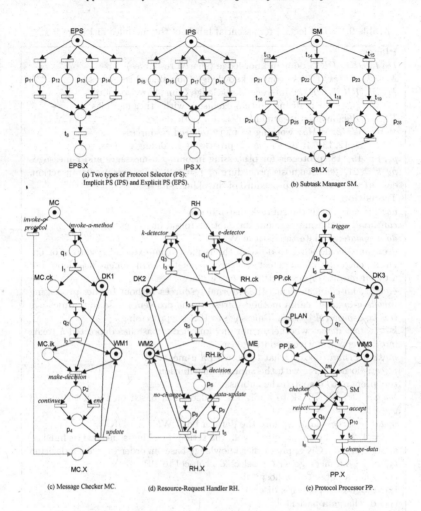

Fig. 9.1 Processes for the five primitive modules of HC-AGENT.

Special modeling and verification techniques for handling resource sharing:

As mentioned in Section 9.1, a multi-agent system shares many resources. For the agent CLINIC, examples of resources include: Domain Knowledge (DK, including specific medical knowledge bases, diagnosis plans for patients, databases of patients' records, etc.), Inference Knowledge (IK, many inference rules which specify inference relations between DK, patient information and WM), Control Knowledge (CK, controlling the execution of IK and DK), Working Memory (WM, a memory used for

Table 9.1   The legend for essential labels of the modules in Figure 9.1.

Places:
DK1, DK2, DK3: domain knowledge (shared resources)
MC.ck, RH.ck, PP.ck: control knowledge (non-shared resources)
MC.ik, RH.ik, PP.ik: Inference knowledge (non-shared resources)
ME: a place for realizing mutual exclusion with other modules.
PLAN: documented treatment plans for patients
WM1, WM2, WM3: working memory (shared resources)
$p_i, i = 11, 12, 13, 14$: protocols for processing incoming p-messages
$p_j, j = 15, , 18$: protocols for processing incoming u-messages and r-messages
$p_k, k = 21, , 28$: diagnosis procedure $q_i, i = 1, 2, , 9$: procedure of an action.
e.g., q1 represents the procedure of 'invoking-a-method'
Transitions:
accept: agree with the subtask distribution
continue: send a message out for another round of processing
data-update: update the data in WM
decision: decide what to do with the request from external resources or environment *e-detector*: detect whether or not the environment has deposited a token into its representative resource place
end; t4; no-change: return the external resources without further processing
invoke-a-method: call a method to resolve the problem raised in a u-message
invoke-a-protocol: pass a u-message to IPS for processing
k-detector: detect whether or not any knowledge base has deposited a token into its representative resource place
make-decision: decide what to do with a u-message
reject: do not agree with the subtask distribution
tm: prepare to dispatch the subtasks
trigger: trigger PP.ck to apply knowledge-bases based on the decision and the PLAN
update; change-data: update the data in both WM and DK
$l_i, i = 1, 2, , 8$: finish an action. e.g., the action 'invoking-a-method' is finished at l1. t_1, t_3, t_6: CK applies other knowledge bases in order to make a decision
t_8, t_9: pass a message with a selected protocol to PP
t_{13}: refer a patient to a hospital
t_{14}: diagnose a patient's disease
t_{15}: discharge a patient
t_{16}: allocate a bed for a patient and diagnose his/her disease
t_{17}: obtain a drug
t_{18}: check a disease
t_{19}: instruct a patient and inform the GP

storing the temporary data generated by CK or the users. The data in WM can be modified or deleted) and places ME for implementing mutual exclusion.

As will be seen in Section 9.2.3, special places are added in the modules' PNPs for representing the resources. As described in Chapter 7, our place-merging technique for handling resource sharing will preserve the desirable properties.

9.2.3 Integrating the modules of the CLINIC (Substep 1.2 in Section 9.2.1)

This subsection applies the operators of PPPA to create the CLINIC from the PNPs obtained in Section 9.2.2.

Step 1.2.1 (Composition and Reduction): This sub-step applies the composition operators and reduction operators on the PNPs of Figure 9.1. The intermediate results are the PNPs symbolically shown in Figure 9.2.

(a) MC and RH are integrated with the operator Choice (Definition 4.2), resulting in the PNP MR (Figure 9.2(a)). This is feasible because MC and RH handle *two mutually exclusive types of messages*, namely, *u-messages* and *r-messages*.

(b) MR and IPS are integrated with the operator Enable (Definition 4.1), resulting in the PNP MRI (Figure 9.2(b)). This is feasible because, after handling a u-message in MC or handling an r-message in RH, the next step is to seek an implicit or explicit protocol in IPS for further processing.

(c) The path $(MR.X, \varepsilon_1, IPS)$ is first reduced by using the operator Reduce-P-Path (Chapter 6) to the single place IPS. EPS and MRI are then integrated by using the operator Choice, resulting in the PNP EMRI (Figure 9.2(c)). The integration is feasible because EPS handles *p-messages* while MRI handles *u-messages* and *r-messages*.

(d) EMRI and PP are first integrated with the operator Enable. Then, the path $(EMRI, \varepsilon_2, MRI)$ is reduced to place AGENT and the path $(EMRI.X, \varepsilon_3, PP)$ (not shown) is reduced to place PP by using Reduce-P-Path. The result is the PNP AGENT shown in Figure 9.2(d). This is feasible because PP is invoked provided that a protocol in EPS or IPS has been selected.

(a) Process for MR, the result of integrating modules MC and RH with the operator Choice.

(b) Process for MRI, the result of integrating components MR and IPS with the operator Enable.

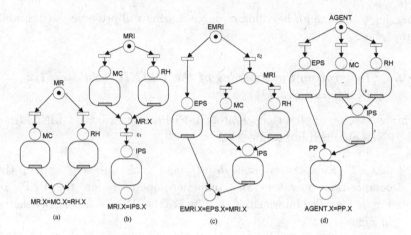

Fig. 9.2   Step 2.1 for constructing the agent system HC-AGENT.

(c) Process for EMRI, the result of reducing path $(MR.X, \varepsilon_1, IPS)$ to IPS with the operator Reduce-P-Path and integrating EPS and MRI with the operator Choice.

(d) Process for AGENT, the result of integrating EMRI and PP with the operator Enable and reducing paths $(EMRI, \varepsilon_2, MRI)$ and $(EMRI.X, \varepsilon_3, PP)$ (not shown) to $AGENT$ and PP, respectively, by the operator Reduce-P-Path.

Step 1.2.2 (Refinement in Chapter 5):   The place SM in Fig 9.1(e) represents a function call to the module Subtask Manager (Figure 9.1(b)). At a more detailed level of specification, this place can be refined with its PNP representation (Figure 9.1(b)). After this refinement, the entire agent is shown as Petri net (N_1, M_1) in Figure 9.3.

Step 1.2.3 (Place merging (Chapter 7)):   This step gives a solution to the resource sharing problem arising in Step 1.2.1 and Step 1.2.2. The shared resources DK and WM are represented in Figure 9.3 by places having the same label with different suffices, such as DK1, DK2 and DK3. Obviously, in an integrated system, since the places in each set represent the same resource, they should be merged into a single place. Such merging is done according to MERGE-N-PLACE (Chapter 7). Note: The result (N, M_0) of this step is not shown because the figure will be too complicated.

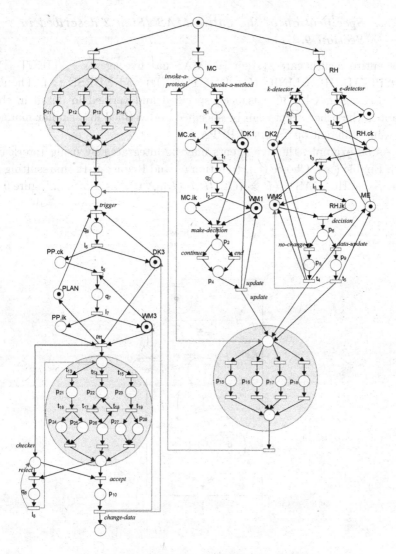

Fig. 9.3   Petri net process (N_1, M_1) for CLINIC.

(N_1, M_1) is the result of refining Figure 9.2(d) with the modules of Figure 9.1.

(N, M_0) is the same as (N_1, M_1) except that the resource places DK and WM have been merged.

9.2.4 *Specification of the entire MAS (Step 2 described in Section 9.2.1)*

The entire health care system HC-MAS has five agents PATIENT (P), HOSPITAL (H), CLINIC (C), DOCTOR (D) and NURSE (N). Though only one agent CLINIC has been specified and verified in detail in this chapter, the other agents can be specified and verified in a similar manner by using different operators of PPPA.

As an example, the five agents may be integrated by using Interleave (||||), Enable (≫), Choice ([]), Iteration (') and Reduce-P-Path, resulting in the system HC-MAS = $(P \gg (C[](H \gg (D|||N))))'$ as shown in Figure 9.4.

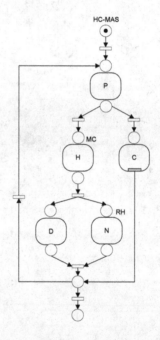

Fig. 9.4   The high level illustration of HC-MAS.

9.3   Verifying the MAS (Step 3 Described in Section 9.2.1)

In this section, it will be shown that all the composite components created in Step 1 and Step 2 are correct. For this example, a component is considered to be *correct* if it is almost live, bounded and almost reversible and terminates properly. Generally, it is up to the designer to select the set of properties for defining the meaning of correctness.

Before giving the details, the steps of verification and the property-preserving characteristics of the operators are summarized in Figure 9.5 and Table 9.2, respectively.

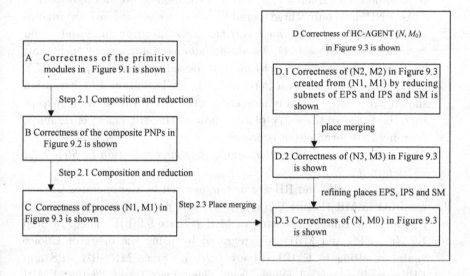

Fig. 9.5   Summary of steps for creating and verifying HC-AGENT.

Table 9.2   Summary of property-preserving characteristics of the operators.

Operator	Constituent Component	Composite Component	Reference
Composition	$b + tp$ $l + tp$ $r + tp$ tp	b 1 r tp	Theorems 4.1, 4.2, 4.3
Place Refinement	$b + tp$ $l + tp + c$ $r + tp$ tp	b $1+c$ $r + c$ tp	Theorem 5.6
Reduce-N-Place	Siphon-Trap Condition + Positive P-invariant	$l + b$	Theorem 7.5
Reduce-Path	$1(resp., b, r)$ tp	$1(resp., b, r)$ tp	Theorems 6.1, 6.2, 6.3
b: bounded, l: almost live, r: almost reversible, tp: terminate properly, c: condition			

A *Correctness of the primitive modules created in Section 9.2.2:*

For CLINIC, the primitive modules are MC, RH, EPS, IPS, SM and PP shown in Figure 9.1. Since their structures are quite simple, it is not hard to see that every one of them is a PNP and is correct. For example, all of them cannot self-start. Another example is the proper termination (Definition 3.7) property of the module MC in Figure 9.1(c). This PNP has only three possible firing sequences: $\sigma_1 = $ *invoke-a-protocol*; $\sigma_2 = $ *invoke-a-method, l1, t1, l2, make-decision, end, update* and $\sigma_3 = $ *invoke-a-method, l1, t1, l2, make-decision, continue, update* such that, after firing each sequence, a token is deposited into the exit place MC.X and, by the definition of a Petri net process (Definition 3.1), execution should terminate. At this moment, initial token distribution (except that the token in the entry place is now in the exit place) is resumed, meaning that termination is proper.

B *Correctness of the PNPs after composition and reduction in Step 1.2.1 of Section 9.2.3:*

In Step 1.2.1, MC and RH are first integrated by the operator Choice, resulting in MR (Figure 9.2(a)). Next, MR and IPS are integrated by the operator Enable, resulting in MRI (Figure 9.2(b)). Then, in Step 1.2.1(c), EPS and MRI are integrated by using the operator Choice again, resulting in EMRI (Figure 9.2(c)). Since MR, RH, IPS and EPS are all correct, it follows from Theorems 4.1 and 4.2 that EMRI is correct. Step 1.2.1(c) also reduces the path $(MR.X, \varepsilon_1, IPS)$ to place *IPS*. Since $^\bullet MR.X \neq \phi, MR.X^\bullet = \{\varepsilon_1\}$ and $M_0(IPS) = 0$, this reduction satisfies Theorem 6.2(1) and Theorem 6.3(16a) and hence preserves the correctness. Hence, EMRI is correct. In Step 1.2.1(d), EMRI and PP are integrated by using the operator Enable and the paths $(EMRI, \varepsilon_2, MRI)$ and $(EMRI.X, \varepsilon_3, PP)$ are reduced to single places. Since both EMRI and PP are correct, by Theorem 4.1, the produced process EMRIP (not shown) after integrating EMRI and PP is correct. In EMRIP, since $MRI^\bullet \neq \phi$, $MRI = \{\varepsilon_2\}$ and $M_0(MRI) = 0$, Theorem 6.2(2a) and Theorem 6.3(16a) are satisfied. Since $^\bullet EMRI.X \neq \phi, EMRI.X^\bullet = \{\varepsilon_3\}$ and $M_0(PP) = 0$, Theorem 6.2(1) is satisfied. Hence, reducing these two paths preserves the correctness.

C *Correctness of the PNPs after the refinement in Step 1.2.2 of Section 9.2.3:*

Since process SM is correct and place SM in PP (therefore in the process AGENT) has input transition tm, it follows from Theorem 5.6(12)(13)(14) that refining place SM with the process SM shown in Figure 9.1(b) preserves the correctness. Hence, the Petri net process (N_1, M_1) in Figure 9.3 is correct.

D *Correctness of the PNPs after place merging in Step 1.2.3 of Section 9.2.3:*

D.1 *Correctness of the PNP (N_2, M_2) after subnet reduction:*

Purely for simplifying the verification procedure, (N_1, M_1) is transformed to (N_2, M_2) by reducing the subnets EPS, IPS and SM in Figure 9.3 each to a single place. (N_2, M_2) is also shown in Figure 9.3 where the subnets of EPS and IPS and SM are now considered as single places. Since (N_1, M_1) and its subnets EPS, IPS and SM are correct, it follows from Theorems 15, 16, 17 and 18 in [SM (1983)] that (N_2, M_2) is correct.

D.2 *Correctness of the PNPs after place merging:*

In Step 1.2.3, two sets of resource places $Q1 = \{DK1, DK2, DK3\}$ and $Q2 = \{WM1, WM2, WM3\}$ in (N_2, M_2) are merged by MERGE-N-PLACE to DK and WM, respectively, resulting in Petri net process (N_3, M_3) (Figure 9.3). In the following, it will be shown that (N_3, M_3) is almost live and bounded by applying Theorem 7.5(17) on the associated process of (N_2, M_2). Furthermore, it can be observed without theoretical support that (N_3, M_3) is almost reversible and terminates properly.

Before applying Theorem 7.5(17), let us make two observations: (a) Theorem 7.5(17) has to be applied on the associated process of (N_2, M_2) instead of (N_2, M_2). This is because Condition (b) in Theorem 7.5(17) can never be satisfied by the latter. In fact, the entry place of a process forms a siphon by itself in P_0 that contains no traps, where $P_0 = \{AGENT, EPS, EN, IPS, PP, PLAN, checker, ME, SM, AGENT.X, MC, p_2, p_4, RH, p_6, p_8, p_9, p_{10}, q_1, q_2, q_3, q_4, q_5, q_6, q_7, q_8, MC.ck, RH.ck, PP.ck, MC.ik, RH.ik, PP.ik\}$. (b) The set of all the minimal siphons of N_2 are: $D_1 = \{DK1, q_1, p_2, p_4\}$, $D_2 = \{DK2, q_3, p_8\}$, $D_3 = \{DK3, SM, q_6, q_8, p_{10}\}$, $D_4 = \{DK3, checker, q_6, q_8, p_{10}\}$, $D_5 = \{WM1, q_2, p_2, p_4\}$, $D_6 = \{WM2, q_5, p_6, p_8, p_9\}$, $D_7 = \{WM3, SM, q_7, q_8, p_{10}\}$, $D_8 = \{WM3, checker, q_7, q_8, p_{10}\}$, $D_9 = \{ME, p_6, p_8, p_9, RH.ik, q_5, RH.ck, q_3, q_4\}$, $D_{10} = \{PP.ck, q_6, PLAN, PP.ik, q_7\}$, $D_{11} = \{EN, q_4\}$, $D_{12} = \{AGENT, AGENT.X, PP, checker, EPS, IPS, MC, p_2, p_4, RH, p_6, p_8,$

$p_9, p_{10}, q_1, q_2, q_3, q_4, q_5, q_6, q_7, q_8, MC.ik, RH.ik, PP.ik, MC.ck,$
$RH.ck, PP.ck\}$ and $D_{13} = \{AGENT, AGENT.X, PP, SM,$
$EPS, IPS, MC, p_2, p_4, RH, p_6, p_8, p_9, p_{10}, q_1, q_2, q_3, q_4, q_5,$
$q_6, q_7, q_8, MC.ik, RH.ik, PP.ik, MC.ck, RH.ck, PP.ck\}$. These
siphons are also traps initially marked by M_2. Furthermore, it
can be shown that any siphon containing $DK1$ ($resp., DK2, DK3$)
also contains $D1$ ($resp., D_2, D_3 or D_4$) and that any siphon containing $WM1$ ($resp., WM2, WM3$) also contains D_5 ($resp., D_6, D_7 or$
D_8). For example, let siphon D contain $WM1$. Then, $l_2 \in \bullet D$ implies $q_2 \in D$ and $update \in \bullet D$ implies $p_4 \in D$ which in turn implies
$p_2 \in D$. Hence, $D5 = \{WM1, q_2, p_2, p_4\} \subseteq D$.

We are now going to show that the two conditions of Theorem 7.5(17) are satisfied for the associated process of (N_2, M_2).
For Condition (b), for example, the siphon D_9, where $D_9 \subseteq P_0$,
contains the trap $S = D_9$ marked by M_2. For the siphon $D =$
$\{WM1, WM2, WM3, SM, ME, q_2, q_5, q_7, q_8, p_2, p_4, p_6, p_8, p_9,$
$p_{10}\}$, Observation(b) in the last paragraph implies that $D \supseteq D_5 \cup$
$D_6 \cup D_7 \supseteq Q_2$. Since for $i = 5, 6, 7$, each D_i is an initially marked
trap, $S = D_5 \cup D_6 \cup D_7$ is also an initially marked trap satisfying
$Q_2 \subseteq S$.

For Condition (a) (i.e., Theorem 7.5(12b)), it is not hard to ob-
serve that each of the primitive PNP in Figure 9.1 is SM-coverable.
By Theorem 4.1(12), 4.2(12), 6.3(12b), 6.5(10) and 5.6(7), after
applying Choice, Enable, Reduce-P-Place, Reduce-subnet and Refinement operators, the resulting PNP (N_2, M_2) is still almost SM-
coverable. Since the minimal SM-component containing any resource place is unique, e.g., the SM-component containing place
$DK1$ is $DK1$ invoke-a-method $q_1 t_1$ $DK1$ make-decision p_2 (continue,
end) p_4 update $DK1$, Theorem 7.5(12b) is satisfied and hence the
associated process of (N_2, M_2) is SM-coverable.

Note: During modeling, the nine paths ($invoke - a -$
$method, q_1, l_1$), (t_1, q_2, l_2), ($k - detector, q_3, l_3$), ($e - detector, q_4, l_4$),
(t_3, q_5, l_5), ($trigger, q_6, l_6$), (t_6, q_7, l_7), ($reject, q_8, l_8$) and ($no -$
$change, p_8, t_4$) shown in Figure 9.1(c)(d)(e) can be collapsed into
nine single transitions invoke-a-method, t_1, k-detector, e-detector,
t_3, trigger, t_6, reject and no-change, respectively. This may not
have any adverse effects on the logic of the modules individually.
However, these transitions will then form self-loops with the places
$DK1, WM1, DK2, EN, WM2, DK3, WM3, SM, checker$ and

$DK2$, respectively. This violates the requirement of the method for handling resource sharing (Theorem 7.5) that the Petri net under concern should be ordinary and pure.

D.3 *Correctness of the agent system CLINIC after place refinement:*
Finally, the places EPS, IPS and SM in (N_3, M_3) are restored by applying Place-Refinement on (N_3, M_3). Since all the places EPS, IPS and SM have input transitions in N_3, by Theorem 5.6(12)(13)(14), the produced Petri net process (N, M_0), i.e., the model for the agent CLINIC, is correct. The architecture for (N, M_0) is shown in Figure 9.3. For clarity, merging of these resource places is not shown.

E *Correctness of the entire HC-MAS in Section 9.2.4:*
Verification for HC-MAS is similar as for CLINIC according to the operators applied to the specification. For the example in Figure 9.4, operators Interleave, Enable, Choice, Iteration and Reduce-P-Path are applied. If all the five agents are correct, by Theorems 4.3, 4.1, 4.2, 3.5, 6.2, and 6.3(16), the entire HC-MAS is also correct.

9.4 Remarks

Due to the great variation in requirements and proneness to modification, it is not feasible to pinpoint a specific architecture and an initial set of primitive modules for the design of a general MAS. Hence, it is more appropriate to adopt a component-based approach so that different MASs may start from different primitive modules and be composed in different ways. This chapter has presented a formal methodology for such a purpose. The methodology has two main features: (1) By applying PPPA, it provides the means for representing the agents' modules as Petri net processes (PNP), integrating simple PNPs into composite ones, and linking several agents into a single multi-agent system. For illustration, only four composition operators, one refinement operator and some reduction operators have been applied in this chapter. (2) A property-preservation requirement is embedded into the algebra for guaranteeing the correctness in every step of the integration process. In fact, every operator of this algebra can preserve about nineteen properties, though only the four most basic ones, namely, liveness, boundedness, reversibility and proper termination have been considered in this chapter. Hence, agent systems with a great variety of properties can be accommodated.

Chapter 10

Application of PPPA to Job-Shop Scheduling Systems

This chapter introduces a PPPA-based system design method for modeling and optimizing the job-shop scheduling systems (JSS). Several operators are presented for handling the combination of JSS modules and resource-sharing. Each operator can obtain the partial schedule and makespan of each component. The optimal scheduling of JSS system can be derived from these theoretical results step by step. With the PPPA-based method, one can not only design a correct system model which is live, bounded, reversible and terminate properly, but also find the optimal scheduling of JSS in a formal way.

10.1 Background and Motivation

Job shop scheduling (JSS) is concerned with the efficient allocation of resources, e.g., machines, Job shop scheduling (JSS) is concerned with the efficient allocation of resources, e.g., machines, materials, robots, workers, AGVs etc, for manufacturing the productions over time. Scheduling occurs whenever we required these shared resources for productions. The objective of scheduling is to find a way to assign these resources so that the production constraints will be satisfied and the performance criterion is maximized or minimized. A typical performance goal is to minimize the makespan, i.e., the time required to complete all the jobs. The JSS problem considered in this chapter is formally stated as follows: There are a finite set of n jobs, each of which consists of a chain of operations, and a finite set of m machines, each of which can handle at most one operation at a time. Each operation needs to be processed during an uninterrupted period of a given length on a given machine. There may be some other resources such as workers responsible for the machines, robots and so on. All that we do is to find a schedule, or an allocation of the operations to time intervals for machines and other resources so that, the makespan is minimized.

Recently, more attentions are paid on two features by researchers, i.e., *Resource-sharing* and *Component-based* architecture in JSS system design. The background is similar to that of a manufacturing system which was introduced in Chapter 8. In the literature, much work involved the research on the above features. We can find the description of the related approaches and a brief review in Chapter 2 of Deniz. K. Terry's PhD thesis [TER (1996)].

As a component of a variety of methodologies developed to generate JSS system schedules, Petri net is very powerful for modeling the concurrent processing, resource-sharing, routing flexibility, sequential steps, concurrent use of multiple resources, machine setup times, job batch sizes and so on in JSS systems [PM (1995); PX (1997); ZV (1999); XZ (1998)]. As for schedule optimization, there are two types of Petri net based methodologies for JSS problems: 1) Heuristic searching algorithms based on reachability graph of Petri net models [SS (1991); LD (1994); SCF (1994)]; and 2) GA algorithms for optimizing scheduling [WXY (2008); LFC (2003); CFL (1998)]. Pertri nets are attracting the attention of many researchers for the design of JSS system and the optimization of the scheduling of the system [XH (2001); GZC (2003); CC (2003)].

However, the following shortages exist for current Petri net based methodologies: 1. Researchers put forward the Petri net models for solving their specified problems. Therefore, the performance of Petri nets cannot be explored efficiently. Although it is good for obtaining better optimization results, researchers cannot apply them for general problems; 2. The real time JSS system is usually quite complex and large, consequently, the Petri net model will be quite large. The explosion problem of Petri nets makes the current Petri net analysis methods not applicable.

In order to overcome the hardship mentioned, this chapter handles JSS problem from two aspects: system design and scheduling optimization. On system design aspect, this chapter applies a component-based approach for JSS system design specification and verification. The basic component model is an extended Petri net process (EPNP) which is similar to the Petri net process (PNP) discussed in Chapter 3. The difference between EPNP and PNP is that EPNP is an extended Petri net while PNP is a general Petri net. Hence, all the results for PPPA will also be true for the JSS design in this chapter. With our component-based approach, the JSS system is correct from the viewpoint that the system contains some desired properties such as liveness, boundedness, reversibility, proper termination, etc.

As for system scheduling optimization, based on EPNP, some information is provided in this chapter on finding the possible scheduling (the firing sequences of the Petri net model) and makespan of the composite components from the constituent components for the integration operators in PPPA. In order to handle resource-sharing problem, we combined the place-merging operator and the integration operators for system modeling. The firing sequences and the upper bound of the makespan for these combined operators are obtained. The algorithm for finding the makespan and its corresponding scheduling is also presented for the combined operators. With the component-based method for system design and schedule optimization, the schedule and makespan of the composite components can be calculated based on their constituent components. In order to reduce the calculation time, traditional algorithms such as GA, Branch-and-bound methods, etc. can be applied together with the algorithms presented in this chapter for optimizing the schedule of JSS systems.

In the remaining part of this chapter, PPPA-based approach is introduced in Section 10.2. For each of the integration operators such as Action Prefix, Iteration, Enable, Choice, Interleave, Parallel, Disable, Disable-resume, Place merging and the mix operators, the related theoretical results concerning the firing sequences and makespan are presented. Then Section 10.3 illustrates the application of PPPA to JSS system design and scheduling optimization with an example. Some conclusive remarks are given at the end.

The results in this chapter are mainly extracted from the literature [HUA (2009)].

10.2　JSS System Design and Makespan Calculation

This section presents several operators in PPPA for the design specification of JSS system. For each operator, the theorem about the makespan calculation is provided. Based on these operators and theorems, we can deduce a larger and more complex module, even for the whole system from the smaller ones step by step, and also calculate the schedule of larger modules.

Definition 10.1. An **Extended Petri net process**(EPNP) $B = (N, p_e, p_x)$, where $N = (P, T, F, W, M_s, C, \tau)$ is an extended Petri net with three additional elements comparing to general Petri nets, i.e., a set of weights W, a set of colors C and a set of time τ, p_e is the entry place and p_x is the exit place of EPNP.

The performance and characteristics, e.g., proper initialization, proper termination, handling resource-sharing, etc. of extended Petri net process (EPNP) is similar to Petri net process (PNP) defined in Definition 3.1, the difference is that in EPNP, N is an extended Petri net while in PNP, N is a general Petri net. Hence, all the theoretical results for PNP are also true for EPNP and can be applied to design a correct JSS system.

Basic model for operations

In job-shop scheduling systems, operations either do not need any resources or need resources such as machines, robots, workers, etc. Hence, the corresponding models for the operations in a JSS have two different forms.

For the first case, we use a transition t to represent an operation; t has a single input place representing the start of the operation and a single output place representing the end of the operation (Figure 10.1(a)). For the second case, one more place representing the resources should be associated to t and form a self-loop. Initially, there are some tokens assigned to the resource place. We use tokens with different appearance forms (e.g, triangle and rectangle ,etc) to represent different resources (Figure 10.1(b)).

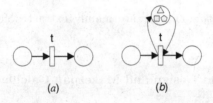

(a)   (b)

Fig. 10.1   The extended Petri net model for operations in JSS.

Based on the relationship of the operations and the control flow of JSS, we use some operators in PPPA to combine these operations in JSS.

In this chapter, the basic model is the extended Petri net process. Although extended Petri net process and the general Petri net process have different structural and behavioral properties and firing rules. Their pictorial representations of these operators are same.

In the following presentation, we use notation $B = (N, p_e, p_x)$ (resp., $B_i = (N_i, p_{ie}, p_{ix})$) to represent an extended Petri net process, where $N = (P, T, F, W, M_s, C,)$ (resp., $N_i = (P_i, T_i, F_i, W_i, M_{is}, C_i, \tau_i)$) is an extended Petri net with an initial static marking $M_s = (0, M_c, 0)$ (Definition 3.4) (resp., M_{is}), p_e and p_x (resp., p_{ie} and p_{ix}) are the entry and exit places of the process, respectively. For simplicity, in the remaining part of this book, we denote the initial marking of B (resp., B_i) as $M_e = p_e + M_c$ (resp., $M_{ie} = p_{ie} + M_{ic}$) instead of $(1, M_c, 0)$, and the exit marking of B (resp., B_i) as $M_x = p_x + M_c$ (resp., $M_{ix} = p_{ix} + M_{ic}$) instead of $(0, M_c, 1)$.

In the following, for each operator, the firing sequence and the potential partial schedule in the composite process will be concluded based on the possible firing sequences of the constituent processes. Since the results are trivial, we omit the details of the proof. The conclusions can be used to find the optimal scheduling solutions.

Theorem 10.1. *Suppose (B', M'_e) is induced from (B, M_e) by applying the operator Action Prefix. If $M_e[\sigma\rangle M_x$ in (B, M_e), then $M'_e[b\sigma\rangle M'_x$ in (B', M'_e). If σ has the minimum makespan in (B, M_e), then $b\sigma$ has the minimum makespan in (B', M'_e).*

Theorem 10.2. *Suppose (B', M'_e) is induced from (B, M_e) by applying the operator Iteration. If $M_e[\sigma\rangle M_x$ in (B, M_e), then $M'_e[\varepsilon_1\sigma\varepsilon_2\sigma\varepsilon_2 \ldots \sigma\varepsilon_2\sigma\varepsilon_3\sigma\rangle M'_x$ in (B', M'_e). If σ has the minimum makespan in (B, M_e), then $\varepsilon_2\sigma\varepsilon_3$ has the minimum makespan in (B', M'_e).*

Theorem 10.3. *Suppose (B, M_e) is induced from (B_1, M_{1e}) and (B_2, M_{2e}) by applying the operator Choice. If $M_{1e}[\sigma_1\rangle M_{1x}$ in (B_1, M_{1e}) and $M_{2e}[\sigma_2\rangle M_{2x}$ in (B_2, M_{2e}), then $M_e[\sigma_1\varepsilon\sigma_2\rangle M_x$ in (B, M_e). If σ_1 has the minimum makespan in (B_1, M_{1e}), σ_2 has the minimum makespan in (B_2, M_{2e}), then $\sigma_1\varepsilon\sigma_2$ has the minimum makespan in (B, M_e).*

Theorem 10.4. *Suppose (B, M_e) is induced from (B_1, M_{1e}) and (B_2, M_{2e}) by applying the operator Enable. If $M_{1e}[\sigma_1\rangle M_{1x}$ in (B_1, M_{1e}) and $M_{2e}[\sigma_2\rangle M_{2x}$ in (B_2, M_{2e}), then $M_e[\varepsilon_1\sigma_1\rangle M_x$ and $M_e[\varepsilon_2\sigma_2\rangle M_x$ in (B, M_e). If σ_1 has the minimum makespan in (B_1, M_{1e}), σ_2 has the minimum makespan in (B_2, M_{2e}), then the firing sequence with the minimum makepan in (B, M_e) is $\min\{\varepsilon_1\sigma_1, \varepsilon_2\sigma_2\}$*

Theorem 10.5. *Suppose (B, M_e) is induced from (B_1, M_{1e}) and (B_2, M_{2e}) by applying the operator Interleave. If $M_{1e}[\sigma_1\rangle M_{1x}$ in (B_1, M_{1e}) and $M_{2e}[\sigma_2\rangle M_{2x}$ in (B_2, M_{2e}), then $M_e[\sigma_1\varepsilon\sigma_2\rangle M_x$ in (B, M_e), where σ is a* *transition sequence consisting of all of the transitions in σ_1 and σ_2 such that the appearance order of the transitions conforms to the order in σ_1 and σ_2 (i.e., the appearance order is preserved). If σ_1 has the minimum makespan in (B_1, M_{1e}) and σ_2 has the minimum makespan in (B_2, M_{2e}), then σ has the minimum makespan in (B, M_e) and $\#\sigma = \max\{\#\sigma_1, \#\sigma_2\}$, where the notation $\#\sigma$ is the makespan value of the firing sequence σ.*

For example, suppose $\sigma_1 = abc$ and $\sigma_2 = de$. Then σ can be any one of the following 10 sequences: *deabc, daebc, dabec, dabce, adebc, adbec, adbce, abdec, abdce, abcde.*

Theorem 10.6. *Suppose (B, M_e) is induced from (B_1, M_{1e}) and (B_2, M_{2e}) by applying the operator Parallel. If $M_{1e}[\alpha_1\tau\beta_1\rangle M_{1x} in (B_1, M_{1e})$ and $M_{2e}[\alpha_2\tau\beta_2\rangle M_{2x}$ in (B_2, M_{2e}), then $M_e[\varepsilon_1\alpha\pi\beta\varepsilon_2\rangle M_x$ in (B, M_e), where α is a transition sequence consist of all of the transitions in α_1 and α_2 such that the appearance order of the transitions in α conforms to the order in α_1 and in α_2, and β is a transition sequence consisting of all of the transitions in β_1 and β_2 such that the appearance order of the transitions in β conforms to the order in β_1 and in β_2. If $\sigma_1 = \alpha_1\tau\beta_1$ has the minimum makespan in (B_1, M_{1e}) and $\sigma_2 = \alpha_2\tau\beta_2$ has the minimum makespan in (B_2, M_{2e}) then $\sigma = \varepsilon_1\alpha\pi\beta\varepsilon_2$ has the minimum makespan in (B, M_e) and $\#\sigma = \max\{\#\alpha_1, \#\alpha_2\} + \#\tau + \max\{\#\beta_1, \#\beta_2\}$.*

Theorem 10.7. *Suppose (B, M_e) is induced from (B_1, M_{1e}) and (B_2, M_{2e}) by applying the operator Disable. If $M_{1e}[\sigma_1\rangle M_{1x}$ in (B_1, M_{1e}) and $M_{2e}[\sigma_2\rangle M_{2x}$ in (B_2, M_{2e}), then $M_e[\varepsilon\sigma\rangle M_x$ in (B, M_e), where σ satisfies: (1) $\sigma = \sigma_1$; (2) $\sigma = \sigma_2$; (3) σ is a transition sequence consisting of the transitions in a firing sequence α of B_1 and σ_2, the appearance order is preserved and $p_{2e}{}^\bullet\bigcap\alpha = \phi$. If σ_1 has the minimum makespan in (B_1, M_{1e}) and σ_2 has the minimum makespan in (B_2, M_{2e}), then σ has the minimum makespan in (B, M_e) and $\#\sigma$ equals to one of $\#\sigma_1$, $\#\sigma_2$, and $\#\alpha + \#\sigma_2$.*

Theorem 10.8. *Suppose (B, M_e) is induced from (B_1, M_{1e}) and (B_2, M_{2e}) by applying the operator Disable-resume. If $M_{1e}[\sigma_1\rangle M_{1x}$ in (B_1, M_{1e}) and $M_{2e}[\sigma_2\rangle M_{2x}$ in (B_2, M_{2e}) then $M_e[\varepsilon_1\sigma\rangle M_x$ in (B, Me), where σ satisfies: (1) $\sigma = \sigma_1$;(2) Suppose $\sigma_1 = \alpha\beta, p_{2e}{}^\bullet\bigcap\alpha = \phi$ and $p_{2e}{}^\bullet\bigcap\beta = \phi$. Then $\sigma = \gamma\beta$, where γ is a transition sequence consisting of transitions in α and $\sigma_2\varepsilon_2$ such that the appearance order is preserved. If σ_1 has the minimum makespan in (B_1, M_{1e}) and σ_2 has the minimum makespan in (B_2, M_{2e}), then σ has the minimum makespan in (B, M_e) and $\#\sigma$ equals to one of $\#\sigma_1$, and $\#\sigma_1 + c\#\sigma_2$, where c is a constant number.*

In JSS, resource-sharing is quite necessary. After the composition of the modules, sharing resource should be considered. Therefore, we apply place merging operator (Figure 10.2) for modeling and analyzing the resource-sharing problem.

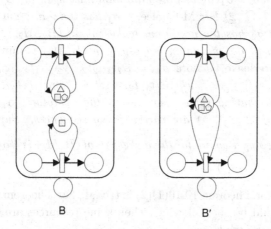

B   B'

Fig. 10.2 $B(Q_i \rightarrow q_i)$: The operator place merging in a single process.

Theorem 10.9. *Let (B, M_e) and (B', M'_e) be defined in Definition 7.1 (Merge-A-Place). If $M_e[\sigma\rangle M_x$ in (B, M_e), then $M'_e[\sigma\rangle M'_x$ in (B', M'_e).*

Note that, by our modeling technology, each transition t representing an operation forms a self-loop with the corresponding resource places. In such a case, Theorem 10.9 is true.

Up to now, we consider several integration operators, including the place merging operators. However, for the place merging operators, we cannot get the makespan of the resultant process from the makespan of the original processes very easily. In order to simplify the computing process and get an exact makespan value, place merging operator will be applied together with other integration operators if there exists resource-sharing in the composition process. For example, two processes are composed by *Enable* operator. If some operations in these two processes share some common resources, *place merging* operator and *Enable* operator will be applied together. The notations *M-Enable*, *M-Choice*, *M-Interleave* etc are used to denote these mixed operators.

Theorem 10.10. *Suppose $M_{1e}[\sigma_1\rangle M_{1x}$ in (B_1, M_{1e}) and $M_{2e}[\sigma_2\rangle M_{2x}$ in (B_2, M_{2e}). (B, M_e) is resulted from B_1 and B_2 by applying mixed operators and $M_e[\sigma\rangle M_x$ in (B, M_e). (1) For operator M-Enable, if σ_1 has the minimum makespan in (B_1, M_{1e}), σ_2 has the minimum makespan in (B_2, M_{2e}), then $\sigma = \sigma_1 \varepsilon \sigma_2$ has the minimum makespan in (B, M_e) and $\#\sigma = \#\sigma_1 + \#\sigma_2 + 1$; (2) For M-Choice, if σ_1 has the minimum makespan in (B_1, M_{1e}) and σ_2 has the minimum makespan in (B_2, M_{2e}), then the firing sequence with the minimum makespan in (B, M_e) is $\min\{\varepsilon_1\sigma_1, \varepsilon_2\sigma_2\}$; (3) For M-Interleave, denote $\sigma_1 = \alpha_1 t_1 \alpha_2 t_2 \ldots \alpha_{k-1} t_{k-1} \alpha_k$ and $\sigma_2 = \beta_1 t'_1 \beta_2 t'_2 \ldots \beta_{m-1} t'_{m-1} \beta_m$, where $t_1, t_2, \ldots, t_{k-1}$ and $t'_1, t'_2, \ldots, t'_{m-1}$ are the transitions sharing some resources in the operators, α_i and $\beta_j (i = 1, 2, \ldots, k, j = 1, 2, \ldots, m)$ are partial firing sequences. Suppose $\#\sigma_1 \geq \#\sigma_2$. Then the upper bound for the makespan of (B, M_e) is $\#\sigma_2 + \sum\limits_{i=0}^{i=k-1} \#t_i$.*

The proof for Theorem 10.10(1)(2) is trivial. For Theorem 10.10(3), the worst case would be the following, when some resources are sharing, it is always process B_2 which waits.

By Theorems 10.5 and 10.9, for the mixed operator *M-Interleave*, the firing sequence σ of (B, M_e) is a transition sequence consisting of all of the transitions in σ_1 and σ_2 such that the appearance order of the transitions conforms to the order in σ_1 and in σ_2. Below, an algorithm is introduced for generating a firing sequence σ of (B, M_e) based on σ_1 and σ_2 with the minimum makespan.

Makespan generating algorithm for M-Interleave operator:

Input: Two firing sequences $\sigma_1 = \alpha_1 t_1 \alpha_2 t_2 \ldots \alpha_{k-1} t_{k-1} \alpha_k$ and $\sigma_2 = \beta_1 t'_1 \beta_2 t'_2 \ldots \beta_{m-1} t'_{m-1} \beta_m$ of (B_1, M_{1e}) and (B_2, M_{2e}), respectively.

Output: a firing sequence σ and the corresponding minimum makespan $\#\sigma$ of (B, M_e).

Step 1: let $\sigma_1 = \alpha_1 t_1 \alpha_2 t_2 \cdots \alpha_{k-1} t_{k-1} \alpha_k$, $\sigma_2 = \beta_1 t'_1 \beta_2 t'_2 \cdots \beta_{m-1} t'_{m-1} \beta_m$, $\sigma = \phi$, and $\#\sigma = 0$.

Step 2: for $i = 1, i < k; j = 1, j < m$

If there exists a leftmost subsequence $\sigma_i = \alpha_i t_i \ldots \alpha_p t_p$ of σ_1 and a leftmost subsequence $\sigma_j = \beta_j t'_j \ldots \beta_q t'_q$ of σ_2 satisfying $\#\sigma_i - \#t_p \leq \#\sigma_j - \#t'_q \leq \#\sigma_i$, then $\sigma = \sigma\sigma_i(\beta_j t'_j \ldots \beta_q) t'_q$ and $\#\sigma = \#\sigma + \#\sigma_i + \#t'_q$.

If there exists a leftmost subsequence $\sigma_i = \alpha_i t_i \ldots \alpha_p t_p$ of σ_1 and a leftmost subsequence $\sigma_j = \beta_j t'_j \ldots \beta_q t'_q$ of σ_2 satisfying $\#\sigma_j - \#t'_p \leq \#\sigma_i - \#t_q \leq \#\sigma_j$, then $\sigma = \sigma\sigma_j(\alpha_i t'_i \ldots \alpha_p)t_p$ and $\#\sigma = \#\sigma + \#\sigma_j + \#t_q$.

Let $\sigma_1 = \sigma_1 - \sigma_i$, $\sigma_2 = \sigma_2 - \sigma_j$, $i = p$ and $j = q$, $i++$, $j++$

Step 3: $\sigma = \max\{\sigma\sigma_1, \sigma\sigma_2\}$ and $\#\sigma = \#\sigma + \max\{\#\sigma_1, \#\sigma_2\}$.

We can apply the above algorithm to generate the firing sequence and compute the makespan for the mixed operators *M-parallel*, *M-disable*, *M-disable-resume*. In the case of more than two composition processes, we can also improve the algorithm easily such that the input is more than two firing sequences and the comparison is between more than two leftmost subsequences.

10.3 JSS Design and Optimization — A Case Study

This section shows how to apply the composition operators in PPPA for designing JSS and finding the optimal scheduling. In order to illustrate the technology more concisely, we will take a real time JSS system as an example.

Suppose 3 jobs are processed in 3 workshops. Each job has 2 or 3 operations which will be processed in the three workshops in a sequence order. In each workshop some machines can be used and each operation only chooses one machine for processing. 4 workers are responsible for these 6 machines in the three workshops. The details of the operation distribution, the processing time of each operation in different machines, and the worker distribution are shown in Table 10.1.

10.3.1 *System design*

For simplicity, in the figures of Petri net model, only those transitions representing the operations are labeled. The resource places R_{ij} in the figures represent machine M_i which are sponsored by worker W_j.

Based on the requirement, Job 1 has three operations. In operation 1, machine M_3 is occupied and which is sponsored by worker W_2. Hence, transition t_1, representing the operation 1, is associated with a resources place R_{32} in the Petri net model (Figure 10.3). For operation 2, it can be processed either in machine M_1 sponsored by worker W_1 or in machine M_2 sponsored by worker W_1 or W_2. Choice operators are applied twice in

Table 10.1   Processing information of jobs.

Job	Operation	Workshop	(machine, time)		W_1	W_2	W_3	W_4
1	1	II	$(M_3, 12)$	M_1	O			
	2	I	$(M_1, 9), (M_2, 10)$	M_2	O	O		
	3	III	$(M_4, 5), (M_5, 6)$ $(M_6, 8)$	M_3		O		
2	1	II	$(M_3, 6)$	M_4			O	
	2	III	$(M_4, 6), (M_5, 9)$ $(M_6, 8)$	M_5			O	O
	3	I	$(M_1, 4), (M_2, 5)$	M_6				O
3	1	I	$(M_1, 5), (M_2, 4)$	Note:"O" means that the worker in the corresponding column is responsible for the machine in the corresponding row.				
	2	III	$(M_4, 8), (M_5, 10)$ $(M_6, 9)$					

this model, one is used for choosing machines (M_1 and M_2), and another is used for choosing workers (W_1 and W_2). Similarly, for operation 3, three machines M_4, M_5 and M_6 will be chosen, and machine M_5 has a choice between the workers W_3 and W_4. Since the three operations are in a sequence order, the three components will be integrated by applying Enable operator (Figure 10.3).

Similarly, the Petri net models for Jobs 2 and 3 can also be obtained (Figure 10.4 and Figure 10.5).

By Theorems 10.3 and 10.4, the three Jobs have the possible firing sequences listed in Figure 10.6. In order for simplicity and clarity, we only list the transitions representing the operations. The processing time and the resources for each operation are assigned to the corresponding transitions. We can find that there are 12 firing sequences for each Petri net model.

Since these three jobs are independent, they can be processed in an Interleave relationship. In the three Petri net models, the resource places with the same labels represent the same resources. Hence, a mixed operator M-interleave will be applied for composing these three modules and obtaining the system model. Since the picture is too complex, we omit it.

Theorem 10.11. *The JSS system design is correct. In this chapter a system is said to be correct if it is almost live, almost bounded, almost reversible and terminates properly. Since each of the primary Petri net modules is correct, by the theorems presented in the previous chapters, the system model*

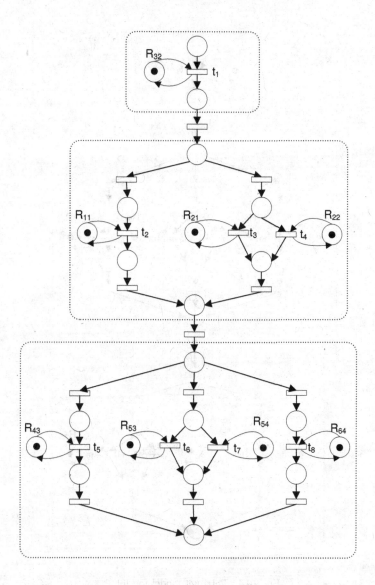

Fig. 10.3 Petri net model for Job 1.

constructed by combining the modules by applying the operators in PPPA is also correct. Hence, the design method can guarantee the correctness of the system.

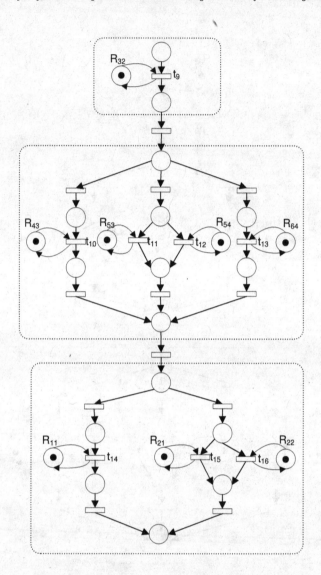

Fig. 10.4   Petri net model for Job 2.

10.3.2   *Scheduling optimization*

By Theorem 10.5 and Makespan generating algorithm for M-Interleave operator, the minimum makespan and the corresponding firing sequence (i.e.,

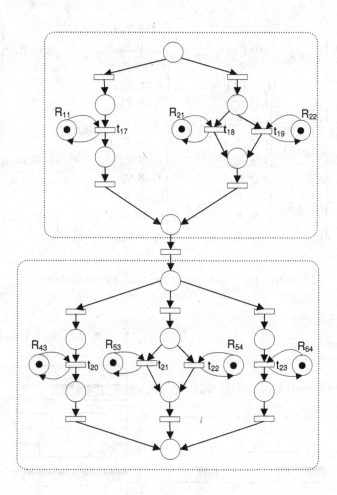

Fig. 10.5   Petri net model for Job 3.

schedule) can be obtained. Theoretically, the algorithm should run totally 12^3 times in order to search for the optimal solution. In practice, quite a lot of firing sequences can be deleted from the list.

In order to illustrate the method, we use two possible schedule examples to run the algorithm. For example, let $\sigma_1 = t_1 t_2 t_5$, $\sigma_2 = t_9 t_{10} t_{14}$, and $\sigma_3 = t_{17} t_{20}$. By Makespan generating algorithm for M-Interleave operator, the schedule is as shown in Figure 10.7(a). Where machines 1, 3 and 4 are used and workers 1, 2 and 3 are needed to sponsor the machines. R_{ij}

$$t_1(12,R_{32}) \left\{ \begin{array}{l} t_2(9,R_{11}) \left\{ \begin{array}{l} t_5(5,R_{43}) \\ t_6(6,R_{53}) \\ t_7(6,R_{54}) \\ t_8(8,R_{64}) \end{array} \right. t_{17}(5,R_{11}) \left\{ \begin{array}{l} t_{20}(8,R_{43}) \\ t_{21}(10,R_{53}) \\ t_{22}(10,R_{54}) \\ t_{23}(9,R_{64}) \end{array} \right. t_{10}(6,R_{43}) \left\{ \begin{array}{l} t_{14}(4,R_{11}) \\ t_{15}(5,R_{21}) \\ t_{16}(5,R_{22}) \end{array} \right. \\ t_3(10,R_{21}) \left\{ \begin{array}{l} t_5(5,R_{43}) \\ t_6(6,R_{53}) \\ t_7(6,R_{54}) \\ t_8(8,R_{64}) \end{array} \right. t_{18}(4,R_{21}) \left\{ \begin{array}{l} t_{20}(8,R_{43}) \\ t_{21}(10,R_{53}) \\ t_{22}(10,R_{54}) \\ t_{23}(9,R_{64}) \end{array} \right. t_9(6,R_{32}) \left\{ \begin{array}{l} t_{11}(7,R_{54}) \left\{ \begin{array}{l} t_{14}(4,R_{11}) \\ t_{15}(5,R_{21}) \\ t_{16}(5,R_{22}) \end{array} \right. \\ t_{12}(7,R_{53}) \left\{ \begin{array}{l} t_{14}(4,R_{11}) \\ t_{15}(5,R_{21}) \\ t_{16}(5,R_{22}) \end{array} \right. \\ t_{13}(8,R_{64}) \left\{ \begin{array}{l} t_{14}(4,R_{11}) \\ t_{15}(5,R_{21}) \\ t_{16}(5,R_{22}) \end{array} \right. \end{array} \right. \\ t_4(10,R_{22}) \left\{ \begin{array}{l} t_5(5,R_{43}) \\ t_6(6,R_{53}) \\ t_7(6,R_{54}) \\ t_8(8,R_{64}) \end{array} \right. t_{19}(4,R_{22}) \left\{ \begin{array}{l} t_{20}(8,R_{43}) \\ t_{21}(10,R_{53}) \\ t_{22}(10,R_{54}) \\ t_{23}(9,R_{64}) \end{array} \right. \end{array} \right.$$

Fig. 10.6   The possible firing sequences for the three jobs.

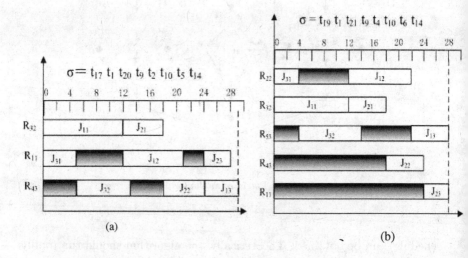

Fig. 10.7   Some possible schedules.

means machine i sponsored by worker j. J_{ij} means Job i's jth operation is being processed in the corresponding time slot. In this case, the number of machines is minimum and the makespan equals 29. In another case, let $\sigma_1 = t_1 t_4 t_6$, $\sigma_2 = t_9 t_{10} t_{14}$, and $\sigma_3 = t_{19} t_{21}$. The schedule is shown in

Figure 10.7(b). In this case, 5 machines are used and 3 workers are needed. The makespan is 28. There are some more schedules with makespan 28. For example, the following four firing sequences correspond to the optimal schedule: $t_1t_{19}t_{20}t_2t_9t_{10}t_7t_{14}$; $t_{18}t_1t_{20}t_9t_3t_{10}t_6t_{14}$; $t_{17}t_1t_{23}t_9t_4t_{10}t_6t_{14}$; $t_{19}t_1t_{23}t_9t_2t_{10}t_7t_{14}$.

In order to reduce the time complexity, an upper bound can be given for the makespan such that, for those partial schedules with the possible makespan exceeding the upper bound, they can be deleted from the firing sequence list. For example, in the above example, the makespan of the system will not exceed 28. Hence, the firing sequences $t_1t_2t_8$, $t_1t_3t_8$, and $t_1t_4t_8$ in the Job 1 model will be deleted from the list since their makespan exceeded 28. As for Job 2 model, before processing, it will wait for 12 time slots because of resources sharing, except the sequence $t_9t_{10}t_{14}$, the makespan for all of other partial schedule will not be less than 28. If an optimal solution is required to be found, only the single firing sequence $t_9t_{10}t_{14}$ will be considered. Hence, very possibly, the algorithm only needs to run no more than 9×12 times.

Theoretically, we can apply our JSS design method to find an optimal scheduling and the idea can refer to [HCW (2007); HUA (2004)]. In practice, for large and complex systems, the main difficulty is to find the optimal scheduling since there are too many possible partial firing sequences. The makespan generating algorithm should run too many times in order to select the optimal one. In order to reduce the time complexity, one of the possible solutions is to combine the branch and bound method [ZCX (1995)] with our algorithm. The idea has been introduced in the above example. Another solution is to combine GA-algorithm [WXY (2008)] with ours, and the idea is as follows: let the partial firing sequences in each module be genes, and the firing sequences of the system be used as the chromosomes. The makespan will be the fitness function and the operators will be newly defined according to the combination operators defined in this book. The branch and bound method, GA method and ours can even be combined together to find the approximation optimal solution in a much more efficient way.

Traditionally, Petri net based methodologies for JSS problem focused on two aspects: one is to model the system with Petri nets and then find the possible schedule by simulation. Another one is based on Petri net model, searching for algorithms for the optimization of the schedule. The common point of current methods is that, the Petri net model is assumed to exist. For a real time complex and large JSS system, generating the

Petri net model is in fact a hard work. How to design a correct Petri net model such that it satisfies the requirement is itself a research topic. Even when the model is given, the simulation and approximation algorithms are not applicable since the state space would be quite large.

In this chapter, a component-based approach for JSS system design and optimization is introduced. It applies the operators in PPPA for specifying and verifying the JSS system such that a correct Petri net model is generated. Based on the approach, we can induce the scheduling of a composite component from the partial scheduling of the constituent components. Hence, the schedule of the whole system can be induced from the original modules step by step.

Theoretically, we can use the component-based approach to find the optimal solution for JSS system in a formal way. For real time systems, due to its complexity, the algorithm may have a very large time complexity. In order to reduce the time complexity and to find an approximation optimal solution, we can apply the traditional optimization technologies such as GA, branch and bound method, hybrid methods, rule-based methods, etc. together with the approach in this book. In order to solve more general JSS problems, we should consider more integration operators, and should also consider the conclusions about the scheduling-preservation and makespan calculation.

Chapter 11

Application of PPPA to Security
Policy Design

One of the most fundamental elements of computer security is security
policy. Current security policy design trends are the composition of components in security systems and interactions between them. Consequently,
in a modular specification and verification of a policy, the composition of
the modules must consistently assure security policies. It is crucial to find
a rigorous and systematic way to predict and assure such critical proper-
ties.This chapter addresses the problem in a formal way. Extended Petri
net process (EPNP) is used to specify and verify security policies in a modular way. Fundamental policy properties, i.e., completeness, termination,
consistency and confluence, are defined in Petri net terminology and some
theoretical results are obtained. This chapter specifies several policy composition operators and presents property preserving results for the policy
correctness verification, according to XACML combiners and property preserving Petri net process algebra (PPPA). As an illustration, the approach
is applied to a large scale complex policy design.

11.1 Background and Motivation

In our information age, the world's economy, communications, entertain-
ment etc. depend on computers which are often connected by networks. In-
estimably valuable information is transmitted through networks, and hence
security policies are required to protect the data (or other resources) from
being processed by any undesirable users (subjects). Consequently, an ur-
gent problem is how to design highly dependable security policies that en-
sure secure access to distributed resources.

As far as policy design is concerned, the following two requirements for
a policy are the main source of difficulty and complexity:

- *Handling resources sharing and cooperation among heterogeneous systems:* Usually, we design a local policy for handling a local request in a local system which usually includes some private resources. In general, however, users' requests for services may be of very diversified nature. For a single local system, it is hardly feasible to contain enough resource information for supporting all kinds of services. One system can not only coordinate various resources but also cooperate with other systems.Consequently, we face the problem of designing a global policy for solving the resources sharing and the cooperation among Heterogeneous systems. This induces the difficult resource-sharing problem into the design of access control policies.
- *Component-based architecture:* At different times, and under different environments, the policies and resources of a system may be modeled, built or owned by different parties. Hence, it is better for a policy to adopt a component-based architecture, in order not to have severe interference in their individual developments. In such an architecture, the policy is considered as loosely-coupled sub-policies. To form a complex global policy, these sub-policies are integrated via various composition operators.

In the recent years, research and development in policies mainly focus on the two features mentioned above.

In a large system, many classes of subjects with different needs process a variety of resources. Different subjects usually have different (even competing) requirements on the use of resources, and their security goals (confidentiality, availability, integrity) may be distinct. Hence, various access requirements have to be consistently authorized and maintained in a single policy. For example, in Chinese wall policy (see Section 11.6.1 for details), when a user requires to access a data DB in a bank COI (conflict of interest) class, based on the policy, it is permitted. In another COI, reading and writing another single data DC are also permitted. But if the same user requires to read DB and write DC, it is prohibited. In this setting, the theory of security policy composition becomes crucially significant. Hence, the idea of the component-based design in in this book can be applied. That is, each simple and original module is first specified independently, then based on the control flow of the system or policy requirement, the modules are composed together into a whole system model. The objective is to deduce the properties of the whole system, based on the properties of the sub-modules, according to some theoretical results about property preservation (i.e., the overall policy preserves the properties of the constituent sub-policies).

During composing policies, conflict resolution arises. The idea of dis-ambiguating among possibly conflicting decisions appear in several works, such as in [JSS (2001); KBB (2003)] and the core of the industrial standard access-control language XACML [MOS (2005)]. However, when the deci-sion conflicts cannot be solved based on the sub-policies decisions but on the activities of the subjects or the systems, XACML combiners are in fact not sufficient. For example, in the Chinese wall policy, although decisions of both sub-policies are Permit, the decision of the composed policy may be Deny.

To address the problems, in this chapter a systematic and formal methodology will be introduced to model security policy and to verify whether required policy properties are assured by the composition of the sub-policies of the systems. The results presented in this chapter are, to the best of our knowledge, among the first efforts on systematic composition and analysis of security policies with Petri nets in the literature.

Motivated by the advantage of Petri nets, much work about applying Petri net for the policy design has appeared in the literature (Section 11.2 gives a brief review). For a specific security policy, the common character-istic of these technologies is that Petri nets are used for the specification, and the reachability-tree related techniques and CPN Tool are applied for the verification. However it is well-known that, when the system is large and complex, these techniques face a state explosion problem. In order to overcome this shortcoming and to strengthen the advantages of the Petri net formalism both in security policy and on software engineering areas, this book presents some pioneer work about applying Petri nets for the security policy design specification and verification in a modular way. It provides the following contributions:

1) It applies the newly defined model for the modular specification, i.e., Extended Petri net Process (EPNP) (Definition 10.1), which is special Petri net with a single entry place and a single exit place working as the module interfaces. In EPNP, colors are assigned to tokens and weights in order to distinguish different types of data and reduce the state space, time constraints are added to the transitions for specifying the duration of executing an operation.

2) Formal definitions for policy-related properties, namely, completeness, termination, consistency and confluence are given in Petri net termi-

nology. These properties, presented in [DOU (2007)] in the rewriting framework, are here adapted to the Petri net approach. Some theoretical results concerning these properties are stated.

3) Some policy composition operators based on Petri nets are specified. For enhancing PPPA, eight simple composition operators based on EPNP are specified which give some hints on applying PPPA to the security policy design. The composition operators in PPPA are mainly useful for those policies that solve conflicts through system actions.

4) When policies are combined, four composition combiners are specified according to XACML to solve decision conflicts according to predefined rules.

5) For each composition operator, the preservation of policy properties is studied. This book is the first one to specify security policy with EPNP and verify the policy properties based on the proof of property preservation, which is one of the most popular verification techniques in software engineering.

It is highly flexible and scalable for our methodology of modeling and verification of security policy. This is because any module (no matter whether or not it is obtained by composing other sub-modules) can be safely replaced with an alternative design without reanalysis of the overall system architecture. Because each module is designed as a correct EPNP model with a specific architecture, the replacement will not change the interfaces and the constraints. It is especially useful for this feature when we design different security policies with the same EPNP architectures (see the specification of the Bank COI in Chinese wall policy in Section 11.6.1 for example).

It is scalable because the overall composition can be analyzed without the interference of internal details of the module design. Verification is separately done by checking whether each sub-module satisfies the constraints of property preservation. Hence, the complexity can be significantly reduced. Furthermore, this methodology is general and can be applied to a large range of security policies design.

The remaining part of this chapter is organized as follows: after giving some related works in Section 11.2, Section 11.3 provides the formal definition of policy properties with Petri net terminology, and some related results concerning the policy properties are given. Section 11.4 is about the modular security policy design technology with a property preserving approach. Eight composition operators based on EPNP are defined and studied. Then in Section 11.5, XACML combiners are specified with EPNP. For each combiner, property preservation results are presented. In Section 11.6, a large scale complex policy design is illustrated based on the methodology mentioned in the previous sections. Some conclusive remarks are given in Section 11.7.

11.2 Related Work

While there exists a huge literature in systems engineering and software engineering on the key idea that complex systems are built by assembling simple components, considering security policies as components and studying their composition are relatively recent research trends.

Policy composition is addressed in [BVS (2002)] through an algebra which is able to define union, intersection, policy templates, among other operations. The work presented in [WJ (2003)] extended this algebra with negative authorizations and non-determinism, an operator for sequential composition is also included. Another alternative for composing access control policies is implemented by the Polymer system [BLW (2005)]. A different approach to composition is taken in [LBO (2006)] for composing policy specifications for web-services security. In [BDH (2007)], the author proposes a set of high-level composition operators coherent with a four-valued logic for policies.

In recent years, term rewriting was applied for the specification and verification of security policy design [DOU (2007); SO (2007)]. The composition of security policies is addressed in [DOU (2007); SO (2008)]. In a rewrite-based specification, policies are expressed by rules similar to natural language: if some conditions are satisfied, then, a given request is evaluated into a decision. For instance it may be granted or denied. Based on the expressive rules, the condition is allowed to granularly specify. Under the condition, decision takes place and these conditions may involve attributes related to subjects or resources. Moreover, control on the rule can be expressed by using the strategic rewriting, and the priorities or choices between possible decisions can also be handled. For instance the specification of XACML combiners is given in [DOU (2007); SO (2008)] in the rewriting context.

It is not a new story that Petri net is used for the specification and verification of the security policy design. In [SMJ (2005)], The consistency of RBAC policies is verified though a colored Petri-net-based (CPN-based) framework. The reachability analysis technique is applied for RBAC policy verification. In [MOR (2000)], CPN is used to specify a real industrial example which is an access control system developed by the Danish security company Dalcotech A/S. Based on the CPN model, the Design/CPN tool is applied for the implementation of automatic code generation. The manuscript [KNO (2000)] defines task-based access control as a dynamic workflow and then specifies the workflow with Petri nets. Literature [ZHL (2006)] and [ZHZ (2006)] model Chinese wall policy and Strict Integrity Policy, respectively, with CPN and apply coverability graph for the verification. [JUS (2003)] uses CPN for the specification of mandatory access control policies and occurrence graph is applied for verification. [DWT (2003)] applies Predicate/Transition net for the modeling and analysis of software security system architectures.

11.3   Petri Net-Based Properties for Security Policy

In this chapter, security policies are built in a modular way from basic modules, specified with extended Petri net processes. Accordingly, the policy properties are now defined on EPNP. The most definitions and theorems in this chapter are extracted from one of our previous work [HH (2010)]. The proof details can be found from the literature.

11.3.1   *Completeness*

A security policy is decision complete (or simply complete) if it computes at least one decision for every incoming request. This property is called totality in [TK (2006)] and [BF (2006)].

Definition 11.1. (completeness)
Suppose a security policy is specified with an EPNP $B = (EPN, p_e, p_x)$. The policy B is complete if for any initial marking M_e, there exists a marking $M_x = M + p_x$ which is reachable from M_e.

Based on the definition, an initial marking represents a request, and the exit marking $M + p_x$ is reachable, implying that the policy will return a decision.

11.3.2 Termination

A security policy terminates if the evaluation of every incoming request terminates.

Definition 11.2. (termination)
Suppose a security policy is specified with an EPNP $B = (EPN, p_e, p_x)$. The policy B is terminating (or strongly terminating) iff B has no infinite firing sequences for any initial marking. B is weakly terminating iff for any initial marking, B has at least one finite firing sequence. If B terminates with an exit marking $M_x = M_0 + p_x$, B is called properly terminating.

If B is weakly terminating, it may have infinite firing sequence(s) but must terminate in some cases (see Figure. 11.1). Strong termination requires that the policy always terminates with a finite number of firing steps; while properly terminating requires the policy to terminate (strongly or weakly) and to reach a special exit state. More discussions on proper termination can refer to Chapter 3.

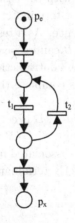

Fig. 11.1   An example of a weakly terminating policy.

The following results allow us to connect these policy properties and the usual notions of boundedness, reversibility, liveness and deadlock-freeness in Petri nets [MUR (1989)].

Theorem 11.1. *Suppose a security policy is specified with an EPNP $B = (EPN, p_e, p_x)$. The following propositions hold:*

1) *If B is almost live, B is complete;*
2) *If B terminates properly, B is complete;*
3) *If B can terminate from any reachable marking and is deadlock-free, then B is complete;*
4) *If B is complete and almost bounded, then it is (strongly or weakly) terminating.*
5) *If B is almost live and bounded, then it is terminating.*
6) *If B is almost live and reversible, then it is properly terminating.*

Proof.

1) If B is almost live, then for any initial marking M_e specifying a request, there exists a reachable marking such that the transition $t \in {}^\bullet p_x$ is firable and $M_x = M + p_x$ is reached after firing the transition t.
2) If B terminates properly, then $M_x = M_0 + p_x$ is a reachable marking and hence B is complete.
3) By contradiction. If B is not complete, then there exists a reachable marking such that either B cannot terminate or it reaches a dead marking. This is in contradiction with the assumption.
4) By contradiction. If B cannot terminate, then either there is a cycle or B has an infinite firing sequence. As a result, either the tokens keep on transferring in the cycle and B cannot be complete or the associated net of B is unbounded. This is in contradiction with the assumption.
5) If B is almost live, then B is complete (based on item 1), and by item 4, B is terminating.
6) Since B is almost live, by item 1, B is complete. That is, $M_x = M + p_x$ is reachable from $M_e = M_0 + p_e$. Since B is almost reversible, $M_0 + p_x$ is reachable from $M + p_x$ in the associated net B_a. Hence $M_x = M + p_x = M_0 + p_x$ in B and B is properly terminating. $\qquad\square$

11.3.3   *Consistency*

A security policy is consistent if it computes at most one access decision for any given input request.

Definition 11.3. (consistency)
Suppose a security policy is specified with an EPNP $B = (EPN, p_e, p_x)$. Then the policy B is consistent iff for any initial marking $M_e = M_0 + p_e$, all reachable markings M_i satisfy that $|M_i(p_x)| \leq 1$ and for any exit markings M_j and M_k, $M_j(p_x) = M_k(p_x)$.

Consistency implies that for any request, the policy returns at most one decision. According to the above definition, all reachable markings can have at most one token ($|M_i(p_x)| \leq 1$) with a unique identical color in place p_x (since $M_j(p_x) = M_k(p_x)$). Note that when the EPNP does not terminate, the decision place p_x will not be marked, so the consistency property is trivially satisfied.

The consistency property can be related to state equations in Petri net theory. In a general Petri net (N, M_0), any reachable marking M satisfies the state equation $M = M_0 + V\mu$, where V is the incidence matrix and μ is the count vector of a firing sequence σ and $M_0[N, \sigma\rangle M$ (see Definition 2.6).

Theorem 11.2. *Suppose a security policy is specified with an EPNP B. Then, B is consistent if*

- *at most one of the following state equations is satisfied:*
  $M_i = M_e + V\mu_i$, *where* $|M_i(p_x)| = 1$

- *and no one of the following state equations is satisfied:*
  $M_i = M_e + V\mu_i$, *where* $|M_i(p_x)| > 1$.

Proof. If the first state equations cannot be satisfied, then the policy cannot make a decision. If at most one of first state equations is satisfied, this implies that there may exist a decision. If the second equation cannot be satisfied, this implies that the case of more than two different decisions is impossible. $\square$

Let us now relate consistency and the notion of confluence defined for Petri nets.

11.3.4 *Confluence*

Definition 11.4. (confluence)
Suppose a security policy is specified with an EPNP $B = (EPN, p_e, p_x)$. The policy B is confluent iff for any initial marking $M_e = M_0 + p_e$ and any two reachable markings $M_i, M_j \in R(B, M_e)$, there exists a reachable marking M_d in B such that $M_d \in R(B, M_i) \cap R(B, M_j)$.

A home space of B, denoted HS, is a set of markings, such that for any $M_i, M_j \in R(B, M_e)$, there exists at least one marking M_d in HS reachable from both M_i and M_j. If a HS contains only one element M_d, then M_d is called a home marking of B. In other words, a home marking is reachable from any marking $M \in R(B, M_e)$.

The confluence property has been studied in the literature [TJM (1988); BRL (2001); LT (2006); MOO (2002)]. It is proved to be a decidable property in Petri net theory. For ordinary Petri nets (with weight 1 on each arc and no self-loop), the confluence can be reduced to confluence of a 2-shallow term rewriting systems [LT (2006)]. The following Theorem is extracted from [LT (2006); MOO (2002)] and the detailed proof can be found in [MOO (2002)].

Theorem 11.3. *Suppose a security policy is specified with an EPNP B.*

1) *If a Petri net has a home marking then it is confluent.*
2) *A safe Petri net (i.e., a PN which satisfies that the number of tokens in any place cannot exceed one for any reachable marking) has a home marking iff it is confluent.*
3) *Any confluent and strongly terminating Petri net has a unique home marking.*

Theorem 11.4. *Suppose a security policy is specified with an EPNP B. If B is consistent and properly terminating, then B is confluent and has a unique home marking.*

Proof.   Since B is properly terminating and consistent, for any request marking $M_e = M_0 + p_e$ there exists a unique exit marking $M_x = M_0 + p_x$ reachable from M_e. From any two reachable markings M_i and M_j, one can reach M_x. Hence B is confluent and M_x is the unique home marking.   □

11.4   Modular Policy Composition Based on EPNP

This section focuses on the composition of the security policies in a modular way. In general, combining security policies may result in inconsistent or non-terminating policies. We explore which syntactic conditions and which operators can guarantee the preservation of these suitable properties for the composition of two policies.

In order to give some hints about how to apply PPPA for the specification of security policy design, we restrict our attention to only four logic related composition operators, namely Enable, Choice, Interleave and Disable, and four application related compositions, namely place merging, transition merging, place refinement and transition refinement, and to the security-policy related properties, i.e., completeness, termination, consistency and confluence. The application of other operators follows similar ideas.

11.4.1 *Logic-based operators for composition*

Theorem 11.5. *Let B be the policy obtained from two sub-policies B_1 and B_2 by applying the composition operator Enable (Definition 4.1, Figure 4.1). Then,*

1) *B is complete iff B_1 and B_2 are complete.*
2) *B is strongly (resp., properly) terminating iff B_1 and B_2 are strongly (resp., properly) terminating; B is weakly terminating iff B_1 and B_2 are weakly terminating, or B_1 is complete and B_2 is weakly terminating.*
3) *If both B_1 and B_2 are consistent, then B is consistent.*
4) *If both B_1 and B_2 are confluent, then B is confluent.*

Proof. For each firing sequence σ in B, it is either a firing sequence in B_1 or a union of a sequence σ_1 in B_1 and a sequence σ_2 in B_2, where $M_{1e}[B_1, \sigma_1\rangle M_{1x}$ in B_1. The remaining part of the proof is trivial. $\quad\square$

Theorem 11.6. *Let the policy B be obtained from two sub-policies B_1 and B_2 by applying the composition operator Choice (Definition 4.2, Figure 4.3). Then,*

1) *B is complete iff B_1 and B_2 are complete;*
2) *B is strongly (resp., weakly, properly) terminating iff B_1 and B_2 are strongly (resp., weakly, properly) terminating;*
3) *B is consistent if B_1 and B_2 are consistent and output the same colored token in the exit places;*
4) *B is not always confluent even if both B_1 and B_2 are confluent. B is confluent if B_1 and B_2 are proper terminating and output the same token in their exit place.*

Proof. After applying Choice operator, the control flow is within one of the sub-policies, so the first two properties are trivial. For Item 3, although both B_1 and B_2 are consistent, they may output different decisions. Hence B is not always consistent unless both B_1 and B_2 always output the same colored token in their exit place. As for Item 4, suppose

$M_i \in R(B_i, M_{ie}), i = 1, 2$. Then $M_i \in R(B, M_e)$. There is no reachable marking $M \in R(B, M_1) \cap R(B, M_2)$ except $M = M_x = M_0 + p_x$. Hence, B is not confluent unless both B_1 and B_2 terminate properly and output the same colored token in their exit place.   □

Theorem 11.7. *Let the policy B be obtained from two sub-policies B_1 and B_2 by applying the composition operator Interleave (Definition 4.3, Figure 4.5). Then,*

1) *B is complete iff B_1 and B_2 are complete;*
2) *B is strongly (resp., properly) terminating iff B_1 and B_2 are strongly (resp., properly) terminating; B is weakly terminating iff B_1 and B_2 are weakly terminating;*
3) *B is consistent iff B_1 and B_2 are consistent;*
4) *B is confluent iff B_1 and B_2 are confluent.*

Proof.

1) B is complete iff transition ε_2 is firable, i.e., both B_1 and B_2 are complete.

2) Since B_1 and B_2 are executed independently, each firing sequence of B is a union of sequences of B_1 and B_2. Item 2 follows easily.

3) If both B_1 and B_2 are consistent, for a request, the outputs of each sub-policy are always the same, by Definitions 4.3 and 11.3, the output of transition ε_2 is unique and B is consistent. On the other hand, if B is consistent, the token color (a 2-dimension vector) in p_x is unique, correspondingly, each entry of the 2-dimension vector is unique, i.e., the token color in places p_{1x} and p_{2x} should be unique, implying both B_1 and B_2 are consistent.

4) For any two reachable markings $M_i = P_{im} + Q_{im}$ in B, where P_{im}, Q_{im} are markings in B_1 and B_2 respectively and $i = 1, 2$, since both B_1 and B_2 are confluent, there exist $M'_1 \in R(B_1, P_{1m}) \cap R(B_1, P_{2m})$ and $M'_2 \in R(B_2, Q_{1m}) \cap R(B_2, Q_{2m})$. Then $M = M'_1 + M'_2 \in R(B, M_1) \cap R(B, M_2)$ and B is confluent.

□

In order to apply Disable operator to policy composition, we give a little modification for the definition of Disable as follows:

Definition 11.5. (disable for policy composition (Figure 11.2))
For two processes $B_i = (P_i, T_i, F_i, W_i, M_{i0}, C_i, P_{ie}, P_{ix})$ $(i = 1, 2)$, their composition by Disable, in notation $B_1 [\rangle B_2$, is defined as the process

$B = (P, T, F, W, M_0, C, p_e, p_x)$, where $P = P_1 \cup P_2 \cup \{p_{1e}, p_{2e}\}$, p_e, p_x are the newly added entry place and exit place, respectively; $T = T_1 \cup T_2 \cup \{t_0, t_c\}$; $F = F_1 \cup F_2 \cup \{(p_e, t_0), (t_0, p_{1e}), (t_0, p_{2e}), (p_{1x}, t_c), (p_{2x}, t_c), (t_c, p_x)\} \cup \{(P_d, T_d), (T_d, p_{2x})\}$, where $P_d \subseteq P_2, T_d \subseteq T_1$; $W = W_1 \cup W_2 \cup \{W(p_e, t_0), W(t_0, p_{1e}), W(t_0, p_{2e}), W(p_{1x}, t_c), W(p_{2x}, t_c), W(t_c, p_x)\} \cup \{W(P_d, T_d), W(T_d, p_{2x})\}$, where $W(t_c, p_x)$ is a 2-dimensional vector $(W(p_{1x}, t_c), W(p_{2x}, t_c))$; $M_0 = M_{10} \cup M_{20}$; $C = C_1 \cup C_2$.

The Disable composition here is similar to Interleave. The difference is that there exist some places P_d in B_2 which are connected to some transitions T_d in B_1 such that once T_d are fired, B_2 is dead and cannot output decisions normally. According to the policy requirement, we may add an additional arc (T_d, p_{2x}) for specifying the decisions of sub-policy B_2 if it is disabled.

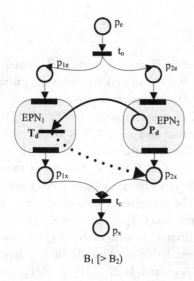

$B_1 [> B_2)$

Fig. 11.2   Policy composition via Disable operator.

Theorem 11.8. *Let the policy B be induced from two sub-policies B_1 and B_2 by applying the composition operator Disable. Then,*

1) Suppose the transitions in T_d are never fired in B. Then the property preservation results are the same as those for the Interleave operator in Theorem 11.7.

2) *Suppose some transitions in T_d are fired and all the transitions in B_2 are disabled in B. Then,*

- *B is complete if B_1 is complete;*
- *B is strongly terminating if B_1 is strongly terminating; B is weakly terminating if B_1 is weakly terminating;*
- *B is consistent if B_1 is consistent;*
- *B is confluent if B_1 is confluent.*

3) *If B_1 and B_2 are both complete (resp., terminating), B is complete (resp., terminating), but in general confluence and consistency are not preserved.*

4) *Suppose B_2 satisfies the following conditions: ${}^{\bullet}(P_d^{\bullet}) = \{p_d\}$, $P_d^{\bullet} = {}^{\bullet}p_{2x}$, and $W(P_d, {}^{\bullet}p_{2x}) = W(P_d, T_d), W({}^{\bullet}p_{2x}, p_{2x}) = W(T_d, p_{2x})$. Then B is consistent (resp., confluent) iff B_1 and B_2 are consistent (resp., confluent).*

Proof.

1) If T_d are never fired in B, the operator is the same as Interleave.
2) Once T_d are fired, B_2 is dead and the remaining flow is occurring in B_1. Hence, B preserves the properties of B_1.
3) Since both B_1 and B_2 are complete, after applying the Disable operator, transition t_c is firable and hence B is complete. Generally, T_d may be fired or not. But in both cases, the length of firing sequences in B cannot exceed the sum length of two firing sequences selected from B_1 and B_2, respectively. Hence, B preserves the termination property. On the contrary, the consistency and confluence properties cannot be preserved in general: since T_d may be fired, B may produce a decision which is different from a decision of B_1. In that case B is obviously not consistent. For a reachable marking M in B, suppose T_d is firable and $M[B, T_d\rangle(P_{1m} + Q_{1m})$ and $M[B, *\rangle(P_{2m} + Q_{2m})$, where Q_{1m} is a dead marking in B_2, while Q_{2m} is a reachable marking in B_2. Then, there does not exist a reachable marking that belongs to $R(P_{1m} + Q_{1m}) \cap R(P_{2m} + Q_{2m})$. Hence, B is not confluent in general.
4) When B_2 satisfies the given conditions, the reachable marking (i.e. the final decision) in B_2, from firing transitions in $P_d^{\bullet}$, is the same as that from firing T_d. Hence, whether or not T_d is fired, the reachable marking states in B_2 are the same. The remaining part of the proof is similar to the proof for Theorem 11.7 and omitted.

□

11.4.2 Policy composition via resources sharing

For cooperation, two or more security components will interoperate for accessing resources from the cooperated domains. In this case, we should design a global policy by composing the local policies for sharing resources.

Example 11.1. (printer accessing policy (PAP))

Given two domains D_1 and D_2, let us assume that each domain has some resources (printers and xeroxing machines) for use. The local policy for accessing resources is that once the requested resources are available, the local user can access them; after being used, the resources should be released. For a global domain D composed from D_1 and D_2, the resources access policy is not changed, that is, each user can request accessing any resource once it is available and release the resource after using. The difference of global policy and local policy is that the resource set is changed.

For EPNP specification, the policy composition is formally defined as follows, the operator is called M-Interleave in Chapter 10:

Definition 11.6. (M-Interleave (Figure 11.3))

For two processes $B_i = (P_i \cup R, T_i, F_i, W_i, M_{i0}, C_i, \tau_i, p_{ie}, p_{ix})$ ($i = 1, 2$), their composition by sharing resources R, where, $M_{10}(R) = M_{20}(R)$, denoted $B_1[R]B_2$, is defined as the EPNP $B = (P \cup R, T, F, W, M_0, C, \tau, p_e, p_x)$, where $P = P_1 \cup P_2 \cup \{p_{1e}, p_{2e}, p_{1x}, p_{2x}\}$, p_e and p_x are newly added interface places; $T = T_1 \cup T_2 \cup \{t_e, t_x\}$; $F = F_1 \cup F_2 \cup \{(p_e, t_e), (t_e, p_{1e}), (t_e, p_{2e}), (p_{1x}, t_x), (p_{2x}, t_x), (t_x, p_x)\}$; $W = W_1 \cup W_2 \cup \{W(p_e, t_e), W(t_e, p_{1e}), W(t_e, p_{2e})\} \cup \{W(p_{1x}, t_x), W(p_{2x}, t_x), W(t_x, p_x)\}$; where $W(t_x, p_x)$ is a 2-dimension vector $(W(p_{1x}, t_x), W(p_{2x}, t_x))$ $M_0 = M_{10} \cup M_{20}$; $C = C_1 \cup C_2$; $\tau = \tau_1 \cup \tau_2$ ($\tau(t_e) = \tau(t_x) = 0$).

Based on the definition, M-Interleave operator is applicable for two EP-NPs with the same set of resource places for sharing, i.e., they have the common resource places R with the same number of initially marked tokens. Hence, for policy composition, the EPNP specification for each local policy may need a modification before applying fusion operator. That is, adding some new resource places (e.g., place r_2 in Figure 11.4(a) is newly added) and modify the number of tokens for some resource places (e.g., the number of tokens in r_1 and r_2 in Figure 11.4(b) are changed).

In order to illustrate the above definition, let us consider Example 11.1. In each local domain, a user may request printing, or copying, or printing

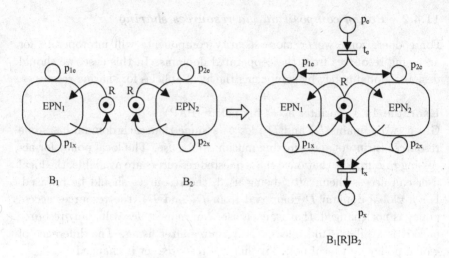

Fig. 11.3   Policy composition via resource sharing.

and copying, or doing nothing. Based on local policy, once the resource is available, the decision for the request is permitted. Let us suppose that the domain D_1 has one printer (r_1) and no xeroxing machine, the domain D_2 has one printer (r_1) and one xeroxing machine (r_2). The EPNP based policy specification is shown in Figure 11.4(a), where a user from D_1 can only successfully request printing (firing t_{51}) or doing nothing (firing t_{91}). Neither copying nor "printing and copying" is possible. In domain D_2, a user can successfully request printing (firing t_{52}), or copying (firing t_{72}), or printing and copying (firing t_{12}), or doing nothing (firing t_{92}) once the requested resource is available.

In this cooperation context, the global domain now has two printers and one xeroxing machine for sharing. Hence, Figure 11.4(a) will be changed to Figure 11.4(b) by modifying their resource places r_1 and r_2. At last, place fusion operator is applied for policy composition as shown in Figure 11.4(c). The specification of the places and transitions are explained in Table 11.1.

Note that resource sharing may result in unsafe interoperation. For instance, in the above example, copying is not permitted in D_1 (transition t_{71} does not appear in Figure 11.4(a)) but is permitted in the global domain (transition t_{71} may be enabled in Figure 11.4(c)). However, whether or not to cooperate is the manager's business and it is out the scope of this paper. Instead, this paper focuses on how to compose the policies and assumes that adding new resources to a local policy is safe. For example, in Example 11.1,

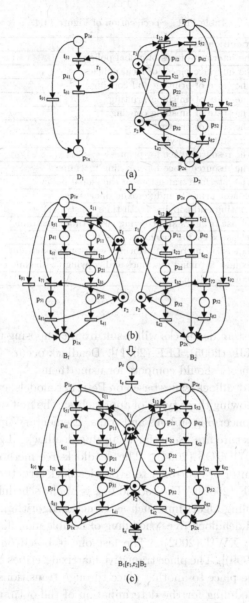

Fig. 11.4 EPNP-based specification for Example 11.1.

the manager decides whether or not to share resources (i.e., to transform
Figure 11.4(a) into Figure 11.4(b)), while our business is to consider how to
specify sharing (i.e., to transform from Figure 11.4(b) into Figure 11.4(c)).

Table 11.1   Specification of Figure 11.4.

node	specification
p_e, p_x	the interface places of global policy
p_{ie}, p_{ix}	the interface places of local policies
p_{1i}	The state of printing and copying
p_{2i}	The state of finishing printing
p_{3i}	The state of finishing copying
p_{4i}	The state of printing
p_{5i}	The state of copying
r_1	The resource place for printers
r_2	The resource place for xeroxing machine
t_e, t_x	The interface transitions of the global policy
t_{1i}	request printers and xeroxing machines
t_{2i}	printing and copying (with time constraint)
t_{3i}, t_{6i}	release printers and copying (with time constraint)
t_{4i}, t_{8i}	release xeroxing machines
t_{5i}	request printers and printing (with time constraint)
t_{7i}	request xeroxing machines and copying (with time constraint)
t_{9i}	request nothing

Generally, system deadlocks will result from composing two systems by M-Interleave [XJF (2005); LEE (2002)]. Deadlock occurs when resources are limited and users should compete for using them.

For handling deadlock issues based on Petri net models, current research considers the following three types of approaches: The first one is relying on the techniques concerning siphons and traps. Elementary siphon invariants in Petri net structure are useful for analyzing deadlock [LZ (2004); HLW (2006); LW (2007); ROS (2004)]. The deadlock-free method is addressed by adding a monitor or controller to avoid deadlock structure [LW (2007); ROS (2004); XJF (2005)]. The second one is about scheduling algorithm. Heuristic scheduling algorithm, such as Genetic Algorithm, is used to get an optimum and deadlock-free scheduling of flexible manufacturing system in [XW (2002); XWU (2002)]. The last one is based on the transitive matrix [LIM (1999)]. The place transitive matrix describes the transferring relation from one place to another place through transitions. The analysis of the cyclic scheduling for the determination of the optimal cycle time is studied, and transitive matrix has efficiently been used to slice off some subnets from the original net in [LEE (2002)]. Deadlock-free conditions are given in [SL (2002); KL (2006)] based on transitive matrix, and an algorithm for finding deadlock is given in [LEE (2004)] using the theory of transitive matrix.

The above mentioned approaches can be applied for designing a deadlock-free policy, and the deadlock-handling problem is outside the scope of this paper, that mainly considers deadlock-free policy composition. Accordingly, we specify a deadlock free local policy as follows: let us suppose that a user requests a resource r_1 for executing an operation t_1. Before releasing r_1, the same user will request another resource r_2 for executing another operations t_2 (Figure 11.5(a)). In this case, for the Petri net specification, both resources will be occupied by the user at the very beginning time of requesting r_1 (Figure 11.5(b)). For instance, in the above example of policy composition, for the request of "printing and copying", in order to avoid system deadlock, transition $t_{1i}, i = 1, 2$ requests both printers and xeroxing machines even if printing and copying are not executed at the same time. We call such a way of requesting resources "complete occupying".

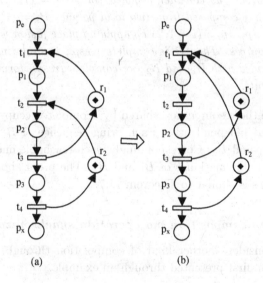

(a)   (b)

Fig. 11.5   Modification of sharing resources.

With "complete occupying" in each local policy, each resource will be requested orderly without circular waiting and the composed system is deadlock free. However, the time duration of handling a request may be delayed. For instance, in the above printer accessing policy, let us assume that user A from domain D_1 requests printing and copying and user B from domain D_2 requests copying. Let us assume that there is only one printer and one xeroxing machine and in the EPNP specification, the time for printing is 10, for copying is 5, that is, $\tau(t_{21}) = \tau(t_{51}) = \tau(t_{52}) = 10, \tau(t_{31}) = \tau(t_{71}) = \tau(t_{72}) = 5$, and for firing other transitions is 0. Based on the global policy, the minimum time for handling these two requests is 15, that is, while user A is printing, user B can copy simultaneously. However, in the "complete occupying" case, since both the printer and the xeroxing machine are occupied by user A, user B has to wait until user A releases the xeroxing machine. In this case, the total time for handling the requests is 20.

Although "complete occupying" may delay the decision for a request, it prevents a policy from the difficult deadlock-handling problems. Searching for an optimal scheduling algorithm for fastly handling a request is out of scope of this chapter.

For policy composition, we get the following conclusions.

Theorem 11.9. *Let us consider a global policy* $B = (P \cup R, T, F, W, M_0, C, \tau, p_e, p_x)$ *composed from two local policies* $B_i = (P_i \cup R, T_i, F_i, W_i, M_{i0}, C_i, \tau_i, p_{ie}, p_{ix})$ $(i = 1, 2)$ *by applying place fusion with "complete occupying" resources. Then* B *is complete (resp., terminating, consistent, confluent) provided both* B_1 *and* B_2 *are complete (resp., terminating, consistent, confluent).*

Proof.   Since the resources are shared by "complete occupying", B_1 and B_2 are executed independently. Each firing sequence of B is a union of sequences of B_1 and B_2. Correspondingly, each reachable marking in B is a union of reachable markings of B_1 and B_2. The proof about preserving policy properties is similar to Theorem 11.7.   □

11.4.3   *Policy composition via operation synchronization*

Let us now consider another kind of composition through an operation synchronization, first presented through an example.

Example 11.2. (writing accessing policy (WAP))
Let us consider two local policies, each one designed for writing some local documents. Each local policy is as follows: a specific user requests writing a document. Once the document is available, it can be written, then, after writing, the document is returned to its place. In the context of cooperation, due to security reasons, the global policy requires that some special documents in a set D (such as contracts) cannot be signed unless two specific users from different domains sign them together.

For the above example, the global policy should be composed from the two local policies by applying transition fusion. It is formally defined as follows and Figure 11.3 can be referred for understanding the definition if we replace the common resource place set R in the figure with common transition set S.

Definition 11.7. (transition fusion) For two processes $B_i = (P_i, T_i \cup S, F_i, W_i, M_{i0}, C_i, \tau_i, p_{ie}, p_{ix})$ $(i = 1, 2)$, their composition by operation synchronization (applying transition fusion for S), denoted $B_1[S]B_2$, is defined as the process $B = (P, T \cup S, F, W, M_0, C, \tau, p_e, p_x)$, where $P = P_1 \cup P_2 \cup \{p_{1e}, p_{2e}, p_{1x}, p_{2x}\}$, p_e and p_x are newly added interface places; $T = T_1 \cup T_2 \cup S \cup \{t_e, t_x\}$; $F = F_1 \cup F_2 \cup \{(p_e, t_e), (t_e, p_{1e}), (t_e, p_{2e}), (p_{1x}, t_x), (p_{2x}, t_x), (t_x, p_x)\}$; $W = W_1 \cup W_2 \cup \{W(p_e, t_e), W(t_e, p_{1e}), W(t_e, p_{2e})\} \cup \{W(p_{1x}, t_x), W(p_{2x}, t_x), W(t_x, p_x)\}$; where $W(t_x, p_x)$ is a 2-dimension vector $(W(p_{1x}, t_x), W(p_{2x}, t_x))$ $M_0 = M_{10} \cup M_{20}$; $C = C_1 \cup C_2$;

$$\tau(t) = \begin{cases} \tau_i(t), t \in T_i; \\ max\{\tau_i(t)\}, t \in S. \end{cases}$$

To illustrate the definition of transition fusion, let us consider the above Example 11.2. Figure 11.6 gives the specification of the two local policies, where transition t_{1i} requests a document for processing; if some normal documents are requested, t_{2i} is fired for processing documents, if some special documents that belong to D are requested, t_2 is fired; after processing document, t_{3i} is fired for returning the documents $(i = 1, 2)$. After policy composition, the global policy requires that the normal documents are processed by local users, while for processing the special documents in D, two specific users from different domains should be both on the scene (on-line or off-line) and execute the processing operation together. Hence transitions t_2 in both local policies should be fused into a single transition t_2 in the global policy.

For policy composition via transition fusion, we have the following results.

Theorem 11.10. *Let us consider a global policy* $B = (P, T \cup S, F, W, M_0, C, \tau, p_e, p_x)$ *composed from two local policies* $B_i = (P_i, T_i \cup S, F_i, W_i, M_{i0}, C_i, \tau_i, p_{ie}, p_{ix})$ $(i = 1, 2)$ *by applying transition fusion. Then* B *is complete (resp., terminating, consistent, confluent) provided both* B_1 *and* B_2 *are complete (resp., terminating, consistent, confluent).*

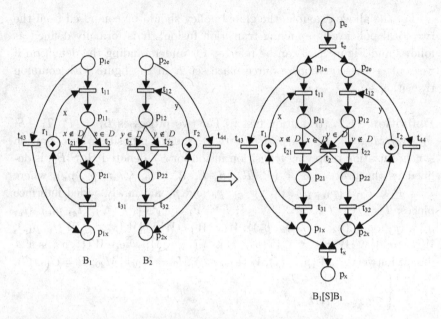

Fig. 11.6   Policy composition via transition fusion.

Proof.   By transition fusion, the control flow in each local policy remains unchanged except that the total time duration of executing the policy is delayed. Each firing sequence of B is a union of sequences of B_1 and B_2. The proof about preserving policy properties is similar to Theorem 11.7.

□

11.4.4   *Policy refinement*

For security requirement, sometimes an we can add an encapsulated policy to an existing policy. The former is named a sub-policy, whereas the latter is named a super-policy. In this section we mainly discuss how to compose a sub-policy and a super-policy by applying place refinement and transition refinement based on EPNP specification.

The formal definition of policy composition via place refinement is similar to Definition 5.5 (Transformation PR).

Based on place refinement, a sub-policy B_2 is inserted into a super-policy B_1 by refining place p_r with B_2. Those input (resp., output) transitions of p_r in B_1 become the input (resp., output) transitions of place p_{2e} (resp., p_{2x}) in B. The tokens of p_r in B_1 will move to place p_{2e} in B. Other parts will not change (Figure 11.7).

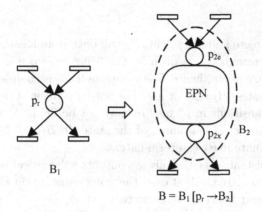

Fig. 11.7   Policy composition via place refinement.

In order to illustrate the application of place refinement for policy composition, let us consider the following example.

Example 11.3. (document accessing policy)
Suppose there are two types of documents distinguished as local document (L) and global document (D). The policy about accessing a document is as follows: for requesting a local document, once the document is available, the request is permitted. As for requesting a global document, for security reason, it should be decided by applying Chinese wall policy (CWP, the detailed specification is introduced in Section 11.6.1). Hence, as a sub-policy, Chinese wall policy will be composed into the super-policy by applying place refinement operator (the details of adding CWP to the super-policy is in Section 11.6).

The following is about the property preservation results for place refinement.

Theorem 11.11. *Suppose a global policy* $B = (P, T, F, W, M_0, C, \tau, p_e, p_x)$ *is composed from two local policies* $B_i = (P_i, T_i, F_i, W_i, M_{i0}, C_i, \tau_i, p_{ie}, p_{ix})$ *($i = 1, 2$), by refining place* $p_r \in P_1$ *into* B_2.

1) If B_1 *is complete,* B_2 *is complete and consistent, then* B *is complete;*
2) If both B_1 *and* B_2 *terminate, so does* B*;*
3) If both B_1 *and* B_2 *are consistent, so is* B*;*
4) If B_1 *is confluent and* B_2 *terminates properly, then* B *is confluent.*

Proof.

1) If B_2 is complete and consistent, B_2 will output an identical token each time it is referred. Consequently, the firing sequence in B_1 part will remain unchange and hence B is complete if B_1 is complete.

2) If B_2 terminates, B_2 may or may not output a token. Correspondingly, in B, the transitions in $p_r^\bullet$ may or may not be fired. As a result, each reachable state of B is a union of the states in B_1 and B_2. Obviously, B will terminate if B_1 can terminate.

3) If B_2 is consistent, B_2 either always outputs an identical token, or never outputs a token. In the first case, the firing sequences in B_1 part remain unchanged and hence B is consistent. If B_2 never outputs a token, then the transitions in $p_{2x}^\bullet$ are never fired in B. As a result, some firing sequences in B_1 part will never appear in B. But those firing sequences that do not include any transition of $p_{2x}^\bullet$ still exist in B. If B_1 is consistent, the control flow in B_1 part will not be changed and hence B is still consistent.

4) If B_2 terminates properly, B_2 will output a single token and resume to its static state M_{20} once it is referred. Each reachable marking of B is a union of the markings of B_1 and B_2. For any two reachable markings $M_i = P_{im} + Q_{im}$ in B, where P_{im}, Q_{im} are markings in B_1 and B_2 respectively and $i = 1, 2$, since B_1 is confluent and B_2 terminates properly, there exist $M_1' \in R(B_1, P_{1m}) \cap R(B_1, P_{2m})$ and $M_{20} \in R(B_2, Q_{1m}) \cap R(B_2, Q_{2m})$. Then $M = M_1' + M_{20} \in R(B, M_1) \cap R(B, M_2)$ and B is confluent. □

If an encapsulated policy is added by transition refinement, we will apply transition refinement operator for policy composition. The property preservation results are shown below and the proof is similar to Theorem 11.11.

Theorem 11.12. *Let a global policy $B = (P, T, F, W, M_0, C, \tau, p_e, p_x)$ be composed from two local policies $B_i = (P_i, T_i, F_i, W_i, M_{i0}, C_i, \tau_i, p_{ie}, p_{ix})$ $(i = 1, 2)$, by applying transition refinement for transition $t \in T_1$.*

1) If B_1 is complete, B_2 is complete and consistent, then B is complete;

2) If both B_1 and B_2 terminate, so does B;

3) If both B_1 and B_2 are consistent, so is B;

4) If both B_1 and B_2 are confluent, so is B.

Based on the composition operators defined in this section, a large security policy system can be specified step by step by composing its different modules. The previous propositions help verifying properties of a large system composed with these operators. However they have shown that the properties of confluence and consistency are not always preserved in such compositions. In the next section, we show how to restore in most cases these properties by a further composition with new EPNP closely related to XACML policy combiners.

11.5 EPNP-Based Specification of XACML Combiners

For a security system, in particular an access control system, the same resource may be requested by different policies and their respective decisions may be different. This question has been largely addressed and a common technique is to use XACML policy combiners to solve the conflicts resulting from applying different policies for the same resources. There are four combiners described as follows:

- *Permit-overrides:* whenever one of the policies answers to a request with a "Permit" decision, the final authorization for the composed policy is "Permit". The policy will generate a "Deny" only in the case where at least one of the sub-policies returns "Deny", and all others return "NotApplicable" or "Indeterminate". When all sub-policy return "NotApplicable", the final output is "NotApplicable". The decision is "Indeterminate" if no sub-policy returns a decision, i.e. when unexpected errors occur in every evaluation attempt.
- *Deny-overrides:* the semantics are similar to *Permit-overrides.* The only difference is to exchange "Permit" and "Deny" in the above description.
- *First-applicable:* the final authorization coincides with the result of the first sub-policy which produces the decision "Permit" or "Deny"; if no sub-policy is applicable, then the decision is "NotApplicable"; if errors occur, then it is "Indeterminate".
- *Only-one-applicable:* the resulting decision will be "Permit" or "Deny" if the single policy that applies to the request generates one of these decisions. The result will be "NotApplicable" if all policies return such decision. The result is "Indeterminate" if more than one policy set returns a decision different from "NotApplicable".

In order to simplify the specification model, we assume in this chapter that there are only two sub-policies and no error occurs in the combiners, so there are only three possible decisions, namely "Permit", "Deny", and

"NotApplicable". Actually it would not be harder but only more technical to handle multiple sub-policies and an additional decision "Indeterminate".

Let us first consider the formal specification of Permit-overrides combiner based on extended Petri net processes. The Petri net structure given in Figure 11.8 is defined as $POC = (p_{3e}, p_{3x}, T_c, F_c, W_c, M_c, C_c)$. Places p_{3e} and p_{3x} have three types of tokens colored with p, d, n respectively, representing the three different decisions. Weights are assigned to each arc. For example, in Figure 11.8, the weight of the arc (p_{3e}, t_{pp}) is $(2, 0, 0)$: this means that firing transition t_{pp} requires at least two tokens colored with p, i.e., both sub-policies return the decision "Permit". After firing transition t_{pp}, the output is $(1, 0, 0)$, meaning that place p_{3x} gets a token colored with p, i.e., the final decision is "Permit".

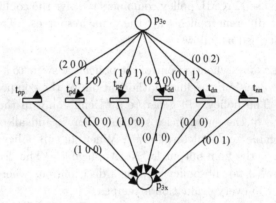

Fig. 11.8   Permit-overrides combiner POC.

For the Deny-overrides combiner (DOC), the specification is similar to the above defined Permit-overrides combiner, just exchanging "Permit" and "Deny", p and d.

The First-applicable combiner $FAC = (\{p_{4e}, p_{4x}\}, T_c, F_c, W_c, M_c, C_c)$ is defined in Figure 11.9, where the entry place p_{4e} represents all possible decisions taken by the sub-policies, while the exit place p_{4x} represents the final decision of the composed policy. T_c is the transition set: for each transition, there is a firing condition assigned to it. For example, the weight $(1, 0, 1)$ means that the decisions of the sub-policy are "Permit" and "NotApplicable" respectively. The condition $\sigma_p \leq \sigma_n$ means that the firing sequence of σ_p is shorter than the sequence σ_n, so the first decision is "Permit". Note that σ_p and σ_n are firing sequences of the two sub-policies. Hence the final

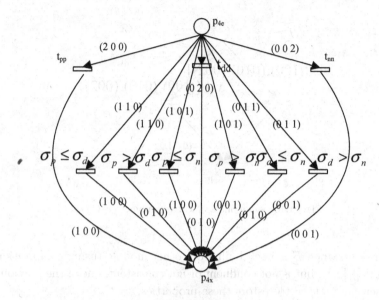

Fig. 11.9   First-applicable combiner FAC.

decision is made based on not only the sub-policies' decisions but also their firing sequences.

For the only-one-applicable combiner, the extended Petri net module $OAC = (\{p_{5e}, p_{5x}\}, T_c, F_c, W_c, M_c, C_c)$ is shown in Figure 11.10. It contains only two transitions: if there exists only one possible decision of "Permit" $(1, 0, 0)$ or "Deny" $(0, 1, 0)$ or there are two decisions of "NotApplicable" $(0, 0, 2)$, the transition t_d can be fired and the place p_{5x} gets a same colored token. Otherwise transition t_n is fired and p_{5x} outputs a decision "NotApplicable".

Theorem 11.13. *POC, DOC, FAC, OAC are EPNP which are complete, strongly terminating, consistent and confluent.*

Proof.   For each of POC, DOC, FAC and OAC, there are only two reachable markings, one is the initial marking and the other is the exit marking. All the firing sequences contain only one transition in T_c. It is obvious that they are complete, strongly terminating, consistent and confluent.   □

As a consequence of Theorems 11.5 and 11.13, if B is complete, strongly terminating, consistent and confluent, and if COM is one of POC, DOC, FAC, OAC, $B >> COM$ will enjoy the same properties.

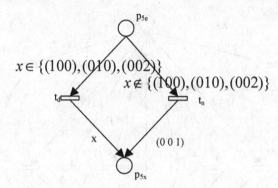

Fig. 11.10   Only-one-applicable combiner OAC.

Moreover, when B is itself built from other modules using composition operators $[], |||, [\rangle$, but is not confluent or not consistent, one of the previous combiner can be used to restore these properties.

Theorem 11.14. *Suppose B is composed from B_1 and B_2 by applying one of Choice, Interleave and Disable operators and COM is any one of the combiners POC, DOC, OAC and FAC.*

1) *$B >> COM$ is complete, strongly terminating if B_1 and B_2 enjoy these two properties.*
2) *Let $B = B_1[]B_2$. $B >> COM$ is consistent if B_1 and B_2 always output the same tokens; $B >> COM$ is confluent if both B_1 and B_2 terminate properly.*
3) *Let $B = B_1|||B_2$ and $COM \in \{POC, DOC, OAC\}$. Then $B >> COM$ is consistent and confluent if B_1 and B_2 enjoy these properties. If $B = B_1|||B_2$ and $COM = FAC$, then $B >> COM$ is confluent. $B >> COM$ is consistent provided that for all reachable markings M_1, M_2 that satisfy $M_{10}[B_1, \sigma_1 \rangle M_1$, $M_{20}[B_2, \sigma_2 \rangle M_2$, $M_1(p_{1x}) > 0$, $M_2(p_{2x}) > 0$, we always have either $\sigma_1 \le \sigma_2$ or $\sigma_2 \le \sigma_1$.*
4) *Let $B = B_1[\rangle B_2$, then $B >> COM$ is consistent provided that 1) COM = FAC; 2) B_1 is consistent and for all reachable markings M_1, M_2 that satisfy $M_{10}[B_1, \sigma_1 \rangle M_1$, $M_{20}[B_2, \sigma_2 \rangle M_2$, $M_1(p_{1x}) > 0$, $M_2(p_{2x}) > 0$, we always have either $\sigma_1 \le \sigma_2$ or $\sigma_2 \le \sigma_1$.*
   $B >> COM$ is confluent if B_1 and B_2 are confluent and satisfy ${}^\bullet(P_d^\bullet) = \{p_d\}$, $P_d^\bullet = {}^\bullet p_{2x}$, and $W(P_d, {}^\bullet p_{2x}) = W(P_d, T_d)$, $W({}^\bullet p_{2x}, p_{2x}) = W(T_d, p_{2x})$.

Proof.

1) By Theorems 11.6, 11.7 and 11.8, B preserves these two properties. By Theorems 11.5 and 11.13, $B >> COM$ also preserves these two properties.

2) If $B = B_1 [] B_2$, when B_1 and B_2 output the same tokens, then the input of COM is the same and hence the final decision is consistent. If both B_1 and B_2 terminate properly, for any two reachable markings, they will reach the exit marking of $B >> COM$ and hence $B >> COM$ is confluent.

3) If $B = B_1 ||| B_2$, since both B_1 and B_2 are consistent, their output is always the same and hence the input for the entry place of COM is always the same, and COM is consistent by Theorem 11.13. By Theorems 11.5 and 11.7, $B >> COM$ is confluent.

4) If $B = B_1 [\rangle B_2$ and the condition is satisfied, then the final decision is the same as that of B_1 and hence consistent. Based on Theorems 11.5, 11.8 and 11.13, B is confluent and so is $B >> COM$.

$\square$

11.6 A Large Scale Complex Policy Design — A Case Study

Let us consider a situation where a user requests access to documents belonging to different competitive companies. Such access is granted or denied on the basis on a Chinese Wall Policy [ZHL (2006)]; if the user has a reading access to a document, he may print it and/or copy it. If he has a writing access, he may modify it, for instance, by signing the document.

The information flow is as follows: a user requests "reading" or "writing" a document according to the Chinese Wall Policy (CWP); in the first case, once the access decision for reading is obtained, the user can continue processing the document by applying the "printer accessing policy" (PAP); in the second case, once the access decision for writing is obtained, the user can process the document by applying the "writing access policy" (WAP).

For cooperation, the two domains are combined. Their printers and xeroxing machines are shared by users from both domains. Generally, they can handle local documents based on their local security policies. But we can assume that some special documents, in a set D, can be processed only if specific users from both domains handle them together.

Based on our approach, a policy design has the following three steps: Specification of primitive modules; composition of sub-policies; and verification of policy correctness. In the following subsections, we introduce the specification and verification of both local and global policies.

11.6.1   *Specification of primitive modules*

The principle of sub-policies PAP and WAP are shown in Section 11.4.2, and Section 11.4.3 respectively. The Petri net based specification of these policy modules are shown in Figure 11.4 and Figure 11.6 respectively. The following is the description of CWP.

Chinese wall policy is about preventing the conflict of interest between clients. Figure 11.11 is an example of Chinese wall policy [ZHL (2006)]. The objects of the database contain the information related to companies; a company dataset (CD) contains objects related to a single company; a conflict of interest (COI) class contains the datasets of companies in competition. For example, the bank COI class contains three competitive companies (i.e., three CDs). The read and write policies in Chinese wall policy are defined as follows [ZHL (2006)].

Read policy: a subject $s \in S$ can read an object $o \in O$ provided that, either there is an object $o' \in O$ such that s has accessed o' and $CD(o') = CD(o)$, or for all objects $o', o' \in PR(s) \Rightarrow COI(o') \neq COI(o)$, where $PR(s)$ is the set of object s has accessed previously.

In other words, a subject s is permitted to read an object o provided that, either s reads the objects all in the same CD, or reads the objects in different COIs. In the same COI, the subject cannot read objects in different CDs.

Write policy: a subject $s \in S$ may write to an object $o \in O$ provided that s is permitted to read o, and for all the objects o', s can read $o' \Rightarrow CD(o') = CD(o)$.

In other words, a subject s is permitted to write an object o only when s can read o and other objects accessible by s are in the same CD with o.

The Bank COI class consists of three CDs, i.e., Bank of America, HSBC Bank and Citibank (Figure 11.11). The data in the same CD will be represented with the same color, and the three CDs are specified with three different colors a, b, and c, respectively (Figure 11.12). Once there is a request, e.g., request for an object in Bank of America, then a token a is put into the entry place and the EPNP model is initially marked. Since requesting two different CD is not permitted, the entry place is initially marked with only one token. For simplicity, we denote the marking and weight in the EPNP with colors instead of vectors. For example, in

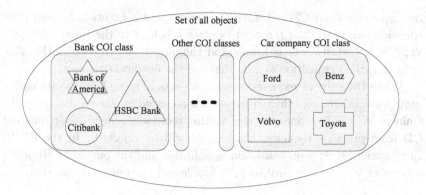

Fig. 11.11   An example for Chinese wall policy.

Figure 11.12, we use a to replace vector $(a, 0, 0)$ and use x, y and z to denote any one of the colors.

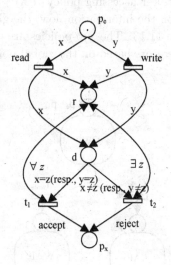

Fig. 11.12   The specification of one COI class of the Chinese wall policy.

When the subject requests to read objects in some CD, the transition *read* is considered. Place r is a record place: once the subject had a request, the corresponding data will have a record in this place. Since firing transition t_1 and t_2 will not consume the tokens in place r, it is possible to specifically record in this way. Suppose a subject had previously requested some objects. Based on the Chinese Wall policy, if the new request belongs to the same CD as before, the new request is granted. In the EPNP specification, if all data recorded in place r belong to the same color, i.e., $\forall z, x = z$, transition t_1 is firable and the output is "Accept"; otherwise, $\exists z, x \neq z$, then transition t_2 is firable and the output is "Reject".

When the subject requests to write objects in some CD, the transition *write* is considered and the request data is recorded in place r. Based on the Chinese Wall policy, once the new writing-requested object is in a different CD from previous requests, the request will be rejected. In the EPNP specification, if $\exists z \neq y$, transition t_2 is firable and the output is "Reject"; otherwise, $\forall z, y = z$, transition t_1 is firable and the output is "Accept".

11.6.2 *Composition of sub-policies*

For handling a local document (including the common documents D), both local policy and global policy contain some sub-policies, that are Chinese Wall policy (CWP), printer accessing policy (PAP) and writing accessing policy (WAP). Based on the information flow, the global EPNP specification is shown in Figure 11.13. The sub-policies are combined by applying Enable and Choice operators based on the global policy requirement.

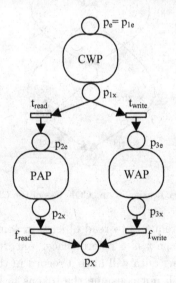

Fig. 11.13   Abstracted specification of the policy (for both local and global use).

11.6.3 *Local policy specification and verification*

The detailed EPNP specification of the local policy is shown in Figure 11.14. Let us explain it.

For a local request, the EPNP-based specification of CWP for one COI is presented in above subsection.

In case of a reading request, transition t_{read} is fired and, provided the decision of CWP was "(read, accept)", the user can request copying the document (firing transition t_{71}), or printing the document (firing transition t_{51}), or printing and copying (firing transition t_{11}). Once there exists an available xeroxing machine (i.e. when r_2 is marked), the user can copy. After copying, transition t_{81} is fired and the xeroxing machine is released; once there exists an available printer ((i.e. when r_1 is marked), the user can print. After printing, transition t_{61} is fired and the printer is released; once there exist available printers and xeroxing machines, the user can print and copy; then transitions t_{31} and t_{41} are fired and both resources are released. If the decision of CWP was "(read, reject)", transition t_{91} is fired. The final transition f_{read} just outputs "OK", meaning that the request has been handled.

In case of a writing request, transition t_{write} is fired and, provided the decision of CWP was "(write, accept)", the user can write the document. When the document is available (i.e. when place r_3 is marked), t_{13} is firable and the user can process writing the document (t_{23} corresponds to the operation of writing); after writing, the document is released and updated (t_{33} is fired). If the decision of CWP was "(write, reject)", transition t_{93} is fired. The final transition f_{write} just outputs "OK" , meaning that the request has been handled.

The local policy (Figure 11.14) consists of three sub-policies, namely CWP, PAP, WAP which are combined by applying Enable and Choice operators. Based on Theorems 11.5 and 11.6, the abstract EPNP model shown in Figure 11.14 is correct, i.e., complete, terminating, consistent and confluent. It is easy to verify that all three sub-policies are correct. At the same time, both PAP and WAP terminate properly. Although CWP does not terminate properly (place r is unbounded), the control flow is similar to that of a properly terminating process. Hence, after adding the three sub-policies to the abstract EPNP by applying place refinement, the resulting local policy (Figure 11.14) is correct (based on Theorem 11.11).

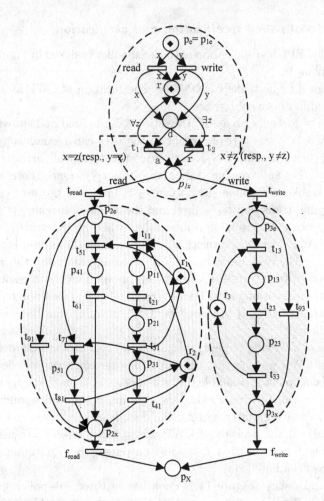

Fig. 11.14   EPNP specification of the local policy.

11.6.4   *Global policy specification and verification*

For the global policy specification (Figure 11.15), based on the policy requirement, CWP, PAP and WAP should apply the global sub-policies. The EPNP specification is similar to local sub-policies with some differences.

For CWP, in an interoperation domain, the policy has to deal with multiple COIs. Hence, it differs from the local policy by changing the conditions of firing transitions t_1 and t_2. Based on Chinese Wall policy, for a new request "reading $x \in (CD \in COI_i)$", there are two cases for accepting the request: the first is when all its recorded data z belong to the same CD as x, that is both x and z have the same color ($x = z$); another case is when x belongs to a COI different from its records, i.e., $x \in COI_i, z \in COI_j$. Otherwise, i.e., $\exists z \neq x, x, z \in COI_i$, the request is rejected. As for a request "writing y", only when all the records and x belong to the same CD, i.e., $\forall z, y = z$, the request is accepted. Otherwise, i.e., $\exists z \neq y$, the request is rejected.

For PAP, the difference is sharing resources r_1 and r_2, which can be specified by place fusion with complete occupying resources.

As for WAP, if the document does not belong to D, transition t_{23} (resp., t_{24}) can be fired just as processing local documents. If the document belongs to D, then the document should be processed by two users from different domains. Hence, the fused transition t_2 is fired and the document is processed by two users together at the same time. Here transition t_2 represents the document processing, which is assigned with time constraints. Transition t_{43} (resp., t_{44}) is used for specifying bypass: when a local document is processed, the cooperating domain can be bypassed directly by firing transition t_{43} or t_{44}.

For the global policy (Figure 11.15), it consists of five sub-policies, i.e., one CWP with multiple COI records, two PAPs, and two WAPs. The two PAPs are combined by applying M-Interleave, and the two WAPs are combined by applying transition fusion. Then, based on the super-policy, i.e., the abstract EPNP (Figure 11.13), all the global sub-policies are added by applying place refinement operator. It is easy to verify that each primitive policy is correct. By Theorems 11.9 and 11.10, the two global sub-policies PAP and WAP are both correct. At the same time, both PAP and WAP terminate properly. Although CWP does not terminate properly (place r is unbounded), the control flow is similar to that of a properly terminated process. Hence, after adding the three sub-policies to the abstract EPNP by applying place refinement, the resulted global policy (Figure 11.15) is correct, i.e., complete, terminate, consistent and confluent (based on Theorem11.11.

Note that, in this example, the sub-policies belong to different types. For instance, CWP will output a decision, while PAP and WAP will not. PAP may result in deadlock if resource competition exits, while WAP never. Theoretically, once the user's request of "reading" or "writing" is permitted, the user will be succeed in copying, printing or writing for the document. However, in order to avoid deadlock, it is possible for a user to wait printers and copy machines for a very long time if many new requests keep on coming and they keep on occupying the resources.

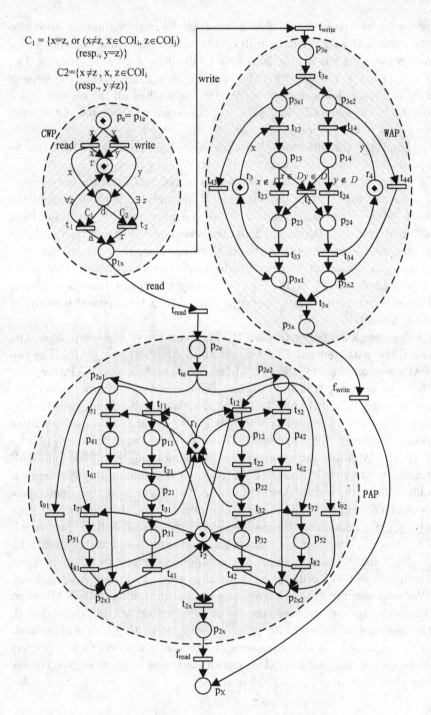

Fig. 11.15   EPNP specification of the global policy.

11.7   Conclusive Remarks

In an extended Petri net, we assign time to transitions for specifying the duration of executing operations, colors are assigned to tokens and weights for distinguishing different data, resources and the precondition of executing operations. Based on the newly defined extended Petri net process, this paper specifies security policies in a modular way. Policy composition operators are specified and property preserving results are stated for verification. This technology is suitable for general security policy design, especially for large and complex security systems. In a real-life software system, the system requirements may be changeable, system action is dynamic and the security policy is complex. Hence, much more policy composition operators should be available for the specification, and much more policy properties should be defined and verified for satisfying the policy requirements. In future work, we will focus on looking for new modeling techniques, composition operators and formal definition and verification of more policy properties.

Because of the complexity and large-scale for a real life system, the policy properties may not always be preserved properly when applying PPPA. For instance, consistency and confluence properties cannot be preserved even by adding a XACML combiner for the Disable operator. In order to design a safe policy, we may try to design a new combiner for composition in order to restore the policy properties for the resultant policy. The combiner may not be based only on the decisions of sub-policies, but possibly also on the activity of the sub-policies. Designing new combiners for restoring the destroyed properties is another interesting future research topic.

Bibliography

[AF (1998)] P. Allen and S. Frost. *Component-Based Development for Enterprise Systems: Applying Select Approach.* Cambridge University Press.

[SZY (1998)] C. Szyperski. *Component Software - Beyond Object-Oriented Programming.* Addison-Wesley.

[SZY (2000)] C. Szyperski. "Component software and the way ahead", *Foundations of Component-Based Systems.* In: Foundations of component-based systems, G.T. Leavens and M. Sitaraman (Eds), Cambridge University Press, pp. 1–20.

[BB (1987)] T. Bolognesi and E. Brinksma. "Introduction to the ISO specification language LOTOS". *Computer Networks and ISDN Systems*, Vol. 14, Issue.1, pp. 25–59.

[LFH (1992)] L. Logrippo, M. Faci and M. Haj-Hussein. "An introduction to LOTOS: learning by examples". *Computer Networks and ISDN Systems*, Vol. 23, Issue 5, 1992, pp. 325–342.

[JL (1992)] R. Janicki and P. E. Lauer. *Specification and analysis of concurrent systems: the COSY approach*, Springer-Verlag, 1992.

[BDK (2001)] E. Best, R. Devillers and M. Koutny. *Petri net algebra*. Springer-Verlag, Berlin, 2001.

[CZ (1994)] T.Y. Cheung and X. Zhu. "A formal-method approach for cyclomatic complexity of distributed software". *Trans. of K.C. Wong Education Foundation Supported Lectures*, K.C. Wong Education Foundation, 1994, pp. 99–118.

[CRM (1994)] T.Y. Cheung, S.Y. Ren and W.M. Mak. "A tool for accessing the quality of distributed software designs specified in LOTOS". *Proc. 5th Annual Working Conference of Southeast Asian Regional Computer Confederation - Software Engineering Education*, 1994, pp. 31–44.

[DES (1998a)] J. Desel. "Place/Transition Petri nets". Advances in Petri Nets, LNCS 1491, Springer-Verlag, 1998, pp. 122–173.

[DES (1998b)] J. Desel. "Linear algebraic techniques for place/transition nets". Advances in Petri Nets, LNCS 1491, Springer-Verlag, 1998, pp. 257–308.

[REI (1985)] W. Reisig. *Petri Nets – An Introduction.* Springer-Verlag, 1985.

[MUR (1989)] T. Murata. "Petri nets: Properties, analysis and applications". *Proceedings of the IEEE*, Vol. 77, No. 4, 1989, pp. 541–580.

[STV (1998)] M. Silva, E. Teruel, R. Valette and H. Pingaud. "Petri nets and production systems". *Advances in Petri Nets*, LNCS 1492, Springer-Verlag, 1998, pp. 85–124.

[DE (1995)] J. Desel and J. Esparza. *Free Choice Petri Nets*. Cambridge University Press, 1995.

[BES (1987)] E. Best. "Structure theory of Petri nets: the free choice hiatus". *Advances in Petri Nets*, LNCS 254, Springer-Verlag, 1987, pp. 168–206.

[CCS (1991)] J. Campos, G. Chiola and M. Silva. "Properties and performance bounds for closed free choice synchronized monoclass queueing networks". *IEEE Transaction on Automatic Control*, Vol. 36, No. 12, 1991, pp. 1368–1381.

[ESP (1990)] J. Esparza. "Synthesis rules for Petri nets, and how they lead to new results". *CONCUR 1990*, LNCS 458, Springer-Verlag, 1990, pp. 182–198.

[DES (1992)] J. Desel. "A proof of rank theorem for extended free choice nets". *ICATPN 1992*, LNCS 616, Springer-Verlag, 1992, pp. 134–153.

[DES (1993)] J. Desel. "Regular marked Petri nets". *Graph-Theoretic Concepts in Computer Science 1993*, LNCS 790, Springer-Verlag, 1993, pp. 264–275.

[TS (1994)] E. Teruel and M. Silva. "Well-formedness of equal conflict systems". *ICATPN 1994*, LNCS 815, Springer-Verlag, 1994, pp. 491–510.

[TS (1996)] E. Teruel and M. Silva. "Structure theory of equal conflict systems". *Theoretical Computer Science*, Vol. 153, No. 1-2, 1996, pp. 271–300.

[RTS (1995)] L. Recalde, E. Teruei and M. Silva. "One well-formedness analysis: the case of deterministic systems of sequential processes". *Structure in Concurrency Theory 1995, Workshops in Computing*, Springer, 1995, pp. 279–293.

[RTS (1996)] L. Recalde, E. Teruei and M. Silva. "{SC}*ECS: A class of modular and hierarchical cooperating systems". *ICATPN 1996*, LNCS 1091, Springer-Verlag, 1996, pp. 440–459.

[STC (1998)] M. Sliva, E. Teruel and J.M. Colom. "Linear algebraic and linear programming techniques for the analysis of place/transition net systems". *Advances in Petri Nets 1998*, LNCS 1491, Springer-Verlag, 1998, pp. 309–373.

[JCL (2002)] L. Jiao, T.Y. Cheung and W.M. Lu. "Characterizing liveness of Petri nets in terms of siphons". *ICATPN 2002*, LNCS 2360, Springer-Verlag, 2002, pp. 203–216.

[CX (1997)] F. Chu and X. Xie. "Deadlock analysis of Petri nets using siphons and mathematical programming". *IEEE Transactions on Robotics and Automation*, Vol. 13, No. 6, 1997, pp. 793–804.

[DAI (1995)] A. A. Desrochers and R. Y. AI-Jaar. *Applications of Petri Nets in Manufacturing Systems: Modeling, Control and Performance Analysis*. IEEE Press, 1995.

[HJC (2003)] H. Huang, L. Jiao and T. Y. Cheung. "Property-preserving composition of augmented marked graphs that share common resources". *Proc.*

IEEE International Conference on Robotics and Automation, IEEE. 2003, pp. 1446–1451.

[CHE (2002)] K.S. Cheung. "Use-case-driven system design: a synthesis methodology based on labelled Petri nets". *PhD thesis*, Dept. of Computer Science, City University of Hong Kong, 2002.

[MAK (2001)] W.M. Mak. "Verifying property preservation for component-based software systems (a Petri-net based methodology)". *PhD Thesis*, Dept. of Computer Science, City University of Hong Kong, 2001.

[VAL (1979)] R. Valette. "Analysis of Petri nets by stepwise refinements". *Journal of Computer and System Sciences*, Vol. 18, Issue 1, 1979, pp. 35–46.

[BA (1971)] J. Bruno and S.M.Altman. "A theory of asynchronous control net-works". *IEEE Transactions on Computing*, Vol. C-20, Issue 6, 1971, pp. 629–638.

[MK (1980)] T. Murata and J. Y. Koh. "Reduction and expansion of live and safe marked graphs". *IEEE Trans. on Circuits and Systems*, Vol. 27, Issue 1, 1980, pp. 68–70.

[SM (1983)] I. Suzuki and T. Murata. "A method for stepwise refinement and abstraction of Petri nets". *Journal of Computer and System Sciences*, Vol. 27, Issue 1, 1983, pp. 51–76.

[VOG (1987)] W. Vogler. "Behavior preserving refinement of Petri nets". *Graph-Theoretic Concepts in Computer Sciences*, LNCS 246, Springer-Verlag, 1987, pp. 82–93.

[BC (1990)] L. Bernardinello and F. de Cindio. "A survey of basic net models and modular net classes". *Advances in Petri nets 1992*, LNCS 609, Springer-Verlag, 1992, pp. 304–351.

[MUL (1985)] K. Muller. "Constructible Petri nets", Elerktr. Inf. Kybern. 21, 1985.

[AAL (1997)] W. M. P. van der Aalst. "Verification of workfolk nets". *ICATPN 1997*, LNCS 1248, Springer-Verlag, 1997, pp. 407–426.

[AAL (1998)] W. M. P. van der Aalst. "The application of Petri nets to workfolk management". *Journal of Circuits, Systems and Computers*, Vol. 8, No. 1, 1998, pp. 21–66.

[HEN (1987)] M. Hennessy. "Axiomatizing finite concurrent processes". *Technical Report 4/87*, Dept. Sci. Univ. of Sussex, Brighon, 1987.

[AH (1988)] L. Aceto and M. Hennesy. "Towards action-refinement in process algebras". *Technical Report 3/88*, Dept. Sci. Univ. of Sussex, Brighon, 1988.

[NEL (1989)] M. Nielsen, U. Engberg and K. Larsen. "Fully abstract models for a process language with refinement". *Proc. Linear Time, Branching Time and Partial Order in Logics and Models for Concurrency 1988*, LNCS 354, Springer-Verlag, 1988, pp. 523–548.

[ACE (1990)] L. Aceto. "Full abstractions for series-parallel pomsets". *Technical Report 1/90*, Dept. Sci. Univ. of Sussex, Brighon, 1990.

[NPW (1981)] M. Nielsen, G. D. Plotkin and G.Winskel. "Petri nets, event structures and domains, part I". *Theory of Computer Science*, Vol. 13, Issue. 1, 1981, pp. 85–108.

[GG (1989)] R. J. Glabbeek and U. Goltz. "Equivalence notions for concurrent systems and refinement of actions". *Mathematical Foundations of Computer Science 1989*, LNCS 379, Springer-Verlag, 1989, pp. 237–248.

[GW (1989)] R. J. Glabbeek and W. P. Weijland. "Refinement in branching time semantics". *Technical Report CS-R8922*, CWI, Amsterdam, 1989.

[GLA (1990)] R. J. Glabbeek and U. Goltz. "Equivalences and Refinement". *Semantics of Systems of Concurrent Processes 1990*, LNCS 469, Springer-Verlag, 1990, pp. 309–333.

[GG (1990)] R. J. Glabbeek and U.Goltz. "Refinement of actions in causality based models". *Technical Report*, Arbeitspapiere der GMD 428, 1990.

[VOG (1990b)] W. Vogler. "Bisimulation and action refinement". *Technical Report SFB-Bericht Nr. 342/10/90A*, Inst.Informatik, Techn. Univ. Munchen, 1990.

[GV (1987)] R. J. Glabbeek and F. Vaandrager. "Petri net models for algebraic theories of concurrency". *PARLE Parallel Architectures and Languages Europe 1987*, LNCS 259, Springer-Verlag, 1987, pp. 224–242.

[DEV (1988)] R. Devillers. "On the definition of a bisimulation notion based on partial words". Petri net newsletter 29, 1988, pp. 16–19.

[BDK (1989)] E. Best, R. Devillers, A. Kiehn and L. Pomello. "Fully concurrent bisimulation", T*echnical Report LIT-202*, Univ.Bruxelles, 1989

[VOG (1990a)] W. Vogler. "Failure semantics based on interval semiwords is a conguence for refinement". *Distributed Computering*, Vol. 4, No.3, 1990, pp. 139–162.

[HCM (2004)] H. Huang, T.Y. Cheung and W. M. Mak. "Structure and Behavior Preservation by Petri-net-based Refinements in System Design". *Theoretical Computer Science*, Vol. 328, Issue 3, 2004, pp. 245–269.

[JIA (2008)] L. Jiao. "Refining and Verifying Regular Petri Nets". *International Journal of System Sciences*, Vol. 39, No. 1, 2008, pp. 17–27.

[PRA (1986)] V. Pratt. "Modeling concurrency with partial orders". *International Journal of Parallel Programming*, Vol.15, No. 1, 1986, pp. 33–71.

[CDP (1987)] L. Castellano, G. De Michelis and L. Pomello. "Concurrency vs. interleaving: an instructive example". Bull. EATCS 31, 1987, pp. 12–15.

[BER (1985)] G. Berthelot. "Checking properties of nets using transformations". *Advances in Petri Nets 1985*, LNCS 222, Springer-Verlag, 1985, pp. 19–40.

[LIP (1981)] R. J .Lipton. "Reduction: a method of proving properties of parallel programs". *Journal of ACM*, Vol. 3, No.12, 1981, pp. 561–567.

[KWO (1977)] Y. S. Kwong. "On reduction of asynchronous systems". *Theoretical Computer Science*, Vol. 5, Issue 1, 1977, pp. 25–50.

[KV (1979)] W. Kowalk and R. Valk. "On reduction of parallel programs". *ICALP 1979*, LNCS 71, Springer-Verlag, 1979, pp. 356–369.

[BER (1986)] G. Berthelot. "Transformations and decompositions of nets". *Advances in Petri Nets 1986*, LNCS 254, Springer-Verlag, 1987, pp. 359–376.

[LF (1985)] K.H. Lee and J. Favrel. "Hierarchical reduction method for analysis and decomposition of Petri nets". *IEEE Trans. on Systems, Man and Cybernetics*, Vol. 15, No. 2, 1985, pp. 272–280.

[LF (1987)] K.H. Lee, J. Favrel and P. Baptiste. "Generalized Petri net reduction method". *IEEE Trans. on Systems, Man and Cybernetics*, Vol. 17, No. 2, 1987, pp. 297–303.

[DES (1990)] J. Desel. "Reduction and design of well-behaved concurrent systems". *CONCUR 1990*, LNCS 458, Springer-Verlag, 1990, pp. 166–181.

[ESP (1994)] J. Esparza. "Reduction and synthesis of live and bounded free choice Petri nets". *Information and Computation*, Vol. 114, Issue 1, 1994, pp. 50–87.

[ES (1990b)] J. Esparza and M. Silva. "On the analysis and synthesis of free choice Petri nets". *Advances in Petri Nets 1990*, LNCS 483, Springer-Verlag, 1990, pp. 241–286.

[ES (1991b)] J. Esparza and M. Silva. "Compositional synthesis of live and bounded free choice Petri nets". *CONCUR 1991*, LNCS 527, Springer-Verlag, 1991, pp. 173–187.

[JCL (2004)] L. Jiao, T.Y. Cheung and W.M. Lu. "On liveness and boundedness of asymmetric choice nets". *Theoretical Computer Science*, Vol. 311, Issue 1-3, 2004, pp. 165–197.

[PX (1996)] J. M. Proth and X. Xie. *Petri Net: A Tool for Design and Management of Manufacturing Systems*. Chichester, New York, Wiley, 1996.

[SB (1996)] R. H. Sloan and U. Buy. "Reduction rules for time Petri nets". *Acta Information*, Vol. 33, No. 5, 1996, pp. 687–706.

[WD (1998)]J. Wang and W. Deng. "Component-level reduction rules for time Petri nets with application in C2 systems". *Proc. IEEE International Conference on Systems, Man and Cybernetics 1998*, 1998, pp. 11–14.

[AC (1978)] T. Agerwala and Y. Choed-Amphai. "A synthesis rule for concurrent systems". *DAC 1978*, IEEE Press, 1978, pp. 305–311.

[SM (1990)] Y. Souissi and G. Memmi. "Composition of nets via a communication medium". *Advances in Petri Nets 1990*, LNCS 483, Springer-Verlag, 1990, pp. 457–470.

[NV (1985)] Y. Narahari and N. Viswanadham. "A Petri net approach to the modeling and analysis of flexible manufacturing systems". *Annals of Operations Research*, Vol. 3, No. 8, 1985, pp. 449–472.

[HJC (2005)] H. Huang, L. Jiao and T.Y. Cheung. "Property-Preserving Subnet Reductions for Designing Manufacturing Systems with Shared Resources". *Theoretical Computer Science*Vol. 332, Issue 1-3, 2005, pp. 461–485.

[JHC (2005)] L. Jiao, H. Huang and T.Y Cheung. "Property-preserving Composition by Place Merging". *Journal of Circuits, Systems and Computers*, Vol. 14, No. 4, 2005, pp. 793–812.

[JHC (2008)] L. Jiao, H. Huang and T.Y Cheung. "Handling Resource Sharing Problem Using Property-Preserving Place Fusions of Petri Nets". *Journal of Circuits, Systems, and Computers*, Vol.17, No. 3, 2008, pp. 365–387.

[ZHO (1996)] M. Zhou. "Generalizing parallel and sequential mutual exclusions for Petri net synthesis of manufacturing systems". *IEEE Int. Conf. On Emerging Technologies and Factory Automation 1996*, IEEE, Vol. 1, 1996, pp. 49–55.

[ZV (1999)] M. Zhou and K. Venkatesh. *Modeling, Simulation, and Control of*

Flexible Manufacturing Systems: a Petri Net Approach. World Scientific Publishing Co. Pte. Ltd, 1999.

[BEC (1985)] C. L. Beck. "Modeling and simulation of flexible control structures for automated manufacturing systems". *M.S thesis* and Robotics Inst. Tech. Rep., Carnegie-Mellon Univ. Pittsburgh, PA, 1985.

[KB (1986)] B.H. Krogh and C.L. Beck. "Synthesis of place/transition nets for simulation and control of manufacturing systems". *Proc. IFIP Symposium on Large Scale Systems 1986*, 1986, pp. 661–666.

[KD (1991)] I. Koh and F. DiCesare. "Modular transformation methods for generalized Petri nets and their applications to automated manufacturing systems". *IEEE Trans. on Systems, Man and Cybernetics*, Vol. 21, Issue 6, 1991, pp. 1512–1522.

[JD (1995)] M.D. Jeng and F. DiCeasare. "Synthesis using resource control nets for modeling shared-resource systems". *IEEE Trans. on Robotics and Automation*, Vol. 11, No. 3, 1995, pp. 317–327.

[JEN (1997)] M.D. Jeng. "A Petri net synthesis theory for modeling flexible manufacturing systems". *IEEE Trans. on Systems, Man and Cybernetics – Part B: Cybernetics*, Vol. 27, No. 2, 1997, pp. 169–183.

[PWX (1997)] J.M. Proth, L. Wang and X. Xie. "A class of Petri nets for manufacturing system integration". *IEEE Trans. on Robotics and Automation*, Vol. 13, No. 3, 1997, pp. 317–326.

[KIN (1997)] E. Kindler. "A compositional partial order semantics for Petri net components". *ICATPN 1997*, LNCS 1248, Springer-Verlag, 1997, pp. 235–252.

[PRE (1997)] R. S. Pressman.*Software Engineering, A practioner's Approach*. McGraw- Hill, fourth edition, 1997.

[AAL (2000)] W. M. P. van der Aalst. "Workflow verification: Finding control-flow errors using Petri-net-based techniques". *Business Process Management 2000*, LNCS 1806, Springer-Verlag, 2000, pp. 161–183.

[PAP (1996)] D. Pape, C. Cruz-Neira and M. Czernuszenko. *CAVE user's guide*, Electronic Visualization Laboratory, University of Illinois at Chigago, May, 1997.

[ZMF (1999)] Y. Zhou, T. Murata, T. DeFanti and H. Zhang. "Fuzzy-timing Petri net modeling and simulation of a networked virtual environment-CAVE". *Proc. of the Workshop within the 20th International Conference on Application and Theory of Petri Nets*, Williamsburg, VA, USA, 1999.

[ZD (1993)] M.C. Zhou and F. DiCesare. *Petri Net Synthesis for Discrete Event Control of Manufacturing Systems*. Kluwer Academic Publishers, 1993.

[BCD (1995)] K. Barkaoui, J. M. Couvreur and C. Dutheillet. "On liveness in extended non self-controlling nets". *ICATPN 1995*, LNCS 935, Springer-Verlag, 1995, pp. 25–44.

[VAL (1990)] K. S. Valvanis. "On the hierarchical analysis and simulation of flexible manufacturing systems with extended Petri nets". *IEEE Trans. on System, Man, and Cybernetics*, Vol. 20, Issue 1, 1990, pp. 94–100.

[XU (2003)] H. Xu and S. M. Shatz. "A framework for model-based design of agent-oriented software". *IEEE Trans. on Software Engineering*, Vol. 29, No. 1, 2003, pp. 15–30.

[COL (2003)] J. M. Colom. "The resource allocation problem in flexible manufacturing systems". *ICATPN 2003*, LNCS 2679, Springer-Verlag, 2003, pp. 23–35.

[REV (1998)] S.A. Reveliotis. "Accommodating FMS operational contingencies through routing flexibility". *Proc. IEEE Int. Conf. On Robotics and Automation 1998*, 1998, pp. 573–579.

[SOU (1990)] Y. Souissi. "On liveness preservation by composition of nets via a set of places". *Advances in Petri Nets 1991*, LNCS 524, Springer-Verlag, 1991, pp. 277–295.

[VMS (1988)] J. L. Villarroel, J. Martinez and M. Silva. "GRAMAN: a graphic system for manufacturing system design". *Proc. IMACS Int. Symp. Syst. Model & Simul.*, 1988.

[AV (1988)] J. S. Ahuja and K. P. Valavanis. "A hierarchical modeling methodology for flexible manufacturing systems using extended Petri nets". *Proc. Int. Conf. On Computer Integrated Manufacturing*, 1988, pp. 350–356.

[DG (1984)] A. Datta and S. Ghosh. "Synthesis of a class of deadlock-free Petri nets". *Journal of the ACM*, Vol. 31, No. 3, 1984, pp. 486–506.

[GH (1986)] G. Goos and J. Hartmanis. "Modular synthesis of deadlock-free control structures". *FSTTCS 1986*, LNCS 241, Springer-Verlag, 1986, pp. 288–318.

[ZDD (1989)] M. Zhou, F. DiCesare and A. A. Desrochers. "A top-down modular approach to synthesis of Petri net models for manufacturing systems". *Proc. IEEE Int. Conf. Robotics and Automation 1989*, 1989, pp. 534–539.

[GV (2003)] C. Girault and R. Valk. *Petri Nets for System Engineering – A Guide to Modeling, Verification, and Applications*. Springer, 2003.

[HAM (2003)] N. Hamerrlain. "Refinement of open Protocols for modeling and analysis of complex interactions in multi-agent systems". *CEEMAS 2003*, LNCS 2691, Springer-Verlag, 2003, pp. 423–434.

[BES (2001)] E. Best, R. Devillers and M. Koutny. *Petri Net Algebra*. Springer, 2001.

[CZ (1998)] T. Y. Cheung and W. Zeng, "Invariant-preserving transformations for the verification of place/transition systems". *IEEE Transactions on Systems, Man and Cybernetics – Part A: Systems and Humans*, Vol. 28, No.1, 1998, pp. 114–121.

[HUA (2004)] H. Huang. *Enhancing the property-preserving Petri net process algebra for component-based system design (with application to designing multi-agent systems and manufacturing systems). PhD Dissertation*, City University of Hong Kong, 2004.

[BGV (1990)] W. Brauer, R. Gold and W. Vogler. "A survey of behavior and equivalence preserving refinement of Petri nets". *Advances in Petri nets 1990*, LNCS 483, Springer-Verlag, 1990, pp. 1–46.

[JD (1993)] M. D. Jeng and F. DeCesare. "A review of synthesis techniques for Petri nets with application to automated manufacturing Systems". *IEEE*

Transactions on System, Man, and Cybernetics, Vol. 23, No. 1, 1993, pp. 301–312.

[HCW (2007)] H. Huang, T. Cheung, X. Wang X. "Applications of property-preserving algebras to manufacturing system design". *Journal of Information Science and Engineering*, Vol. 23, No. 1, 2007, pp. 167–181.

[CMR (2003)] L. Cabac, D. Moldt and H. Rolke. "A proposal for structuring Petri net-based agent interaction protocols". *ICATPN 2003*, LNCS 2679, Springer-Verlag, 2003, pp. 102–120.

[KMR (2001)] M. Kohler, D. Moldt and H. Rolke. "Modeling the structure and behavior of Petri net agents". *ICATPN 2001*, LNCS 2075, 2001, pp. 224–241.

[KMR (2003)] M. Kohler, D. Moldt and H. Rolke. "Modeling mobility and mobile agents using nets within nets". *ICATPN 2003*, LNCS 2679, Springer-Verlag, 2003, pp. 121–139.

[CDS (2002)] J. Cuena, Y. Demazeau, A. G. Serrano and J. Treur. *Knowledge Engineering and Agent Technology*. IOS press, Amsterdam. 2004.

[KGR (1996)] D. Kinny, M. Georgeff and A. Rao. "A methodology and modeling technique for systems of BDI agents". *MAAMAW 1996*, LNCS 1038, Springer-Verlag, 1996, pp. 56–71.

[KMJ (1996)] E. A. Kendall, M. T. Malkoun and C. H. Jiang. "A methodology for developing agent based systems". *DAI 1996*, LNCS 1087, Springer-Verlag, 1996, pp. 85–99.

[HL (2005)] B. Horling and V. Lesser. "A survey of multi-agent organizational paradigms". *Knowledge Engineering Review*, Vol. 19, No. 4, 2005, pp. 281–316.

[FW (1997)] M. Fisher and M. Wooldridge. "On the formal specification and verification of multi-agent systems". *International Journal of Cooperative Information Systems*, Vol. 6, No. 1, 1997, pp. 37–65.

[WOO (1998)] M. Wooldridge. "Agents and software engineering". *AI*IA Notizie*, Vol. 6, No. 3, 1998, pp. 31–37.

[BDJ (1997)] F. M. T. Brazier, B. M. Dunin-Keplicz, N. R. Jennings and J. Treur. "DESIRE: modeling multi-agent systems in a compositional formal framework". *International Journal of Cooperative Information Systems*, Vol. 6, No. 1, 1997, pp. 67–94.

[CJT (2001)] F. Cornelissen, C. M. Jonker and J. Treut. "Compositional verification of knowledge-based systems: a case study in diagnostic reasoning". *EKAW 1997*, LNCS 1319, Springer-Verlag, 1997, pp. 65–80.

[JT (2002)] C. M. Jonker and J. Treur. "Compositional verification of multi-agent systems: a formal analysis of pro-activeness and reactiveness". *International Journal of Cooperative Information System*, Vol. 11, Issue 1-2, 2002, pp. 51–92.

[MK (1999)] T. Miyamoto and S. Kumagai. "A multi-agent net model and the realization of software environment". *ICATPN 1999*, LNCS 1639, Springer-Verlag, 1999, pp. 83–92.

[MW (1997)] D. Moldt, F. Wienberg. "Multi-agent-systems based on colored Petri nets". *ICATPN 1997*, LNCS 1248, Springer-Verlag, 1997, pp. 82–101.

[LL (2003)] S. Ling and S. W. Loke. "MIP-Nets: A compositional model of multi-agent interaction", *CEEMAS 2003*, LNCS 2691, Springer-Verlag, 2003, pp. 61–72.

[XVI (2002)] D. Xu, R. Volz, T. Ioerger and J. Yen. "Modeling and verifying multi-agent behaviors using Predicate/Transition nets", *SEKE 2002*, ACM, 2002, pp. 193–200.

[XYD (2003)] D. Xu, J. Yin, Y. Deng and J. Deng. "A formal architecture model for logical agent mobility", *IEEE trans. on Software Engineering*, Vol. 29, No. 1, 2003, pp. 31–45.

[HUA (1995)] J. Huang, N. R. Jennings and J. Fox. "An agent-based approach to health care management", *Journal of Applied Artificial Intelligence*, Vol. 9, No. 4, 1995, pp. 401–420.

[TER (1996)] D. K. Terry. *A Petri net-based on-line scheduling system for a general manufacturing job shop*, PhD Dissertation, Rensselaer Polytechnic Institute, Troy, New York, 1996.

[PM (1995)] J. M. Proth and I. Minis. "Planning and scheduling based on Petri nets: in Petri nets in flexible and agile automation". *In*: Zhou, M. C. (Ed.), Kluwer Academic Publishers, Boston, MA, 1995, pp. 109–148.

[PX (1997)] J. M. Proth and X. Xie. *Petri nets: a tool for design and management of manufacturing systems*, John Wiley & Sons, 1997.

[XZ (1998)] H. H. Xiong and M. Zhou. "Scheduling of semiconductor test facility via Petri nets and hybrid heuristic search". *IEEE Transactions on Semiconductor Manufacturing*, Vol. 11, No. 3, 1998, pp. 384–393.

[SS (1991)] H. Shih and T. Sekiguchi. "A timed Petri net and beam search based on-line FMS scheduling systems with routing flexibility". *Proceedings of IEEE International Conference on Robotic and Automation*, Sacramento, 1991, 2548–2553.

[LD (1994)] D. Y. Lee and F. DiCesare. "FMS scheduling using Petri nets and heuristic search". *IEEE Transactions on Robotics and Automation*, Vol. 10, No. 2, 1994, pp. 123–132.

[SCF (1994)] T. H. Sun, C. W. Cheng and L. C. Fu. "A Petri net based approach to modeling and scheduling for an FMS and a case study". *IEEE Transactions on Industrial Electronics*, Vol. 41, No. 6, 1994, pp. 593–601.

[WXY (2008)] Y. M. Wang, N. F. Xiao and H. L. Yin. "A two-stage genetic algorithm for large size job shop scheduling problems". *The International Journal of Advanced Manufacturing Technology*, Vol. 39, No. 7-8, 2008, pp. 813–820.

[LFC (2003)] S. Y. Lin, L. C. Fu, T. C. Chiang and Y. S. Shen. "Colored timed Petri net and GA based approach to modeling and scheduling for wafer probe center". *Proceedings of the 2003 IEEE. International Conference on Robotics and Automation 1*, 2003, pp. 1434–1439.

[CFL (1998)] Y. Y. Chung, L. C. Fu and M. W. Lin. "Petri net based modeling and GA based scheduling for a flexible manufacturing system". *Proceedings of the 37th IEEE Conference on Decision and Control*, 1998, pp. 4346–4347.

[XH (2001)] L. Xue and Y. Hao. "Petri net based scheduling for integrated circuits manufacturing". *ACTC Electronica Sinca*, Vol. 29, No.8, 2001, pp. 1064–1067.

[GZC (2003)] Y. J. Gua, X. N. Zhang, Y. Cao, R. J. Zhao and T. Q. Lin. "Workshop scheduling based on RCPN and system development on Internet/Intranet". *Mini-micro Systems*, Vol. 24, No. 7, 2003, pp. 1285–1288.

[CC (2003)] J. Chen and F. F. Chen. "Performance modeling and evaluation of dynamic tool allocation in flexible manufacturing systems using colored Petri nets:an object-oriented approach". *The International Journal of Advanced Manufacturing Technology*, Vol. 21, No. 2, 2003, pp. 98–109.

[HUA (2009)] H. Huang. "Component-based design and optimization for jobshop scheduling systems". *International Journal of Advanced Manufacturing Technology*, Vol. 45, Issue 9, 2009, pp. 958–967.

[JSS (2001)] S. Jajodia, P. Samarati, M. L. Sapino and V. S. Subrahmanian. "Flexible support for multiple access control policies". *ACM Trans. Database Syst*, Vol. 26, No. 2, 2001, pp. 214–260.

[KBB (2003)] A. A. E. Kalam, R. E. Baida, P. Balbiani, S. Benferhat, F. Cuppens, Y. Deswarte, A. Miege, C. Saurel and G.Trouessin. "Organization based access control". *Proceedings of IEEE 4th International Workshop on Policies for Distributed Systems and Networks*, 2003, pp. 120–131.

[MOS (2005)] T. Moses. "Extensible access control markup language (XACML)", Version 2.0. *Technical report*, OASIS, 2005.

[DOU (2007)] D. J. Dougherty, C. Kirchner, H. Kirchner and A. Santana de Oliveira. "Modular access control via strategic rewriting". *ESORICS 2007*, LNCS 4734, Springer-Verlag, 2007, pp. 578–593.

[BVS (2002)] P. A. Bonatti, S. D. C. di Vimercati and P. Samarati. "An algebra for composing access control policies". *ACM Trans. Inf. Syst. Secur.* Vol. 5, No. 1, 2002, pp. 1–35.

[WJ (2003)] D. Wijesekera and S. Jajodia. "A propositional policy algebra for access control". *ACM Trans. Inf. Syst. Secur.* Vol. 6, No. 2, 2003, pp. 286–325.

[BLW (2005)] L. Bauer, J. Ligatti and D. Walker. "Composing security policies with polymer". *PLDI 2005*, ACM, pp. 305–314.

[LBO (2006)] A. J. Lee, J. P. Boyer, L. Olson and C. A. Gunter. "Defeasible security policy composition for web services". *FMSE 2006*, ACM, pp. 45–54.

[BDH (2007)] G. Bruns, D. S. Dantas and M. Huth. "A simple and expressive semantic framework for policy composition in access control". *FMSE 2007*, ACM, 2007, pp. 12–21.

[SO (2007)] A. Santana de Oliveira. "Rewriting-based access control policies". *Electronic Notes in Theoretical Computer Science*, Vol. 171, No. 4, 2007, pp. 59–72.

[SO (2008)] A. Santana de Oliveira. "R'e'ecriture et modularit'e pour les politiques de s'ecurit'e". *PhD thesis*, UHP Nancy 1, 2008.

[SMJ (2005)] B. Shafiq, A. Masood, J. Joshi and A. Ghafoor. "A role-based access control policy verification framework for real-time systems". *Proc.*

of the 10th IEEE International Workshop on Object-Oriented Real-Time Dependable Systems, 2005, pp. 13–20.

[MOR (2000)] K. H. Mortensen. "Automatic code generation method based on coloured Petri net models applied on an access control system". *ICATPN 2000*, LNCS 1825, Springer-Verlag, 2000, pp. 367–386.

[KNO (2000)] K. Knorr. "Dynamic access control through Petri net workflows". *Proceedings of 16th Annual Conference on Computer Security Applications* 2000, IEEE, pp. 159–167.

[ZHL (2006)] Z. L. Zhang, F. Hong and J. G. Liao. "Modeling Chinese wall policy using colored Petri nets". *Proceedings of the 6th IEEE International Conference on Computer and Information Technology 2006*, IEEE, 2006, pp. 162–162.

[ZHZ (2006)] Z. L. Zhang, F. Hong and H. Xiao. "Verification of strict integrity policy via Petri nets". *Proceedings of the International Conference on Systems and Networks Communication*, IEEE, 2006, pp. 23.

[JUS (2003)] K. Juszczyszyn. "Verifying enterprise's mandatory access control policies with coloured Petri nets". *Proceedings of the 12th IEEE International Workshops on Enabling Technologies: Infrastructure for Collaborative Enterprises* 2003. IEEE, 2003, pp. 184–189.

[DWT (2003)] Y. Deng, J. C. Wang, J. Tsai and K. Beznosov. "An approach for modeling and analysis of security system architectures". *IEEE Transactions on Knowledge and Data Engineering*, Vol. 15, No. 5, 2003, pp. 1099–1119.

[HH (2010)] H. Huang and H. Kirchner. "Formal Specification and Verification of Modular Security Policy based on Colored Petri Nets". To appear in *IEEE Transactions on Dependable and Secure Computing*.

[TK (2006)] M. C. Tschantz and S. Krishnamurthi. "Towards reasonability properties for access-control policy languages". *ACMAT 2006*, ACM, 2006, pp. 160–169.

[BF (2006)] S. Barker and M. Fern'andez. "Term rewriting for access control". *DBSec 2006*, LNCS 4127, Springer-Verlag, 2006, pp. 179–193.

[TJM (1988)] F. Tiplea, T. Jucan and C. Masalagiu. "Term rewriting systems and Petri nets". *Analele Stiintifice ale Universitatii Al. I. Cuza*, Vol. 34, No. 4, 1988, pp. 305–317.

[BRL (2001)] R. Verma, M. Rusinowitch and D. Lugiez. "Algorithms and reductions for rewriting problems". *Fundamental Informatics*, Vol. 46, No. 3, 2001, pp. 257–276.

[LT (2006)] I. Leahu and F. Tiplea. "The confluence property for petri nets and its applications". *Proceedings of the 8th International Symposium on Symbolic and Numeric Algorithms for Scientific Computing*. IEEE, 2006, pp. 430–436.

[MOO (2002)] R. Melinte, O. Oanea, I. Olga and F. Tiplea. "The home marking problem and some related concepts". *Acta Cybernetica*, Vol. 15, Issue 3, 2002, pp. 104–115.

[XJF (2005)] K. L. Xing, X. J. Jin and Y. Feng. "Deadlock avoidance Petri net controller for manufacturing systems with multiple resource service". *Proc. IEEE Conference on Robotics and Automation*, IEEE, 2005, pp. 4757–4761.

[LEE (2002)] J. K. Lee. "Scheduling analysis with resources share using the transitive matrix based on P-invariant". *Proc. 41st SICE Annual Conference*, 2002, Vol.2, pp. 1359–1364.

[LZ (2004)] Z. Li and M. Zhou. "Elementary siphons of Petri nets and their application to deadlock prevention in flexible manufacturing systems". *IEEE Transactions on System, Man, and Cybernetics*, Vol. 34, Issue 1, 2004, pp. 38–51.

[HLW (2006)] H. Hu, Z. Li and A. R. Wang. "On the optimal set of elementary siphons in Petri nets for deadlock control in fms". *Proc. 2006 IEEE International Conference on Networking, Sensing and Control*, IEEE, 2006, pp. 244–247.

[LW (2007)] Z. Li and N. Wei. "Deadlock control of flexible manufacturing systems via invariant-controlled elementary siphons of Petri nets". *The International Journal of Advanced Manufacturing Technology*, Vol. 33, No. 1, 2007, pp. 24–35.

[ROS (2004)] E. Roszkowska. "Supervisory control for deadlock avoidance in compound processes". *IEEE Transactions on System, Man, and Cybernetics*, Vol. 34, No. 1, 2004, pp. 52–64.

[XW (2002)] G. Xu and Z. M. Wu. "A kind of deadlock-free scheduling method based on Petri net". *Proc. IEEE HASR'02*, IEEE, 2002, pp. 195–200.

[XWU (2002)]G. Xu and Z. M. Wu. "Deadlock-free scheduling method using Petri net model analysis and GA search. *Proc. 2002 International Conference on Control Application*, IEEE, Vol. 2, 2002, pp. 1153–1158.

[LIM (1999)] J. Liu, Y. Itoh, I. Miyazawa and T. Sekiguchi, "A research on Petri net properties using transitive matrix". *Proc. IEEE International Conference on System, Man, and Cybernetics*, IEEE, 1999, pp. 888–893.

[SL (2002)] Y. J. Song and J. K. LEE. "Deadlock analysis of Petri nets using the transitive matrix". *Proc. 41st SICE Annual Conference*, IEEE, 2002, pp. 689–694.

[KL (2006)] S. Kim, S. Lee and J. Lee. "Deadlock analysis of Petri nets based on the resource share places relationship. *IMACS multiconference on CESA*, 2006, pp. 59–64.

[LEE (2004)]J. Lee, "Deadlock find algorithm using the transitive matrix". *Proc. CIE 2004*, 2004, pp. 24–26.

[ZCX(1995)] M. Zhou, H. Chiu and H. Xiong, "Petri net scheduling of FMS using branch and bound method". *Industrial Electronics, Control, and Instrumentation. Proc. 1995 IEEE IECON*, 1995, pp. 211–216.